A

TREATISE

ON

PLANE AND SPHERICAL

TRIGONOMETRY.

BY

WILLIAM CHAUVENET, A.M.

PROFESSOR OF MATHEMATICS IN THE UNITED STATES NAVY; PROFESSOR OF ASTRONOMY, NAVIGATION, AND NAUTICAL SURVEYING IN THE UNITED STATES NAVAL ACADEMY; FELLOW OF THE AMERICAN ACADEMY OF ARTS AND SCIENCES; MEMBER OF THE AMERICAN PHILOSOPHICAL SOCIETY, ETC.

Fourth Edition.

PHILADELPHIA:
LIPPINCOTT, GRAMBO & CO.
1855.

STEREOTYPED BY L. JOHNSON AND CO.
PHILADELPHIA.

PREFACE.

I HAVE in this treatise endeavored to arrange a course of trigonometrical study sufficiently extensive to enable the student to comprehend readily any applications of trigonometry he may meet with in the works of the best modern mathematicians. With this object, some topics have been introduced which are not usually found in works devoted specially to this subject.

Among those topics, the most important is the solution of the general spherical triangle, or the triangle whose sides and angles are not limited, according to the usual practice, to values less than 180°. The advantage of introducing such triangles into astronomical investigations is sufficiently shown in the applications made of them in the works of BESSEL and other German mathematicians; and especially in the *Theoria Motus Corporum Cœlestium* of GAUSS, who was the first to suggest their employment.

The subject of Finite Differences of triangles, plane and spherical, occupies a large space in Cagnoli's treatise, but has not been admitted into more recent works. It here occupies only a few pages, but no important result of

Cagnoli's Table has been omitted, while a number of the formulæ are much simpler than the corresponding ones given by him.

Although my plan embraces a much more extensive course than is contained in the text-books commonly used, I have studiously kept in view the wants of academic and collegiate classes; and have so arranged the work that a selection of subjects of immediate importance may be readily made. The more elementary portions are printed in a larger type, and are intended to form, independently of the matter in the smaller type, a connected treatise which may be studied as though it were in a separate volume.

Those who may afterwards wish to extend their knowledge will appreciate the advantage of having the higher departments of the subject treated in connection with those fundamental ones to which they are most intimately related.

W. C.

U. S. Naval Academy,
Annapolis, Md., May 1, 1850.

NOTE TO THE FOURTH EDITION.

In this edition, besides a number of minor changes, and the correction of some typographical errors, a very important modification has been made in the solution of the equation $\tan x = p \tan y$ by series (p. 145), which was given in former editions in the usual form as stated by all writers on trigonometry. This form was discovered to lack generality, and consequently to fail in certain applications, in consequence of the omission of the arbitrary term $n\pi$ now introduced. Several subsequent investigations, depending on this, have in like manner been rectified.

W. C.

U. S. Naval Academy, *April* 1, 1854.

CONTENTS.

PART I.

PLANE TRIGONOMETRY.

CHAPTER I.

CHAPTER II.

CHAPTER III.

CHAPTER IV.

CHAPTER V.

CHAPTER VI.

CHAPTER VII.

PART II.

SPHERICAL TRIGONOMETRY.

CHAPTER I.

CHAPTER II.

CHAPTER III.

CHAPTER IV.

CHAPTER V.

CHAPTER VI.

CHAPTER VII.

CHAPTER VIII.

PART I.

PLANE TRIGONOMETRY.

CHAPTER I.

MEASURES OF ANGLES AND ARCS.

1. TRIGONOMETRY is that branch of Mathematics which treats of methods of subjecting angles and triangles to numerical computation.

2. PLANE TRIGONOMETRY treats of methods of computing plane angles and triangles.

It embraces the investigation of the relations of angles in general, a branch of the science not necessarily connected with the elementary solution of triangles, and which has been distinguished as the *Angular Analysis*.

3. By the *solution* of a triangle, in trigonometry, is meant the *computation* of unknown parts of the triangle from given ones.

The triangle has six parts; three angles and three sides. It is shown in geometry, that when any three of these parts are given, provided one of them is a side, the triangle may be constructed, and the unknown parts found by mechanical measurement.

In the same cases, by trigonometry, we compute the unknown parts from the three given ones, without resorting to construction and measurement: a method of inferior accuracy, on account of the unavoidable imperfections of the instruments employed, and the difficulty of distinguishing with the eye the smallest subdivisions of lines and angles.

But here also the case is excluded in which the three angles are given without a side, because there may be an indefinite number of plane triangles, whose angles are equal to the same three given ones, as

Fig. 1.

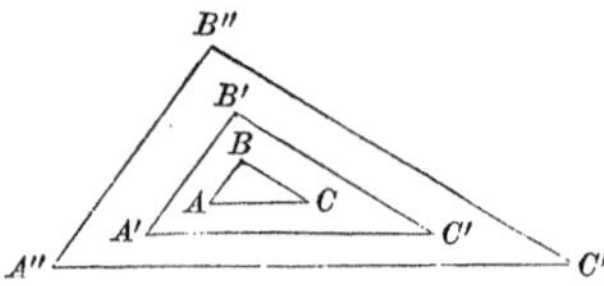

in Fig. 1. the triangles ABC, $A'B'C'$, &c. In this case, all these triangles are similar, and their sides are proportional; or the ratio of AB to AC is equal to the ratio of $A'B'$ to $A'C'$, &c.; so that the *ratios* of the sides to each other are fixed or determinate, although the absolute lengths of these sides are indeterminate.

4. Now, in order to subject a triangle to computation, we must first express the sides and angles by numbers. For this purpose proper units of measure must be adopted.

The unit of measure for the sides of plane triangles is a straight line, as an inch, a foot, a mile, &c.; and the number expressing a side is the number of units of the adopted kind that the side contains.

5. The units by which angles are expressed are, the *degree*, *minute*, and *second;* distinguished by the characters $^\circ\ '\ ''$.

A *degree* is an angle equal to $\frac{1}{90}$ of a right angle; or a degree is $\frac{1}{360}$ of the whole angular space about a point, or $\frac{1}{360}$ of four right angles. Thus, Fig. 2, if the angular space about O is divided into 360 equal parts, of which AOB is one, then AOB is one degree. The right angle will be expressed by 90°; two right angles by 180°, and the whole angular space about a point by 360°.

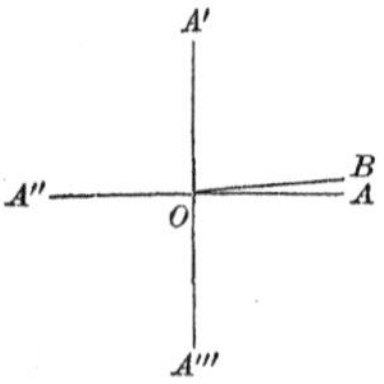

Fig. 2.

A *minute* is an angle equal to $\frac{1}{60}$ of a degree. Therefore, $1^\circ = 60'$; and a right angle $= 90 \times 60' = 5400'$.

A *second* is an angle equal to $\frac{1}{60}$ of a minute. Therefore, $1' = 60''$; $1^\circ = 60 \times 60'' = 3600''$; and a right angle $= 90 \times 60 \times 60'' = 324000''$.

Angles less than seconds are sometimes expressed by *thirds*, *fourths*, *fifths*, &c., marked $'''\ {}^{\text{IV}}\ {}^{\text{V}}$, &c.; a third being $\frac{1}{60}$ of a second; a fourth, $\frac{1}{60}$ of a third; &c. But the more convenient method is to express them as decimal parts of a second; thus $\frac{1}{7}$ of a right angle will be either

$$12^\circ\ 51'\ 25''\ 42'''\ 51^{\text{IV}}, \text{ \&c.}$$

or more conveniently

$$12^\circ\ 51'\ 25''.714, \text{ \&c.}$$

6. The above division of angles is called *sexagesimal*, from the divisor 60 employed in the subdivision of the degree. The *centesimal* division, however, would be preferable in all cases, but cannot now be generally introduced without, at the same time, changing the arrangement of all our tables, the graduation of astronomical and

other instruments, charts, &c. Nevertheless, the attempt has been made in France, and several standard works exist in the French language, in which it is employed throughout.

In the centesimal or French division, the right angle is divided into 100 degrees; the degree into 100 minutes; the minute into 100 seconds, &c. The reduction of these denominations from one to the other requires only a change in the position of the decimal point; thus, in this system 60° 75′ 84″·8 is the same as 607584″·8 or 60°·75848 or 0^q·6075848, the symbol q denoting a quadrant or right angle.

To convert centesimal into sexagesimal degrees, since 100° dec. = 90° sex. *deduct one tenth from the number of centesimal degrees.*

Example. Required the number of sex. degrees in 85° 47′ 43″ dec.

85°·4743 cent.
Deduct $\frac{1}{10}$ = 8 ·54743

76°·92687 sex. degrees and dec. parts.
55′·6122
36″·732

or 76° 55′ 36″·732 sexagesimal.

To convert sexagesimal into centesimal degrees, since we must take $\frac{10}{9}$ of the sex., *divide by* 9 *and move the decimal point one place to the right.*

Example. Required the number of centesimal degrees in 76° 55′ 36″·732 sex. Reducing the minutes and seconds to the decimal of a degree, we have

76°·92687 sex.
$\frac{10}{9}$ of which is 85°·4743 cent.
or 85° 47′ 43″ centesimal.

To distinguish the degrees of the centesimal from those of the sexagesimal division, the former are frequently called *grades,* and are denoted by the character ^{g} instead of °; thus the preceding angle would be 85^g 47′ 43″.

Measures of Arcs.

7. Since the angles at the center of a circle are proportional to the arcs of the circumference intercepted between their sides, these arcs may be taken as the measures of the angles, and we may express both the arc and the angle by the number of *units of arc* intercepted on the circumference.

The units of arc are also the degree, minute, and second. They are the arcs which subtend angles of a degree, a minute, and a second, respectively, at the center. A degree of arc is thus always $\frac{1}{360}$ of the circumference, whatever the radius of the circle may be; and we obtain the same numerical expression of an angle, whether we refer it directly to the angular unit, or to the corresponding unit of arc. The right angle AOA', Fig. 3, and its measure, the quadrant AA', are therefore both expressed by 90°; the semicircumference by 180°, and the whole circumference by 360°.

Fig. 3.

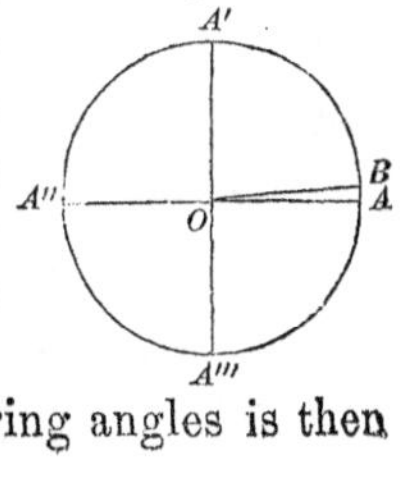

8. The radius of the circle employed in measuring angles is then

arbitrary, and we may assume for it such a value as will most simplify our calculations. This value is *unity;* that is, the linear unit employed in expressing the sides of our triangles, or other lines considered. This value will be generally used throughout this treatise.

9. *To find the length of an arc of a given number of degrees, minutes,* &c.

The semi-circumference of a circle whose radius is unity is known to be 3·14159265; or, the radius being R, the semi-circumference is 3·14159265R. Hence

		When $R=1$
Arc $180°$	$=3{\cdot}14159265\,R$	$=3{\cdot}14159265$
" $1°$	$=0{\cdot}017453293\,R$	$=0{\cdot}017453293$
" $1'$	$=0{\cdot}0002908882\,R$	$=0{\cdot}0002908882$
" $1''$	$=0{\cdot}000004848137\,R$	$=0{\cdot}000004848137$

An arc x therefore, in the circle whose radius is unity, being expressed in degrees, or minutes, or seconds, we find its length by the formulæ

$$\begin{aligned}\text{Arc } x &= 0{\cdot}017453293\,x^{\circ}\\ &= 0{\cdot}0002908882\,x'\\ &= 0{\cdot}000004848137\,x''.\end{aligned}$$

As these factors for finding the length of an arc are often used, it is convenient to have their logarithms prepared.* Thus

$$\begin{aligned}\text{Arc } x &= [8{\cdot}2418774]\,x^{\circ}\\ &= [6{\cdot}4637261]\,x'\\ &= [4{\cdot}6855749]\,x''\end{aligned}$$

in which the rectangular brackets are used to express that the *logarithm* of the factor is given instead of the factor itself.

EXAMPLE. What is the length of the arc $x=38°\ 17'\ 48''$, the radius being $=1$.

$38°\ 17'\ 48''=137868''$	log.	5·1394635
	Log. factor for seconds	4·6855749
$x=0{\cdot}6684031$	log. x	9·8250384

10. *To find the number of degrees, &c. in an arc equal to the radius.*

We have, from the preceding article,

---

* The logarithms in the examples of this work will be taken from *Stanley's Tables,* (published in New Haven, by Durrie and Peck,) the best tables of seven-figure logarithms yet published in this country.

$$R = \frac{180°}{3 \cdot 14159265} = 57° \cdot 2957795$$
$$= 3437' \cdot 74677 = 206264'' \cdot 806$$

11. The angle at the center measured by an arc equal to the radius, is often taken as the unit of angular measure, as this angle will be of an invariable magnitude, whatever is the length of the radius. If x is the number of such units in a given angle, the number of degrees, &c., in it will be found by multiplying by the value of the radius in degrees, &c., found in the preceding article. Thus,

$$x° = x R° = 57° \cdot 2957795\, x = [1 \cdot 7581226]\, x$$
$$x' = x R' = 3437' \cdot 74677\, x = [3 \cdot 5362739]\, x$$
$$x'' = x R'' = 206264'' \cdot 806\, x = [5 \cdot 3144251]\, x$$

Reciprocally, the angle being given in degrees, &c., we reduce it to the unit radius, by dividing by $R°$, R', or R'', thus,

$$x = \frac{x°}{R°} = \frac{x'}{R'} = \frac{x''}{R''}$$

which is evidently the same as multiplying by the factors of Art. 9.

It appears, then, that an angle is expressed in the unit of this article by the length of the arc which measures the angle in the circle whose radius is unity. Hence, an angle thus expressed is said to be given *in arc*. If we put (as is usual)

$$\pi = 3 \cdot 14159265 \cdots$$

π is the circular measure of two right angles, or it is the expression of two right angles in arc. In trigonometry it is therefore common to employ π to denote an angular magnitude of 180°; $\frac{\pi}{2}$ a right angle; $2\,\pi$ four right angles, &c.

12. The *complement* of an angle or arc is the remainder obtained by subtracting the angle or arc from 90°.

The *supplement* of an angle or arc is the remainder obtained by subtracting the angle or arc from 180°.

Thus the complement of 30° is 60°; the supplement of 30° is 150°.

Two angles or arcs are complements of each other when their sum is 90°. They are supplements of each other when their sum is 180°.

13. According to these definitions, the complement of an arc that exceeds 90° is negative. Thus the complement of 120° is $90° - 120° = -30°$. In like manner the supplement of 200° is $180° - 200° = -20°$.

CHAPTER II.

SINES, TANGENTS, AND SECANTS. FUNDAMENTAL FORMULÆ.

14. Having expressed the sides and angles of triangles by numbers, we are next to find such relations between them as shall enable us to combine these two different species of quantity in computation.

As every oblique triangle may be resolved into two right triangles by dropping a perpendicular from one of the angles upon the opposite side, the solution of all triangles is readily made to depend upon that of right triangles. Let us therefore consider a series of right triangles, ABC, $AB'C'$, $AB''C''$, &c., Fig. 4, which have a common angle A. The angles at B, B', B'', being also equal, the triangles are similar; and by geometry

Fig. 4.

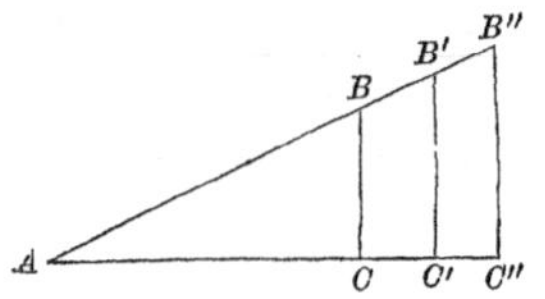

$$BC : AB = B'C' : AB' = B''C'' : AB''$$

or by the definitions of ratio and proportion,

$$\frac{BC}{AB} = \frac{B'C'}{AB'} = \frac{B''C''}{AB''}$$

In like manner it follows that

$$\frac{BC}{AC} = \frac{B'C'}{AC'} = \frac{B''C''}{AC''}$$

and

$$\frac{AB}{AC} = \frac{AB'}{AC'} = \frac{AB''}{AC''}$$

Hence it appears that *the ratios of the sides to each other are the same in all right triangles having the same acute angle;* and, therefore, if these ratios are known in any one of these triangles, they will be known in all of them.

These ratios, then, depending on the value of the angle alone, without regard to the absolute lengths of the sides, may be considered as indices of the angle, and have received special names, as follows:

15. *The* SINE *of the angle is the quotient of the opposite side divided by the hypotenuse.*

Thus, in the right triangle ABC, Fig. 5, if we designate the sides by the small letters a, b, c, we shall have, (whatever the *absolute* length of the sides)

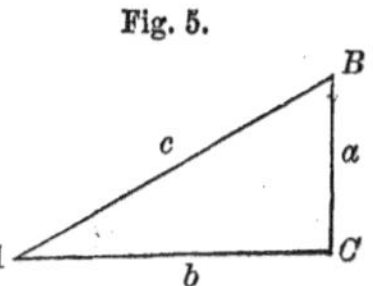

$$\sin A = \frac{a}{c}, \quad \sin B = \frac{b}{c}$$

16. *The* TANGENT *of the angle is the quotient of the opposite side divided by the adjacent side.*

Thus
$$\tan A = \frac{a}{b}, \quad \tan B = \frac{b}{a}$$

17. *The* SECANT *of the angle is the quotient of the hypotenuse divided by the adjacent side.*

Thus
$$\sec A = \frac{c}{b}, \quad \sec B = \frac{c}{a}$$

18. *The* COSINE, COTANGENT, *and* COSECANT *of an angle, are respectively the* SINE, TANGENT, *and* SECANT *of the complement of the angle.*

Since the sum of the two acute angles of a right triangle is one right angle, or 90°, they are, by Art. 12, complements of each other; therefore, according to the preceding definitions, we shall have

$$\left.\begin{array}{ll} \sin A = \cos B = \frac{a}{c} & \cos A = \sin B = \frac{b}{c} \\ \tan A = \cot B = \frac{a}{b} & \cot A = \tan B = \frac{b}{a} \\ \sec A = \operatorname{cosec} B = \frac{c}{b} & \operatorname{cosec} A = \sec B = \frac{c}{a} \end{array}\right\} \quad (1)$$

19. Since $\frac{c}{a}$ is the reciprocal of $\frac{a}{c}$, it follows from the first and last of these equations, that the sine and cosecant of the same angle are reciprocals; and from the other equations, also, that the cosine and secant, the tangent and cotangent are reciprocals. That is,

$$\left.\begin{array}{ll} \sin A = \frac{1}{\operatorname{cosec} A} & \operatorname{cosec} A = \frac{1}{\sin A} \\ \cos A = \frac{1}{\sec A} & \sec A = \frac{1}{\cos A} \\ \tan A = \frac{1}{\cot A} & \cot A = \frac{1}{\tan A} \end{array}\right\} \quad (2)$$

or more briefly,

$$\sin A \operatorname{cosec} A = \cos A \sec A = \tan A \cot A = 1 \qquad (3)$$

SINES, &c. OF ARCS.

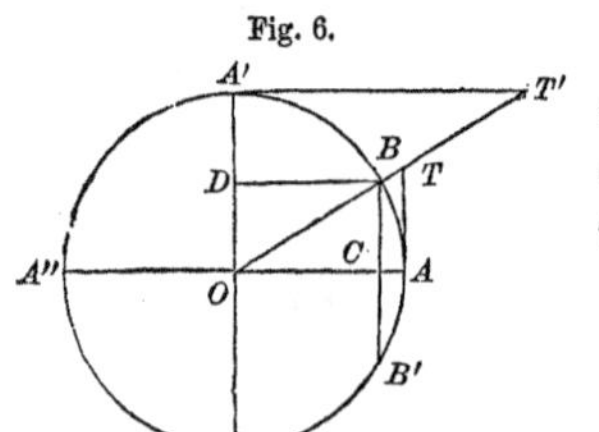

20. *The sine, tangent, and secant of an arc are respectively the sine, tangent, and secant of the angle at the center measured by that arc.* Thus, Fig. 6,

$$\sin AB = \sin AOB = \frac{BC}{OB}$$

The sine of an arc, therefore, does not depend upon the absolute length of the arc, but upon the ratio of the arc to the whole circumference, (Art. 7.) It follows that the relations (2) and (3) are also applicable when A expresses an arc.

21. If the radius $= 1$, all the trigonometric functions above defined may be represented in or about the circle by straight lines. Representing the arc AB, or angle AOB, by x, we have, when $OA = OB = 1$,

$$\sin x = \frac{BC}{OB} = \frac{BC}{1} = BC$$

$$\tan x = \frac{AT}{OA} = \frac{AT}{1} = AT$$

$$\sec x = \frac{OT}{OA} = \frac{OT}{1} = OT$$

and from the arc $A'B = 90° - x$ we find in the same way

$$\cos x = BD = OC$$
$$\cot x = A'T'$$
$$\operatorname{cosec} x = OT'$$

Therefore, in the circle whose radius is unity, *the sine of an arc, or of the angle at the center measured by that arc, is the perpendicular let fall from one extremity of the arc upon the diameter passing through the other extremity.*

The trigonometric tangent is that part of the tangent drawn at one extremity of the arc, which is intercepted between that extremity and the diameter (produced) *passing through the other extremity.*

The secant is that part of the produced diameter which is intercepted between the center and the tangent.

The cosine is the distance from the center to the foot of the sine.

In a circle of any other radius than unity, the trigonometric functions of an arc will be equal to the lines drawn as above, divided by that radius.

The properties here stated have heretofore been used by most writers upon trigonometry as definitions, but without limiting the radius to unity; and it is evidently from this mode of viewing these functions that they have derived their names.

22. Besides the functions already defined, others have been occasionally employed to facilitate particular calculations, as the *versed sine*, which in the circle is the portion of the diameter intercepted between the extremity of the arc and the foot of the sine; thus, Fig. 6, the versed sine of AB is AC, or the radius being $= 1$,

$$\text{versin}\, x = 1 - \cos x \tag{4}$$

by means of which formula we may always substitute versed sines for cosines, and reciprocally.

The *coversed sine* (covers.) is the versed sine of the complement, and *suversed sine* (suvers.) is the versed sine of the supplement.

The *chords* of arcs have also been used, and may be substituted for sines by the formula

$$\text{ch}\, x = 2 \sin \tfrac{1}{2} x \tag{5}$$

which is evident from Fig. 6, where if the arc $BB' = x$, we have chord $BB' = 2\,BC = 2 \sin AB$.

23. From what has now been stated, the student will perceive that angles are to be subjected to computation by means of the quantities sine, cosine, &c., commonly designated by the comprehensive term *trigonometric functions*.* It becomes necessary, therefore, for the computer to know the values of these functions for any given value of the angle. The *trigonometric tables* contain these values for every minute, and sometimes for every second, from 0° to 90°; and with these tables all the numerical computations of trigonometry are carried on. In practice, then, we are not required to compute the functions themselves, and we shall therefore defer the methods for that purpose to a subsequent part of this work, and proceed at once with the investigation of the formulæ and methods by which these tables are rendered available.

Fundamental Formulæ.

24. *Given the sine of an angle, to find the cosine.*

From the right triangle ABC, Fig. 7, we have by geometry $a^2 + b^2 = c^2$

Dividing by c^2, this equation becomes

$$\frac{a^2}{c^2} + \frac{b^2}{c^2} = 1$$

Fig. 7.

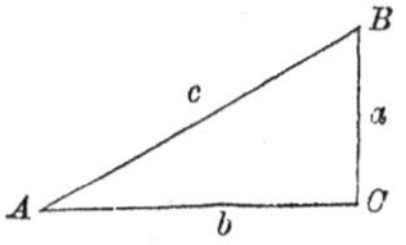

* Also *trigonometric lines*, from the properties explained in Art. 21.

or, by the definitions of sine and cosine (1),

$$\sin^2 A + \cos^2 A = 1 \qquad (6)$$

in which the notation $\sin^2 A$ signifies "the square of the sine of A." From this formula, if the sine is given, we find

$$\cos^2 A = 1 - \sin^2 A = (1 + \sin A)(1 - \sin A)$$

$$\cos A = \sqrt{(1 - \sin^2 A)} = \sqrt{[(1 + \sin A)(1 - \sin A)]} \qquad (7)$$

and if the cosine is given, we find

$$\sin A = \sqrt{(1 - \cos^2 A)} = \sqrt{[(1 + \cos A)(1 - \cos A)]} \qquad (8)$$

25. *Given the sine and cosine of an angle, to find the tangent.*

By (1) we have

$$\tan A = \frac{a}{b}$$

also

$$\frac{\sin A}{\cos A} = \frac{a}{c} \div \frac{b}{c} = \frac{a}{b}$$

therefore

$$\tan A = \frac{\sin A}{\cos A} \qquad (9)$$

And since the cotangent is the reciprocal of the tangent,

$$\cot A = \frac{\cos A}{\sin A} \qquad (10)$$

26. *Given the tangent of an angle, to find the secant.*

The right triangle ABC, Fig. 7, gives

$$c^2 = b^2 + a^2$$

Dividing by b^2, this becomes

$$\frac{c^2}{b^2} = 1 + \frac{a^2}{b^2}$$

or, by the definitions of secant and tangent (1),

$$\sec^2 A = 1 + \tan^2 A \qquad (11)$$

This formula applied to the complement of A gives

$$\operatorname{cosec}^2 A = 1 + \cot^2 A \qquad (12)$$

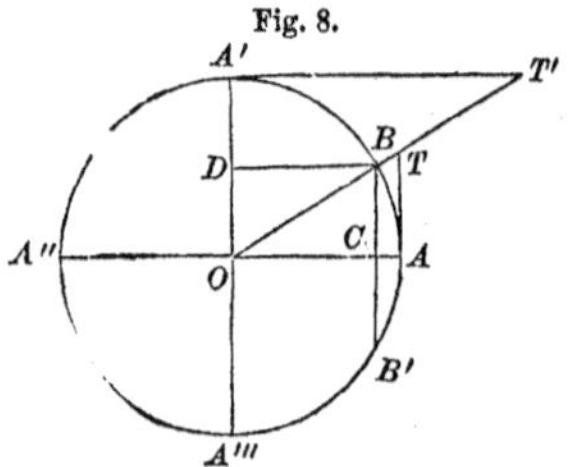

Fig. 8.

27. The preceding formulæ are also directly obtained from Fig. 8. If the angle AOB, or the arc AB, be denoted by x, the right triangle OBC, gives

$$BC^2 + OC^2 = OB^2$$

or remembering that the radius is unity, by Art. 21,

$$\sin^2 x + \cos^2 x = 1 \qquad (13)$$

The triangle OBC gives by the definition, Art. 16,

$$\tan AOB = \frac{BC}{OC}$$

or

$$\tan x = \frac{\sin x}{\cos x} \tag{14}$$

Since the angle BOD is the complement of BOC, $\tan BOD = \cot x$, and the triangle BOD gives

$$\tan BOD = \frac{BD}{OD} = \frac{OC}{BC}$$

or

$$\cot x = \frac{\cos x}{\sin x} \tag{15}$$

In a similar manner the triangles AOT, $A'OT'$ give

$$\sec^2 x = 1 + \tan^2 x \tag{16}$$

$$\operatorname{cosec}^2 x = 1 + \cot^2 x \tag{17}$$

28. The following equations are easily demonstrated by combining (13), (14), (15), (16), (17), and employing the property of the reciprocals (2). They are of frequent use.

$$\sin x = \frac{1}{\operatorname{cosec} x} = \tan x \cos x = \frac{\tan x}{\sec x} = \frac{\cos x}{\cot x} \tag{18}$$

$$\cos x = \frac{1}{\sec x} = \cot x \sin x = \frac{\cot x}{\operatorname{cosec} x} = \frac{\sin x}{\tan x} \tag{19}$$

$$\sin x = \sqrt{(1 - \cos^2 x)}, \qquad \cos x = \sqrt{(1 - \sin^2 x)} \tag{20}$$

$$\sec x = \sqrt{(1 + \tan^2 x)}, \qquad \operatorname{cosec} x = \sqrt{(1 + \cot^2 x)} \tag{21}$$

$$\tan x = \sqrt{(\sec^2 x - 1)}, \qquad \cot x = \sqrt{(\operatorname{cosec}^2 x - 1)} \tag{22}$$

$$\sin x = \frac{\tan x}{\sqrt{(1 + \tan^2 x)}} = \frac{1}{\sqrt{(1 + \cot^2 x)}} \tag{23}$$

$$\cos x = \frac{\cot x}{\sqrt{(1 + \cot^2 x)}} = \frac{1}{\sqrt{(1 + \tan^2 x)}} \tag{24}$$

$$\tan x = \frac{\sin x}{\sqrt{(1 - \sin^2 x)}} = \frac{\sqrt{(1 - \cos^2 x)}}{\cos x} \tag{25}$$

$$\cot x = \frac{\cos x}{\sqrt{(1 - \cos^2 x)}} = \frac{\sqrt{(1 - \sin^2 x)}}{\sin x} \tag{26}$$

29. *To find the sine, &c. of* 30° *and* 60°.

In Fig. 8, let the arc $AB = 30°$, and $BB' = 2\,AB = 60°$. By Art. 21, $\sin AB = BC$, and by geometry the chord of 60°, or of one-sixth of the circumference, is equal to the radius $= 1$; therefore

$$2 \sin 30° = 2\,BC = BB' = 1$$

whence

$$\sin 30° = \tfrac{1}{2} = \cos 60° \tag{27}$$

and by (7)

$$\cos 30° = \sqrt{[(1+\tfrac{1}{2})(1-\tfrac{1}{2})]} = \sqrt{(\tfrac{3}{2}\times\tfrac{1}{2})}$$

whence

$$\cos 30° = \tfrac{1}{2}\sqrt{3} = \sin 60° \tag{28}$$

Then, by (9) and (2),

$$\tan 30° = \frac{\sin 30°}{\cos 30°} = \frac{\tfrac{1}{2}}{\tfrac{1}{2}\sqrt{3}} = \frac{1}{\sqrt{3}} = \cot 60° \tag{29}$$

$$\cot 30° = \frac{1}{\tan 30°} = \sqrt{3} = \tan 60° \tag{30}$$

$$\sec 30° = \frac{1}{\cos 30°} = \frac{2}{\sqrt{3}} = \operatorname{cosec} 60° \tag{31}$$

$$\operatorname{cosec} 30° = \frac{1}{\sin 30°} = 2 = \sec 60° \tag{32}$$

30. *To find the sine, &c. of* 45° Since 45° is the complement of 45°, we have

$$\sin 45° = \cos 45°$$

whence by (13), putting $x = 45°$,

$$\sin^2 45° + \cos^2 45° = 2\sin^2 45° = 2\cos^2 45° = 1$$

$$\sin^2 45° = \cos^2 45° = \tfrac{1}{2}$$

$$\sin 45° = \cos 45° = \sqrt{\tfrac{1}{2}} = \tfrac{1}{2}\sqrt{2} \tag{33}$$

$$\tan 45° = \cot 45° = \frac{\sin 45°}{\cos 45°} = 1 \tag{34}$$

$$\sec 45° = \operatorname{cosec} 45° = \frac{1}{\sin 45°} = \sqrt{2} \tag{35}$$

Fig. 9.

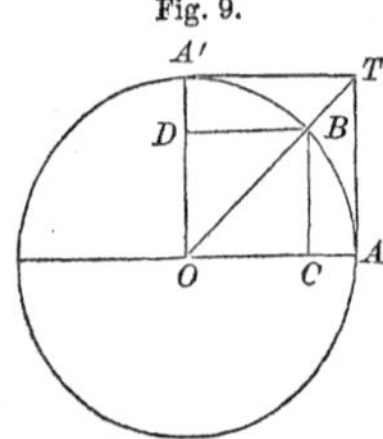

These values are readily verified in the circle, Fig. 9, where $OATA'$ is a square described upon the radius. The diagonal OT bisects the right angle, whence $AOT = 45°$, and $\tan 45° = AT = OA = 1$; $\cot 45° = A'T = 1$; $\sin 45° = BC = OC = \cos 45°$, &c.

31. *The sines and cosines of two angles being given, to find the sine and cosine of the sum, and the sine and cosine of the difference of those angles.*

Fig. 10. Fig. 11.

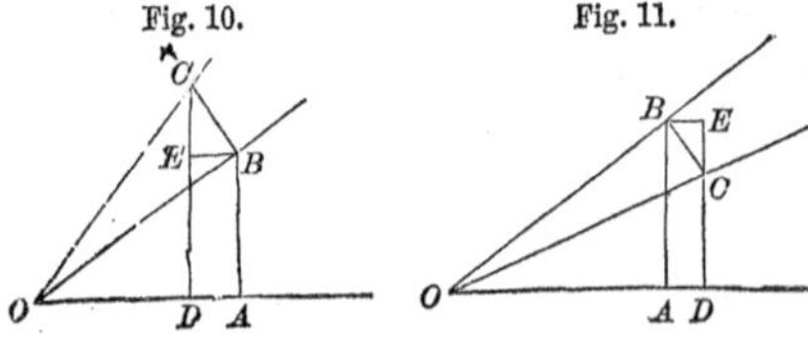

Let the two angles be AOB and BOC, Figs. 10 and 11. At any point B in the line OB draw BC perp. to OB. Draw BA and CD perp. to OA, and BE perp. to CD.

Then the triangles BCE and BOA are mutually equiangular, the three sides of the one being perp. to the three sides of the other respectively; therefore the angle $BCE = AOB$.

Let $x = AOB = BCE$
$y = BOC$

Then, Fig. 10, $x + y = COD$
Fig. 11, $x - y = COD$

and in

Fig. 10, $\sin(x+y) = \frac{CD}{CO} = \frac{BA + CE}{CO} = \frac{BA}{CO} + \frac{CE}{CO}$

Fig. 11, $\sin(x-y) = \frac{CD}{CO} = \frac{BA - CE}{CO} = \frac{BA}{CO} - \frac{CE}{CO}$

and in both figures

$$\frac{BA}{CO} = \frac{BA}{BO} \times \frac{BO}{CO} = \sin x \cos y$$

$$\frac{CE}{CO} = \frac{CE}{CB} \times \frac{CB}{CO} = \cos x \sin y$$

which being substituted in the above expressions of $\sin(x+y)$ and $\sin(x-y)$ give

$$\sin(x+y) = \sin x \cos y + \cos x \sin y \qquad (36)$$
$$\sin(x-y) = \sin x \cos y - \cos x \sin y \qquad (37)$$

Again in

Fig. 10, $\cos(x+y) = \frac{OD}{OC} = \frac{OA - EB}{OC} = \frac{OA}{OC} - \frac{EB}{OC}$

Fig. 11, $\cos(x-y) = \frac{OD}{OC} = \frac{OA + EB}{OC} = \frac{OA}{OC} + \frac{EB}{OC}$

and in both figures,

$$\frac{OA}{OC} = \frac{OA}{OB} \times \frac{OB}{OC} = \cos x \cos y$$

$$\frac{EB}{OC} = \frac{EB}{BC} \times \frac{BC}{OC} = \sin x \sin y$$

therefore

$$\cos(x+y) = \cos x \cos y - \sin x \sin y \qquad (38)$$
$$\cos(x-y) = \cos x \cos y + \sin x \sin y \qquad (39)$$

and (36), (37), (38), and (39) are the required formulæ.

These may be considered as the fundamental formulæ of the trigonometric analysis, and will form the basis of our subsequent investigations. They are equally applicable to arcs represented by x and y (Art. 20).

CHAPTER III.

TRIGONOMETRIC FUNCTIONS OF ANGULAR MAGNITUDE IN GENERAL.

32. THE definitions of sine, &c. given in the preceding chapter apply only to acute angles, since the angle is there assumed to be one of the oblique angles of a right triangle. But we shall now take a more general view of angular magnitude and of the functions by means of which it is subjected to computation.

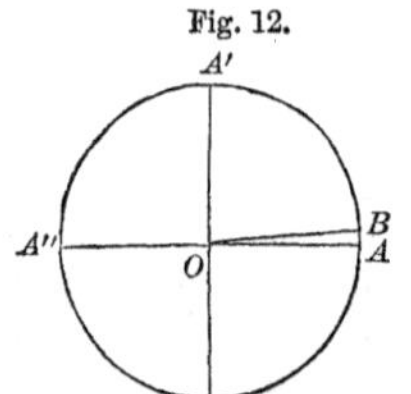

Fig. 12.

If, Fig. 12, we suppose the line OA to revolve from the position OA to OA' in the direction of the arc AA' (or from right to left), it will describe an angular magnitude of 90°; when it arrives at OA'' it will have described an angular magnitude of 180°; at OA''', 270°; and at OA again, 360°. If it now continue its revolution, when it arrives at OA' again, it will have described an angular magnitude of $360° + 90°$, or 450°; and thus we may readily conceive of an angular magnitude of any number of degrees. In like manner we may have arcs equal to or greater than one, two, or more circumferences.

To obtain trigonometric functions for angles and arcs thus generally considered, we shall avail ourselves of the fundamental formulæ established in the preceding chapter; first deducing their values analytically, and then explaining their geometrical signification.

33. *To find the sine, &c. of* 0° *and* 90°. In (37) and (39) let $x = y$; the first members become $\sin(x - x) = \sin 0°$, and $\cos(x - x) = \cos 0°$; and by (13) they are reduced to

$$\begin{aligned} \sin 0° &= \sin x \cos x - \cos x \sin x = 0 \\ \cos 0° &= \cos^2 x + \sin^2 x = 1 \end{aligned}$$

and since 0° and 90° are complements of each other, Art. 12,

$$\sin 0° = \cos 90° = 0 \tag{40}$$

$$\cos 0° = \sin 90° = 1 \tag{41}$$

from which by (9) and (2)

$$\tan 0° = \cot 90° = \frac{\sin 0°}{\cos 0°} = \frac{0}{1} = 0 \tag{42}$$

$$\cot 0^\circ = \tan 90^\circ = \frac{1}{\tan 0^\circ} = \frac{1}{0} = \infty \tag{43}$$

$$\sec 0^\circ = \operatorname{cosec} 90^\circ = \frac{1}{\cos 0^\circ} = \frac{1}{1} = 1 \tag{44}$$

$$\operatorname{cosec} 0^\circ = \sec 90^\circ = \frac{1}{\sin 0^\circ} = \frac{1}{0} = \infty \tag{45}$$

34. *To find the sine, &c. of* 180°. In (36) and (38) let $x = y = 90^\circ$; these equations become by means of the preceding values

$$\sin 180^\circ = 1 \times 0 + 0 \times 1 = 0 \tag{46}$$

$$\cos 180^\circ = 0 \times 0 - 1 \times 1 = -1 \tag{47}$$

whence by (9) and (2)

$$\tan 180^\circ = \frac{0}{-1} = 0 \qquad \cot 180^\circ = \frac{1}{0} = \infty \tag{48}$$

$$\sec 180^\circ = \frac{1}{-1} = -1 \qquad \operatorname{cosec} 180^\circ = \frac{1}{0} = \infty \tag{49}$$

35. *To find the sine, &c. of* 270°. In (36) and (38) let $x = 180^\circ$, $y = 90^\circ$, then

$$\sin 270^\circ = 0 \times 0 + (-1) \times 1 = -1 \tag{50}$$

$$\cos 270^\circ = (-1) \times 0 - 0 \times 1 = 0 \tag{51}$$

$$\tan 270^\circ = \frac{-1}{0} = \infty \qquad \cot 270^\circ = \frac{1}{\infty} = 0 \tag{52}$$

$$\sec 270^\circ = \frac{1}{0} = \infty \qquad \operatorname{cosec} 270^\circ = \frac{1}{-1} = -1 \tag{53}$$

36. *To find the sine, &c. of* 360°. In (36) and (38) let $x = y = 180^\circ$; then

$$\sin 360^\circ = 0 \times (-1) + (-1) \times 0 = 0 \tag{54}$$

$$\cos 360^\circ = (-1) \times (-1) - 0 \times 0 = 1 \tag{55}$$

the same values as for 0°, whence it follows that all the trig. functions of 360° are the same as those of 0°.

The same process continued will give for 450° (= 360° + 90°), the same trig. functions as those of 90°; for 540° the same functions as for 180°, &c.

37. The preceding values now furnish us at once with the values of the functions for all possible values of the angle. In (36) and (38) let $x = 0°$, they are reduced to

$$\sin y = \sin 0° \cos y + \cos 0° \sin y = \sin y$$

$$\cos y = \cos 0° \cos y - \sin 0° \sin y = \cos y$$

which are simply identical equations, and reveal no new property. But if in (37) and (39) we put $x = 0°$, we have, after substituting the functions of 0°,

$$\sin(-y) = -\sin y \qquad \cos(-y) = \cos y \tag{56}$$

whence by (9) and (2)

$$\tan(-y) = \frac{\sin(-y)}{\cos(-y)} = \frac{-\sin y}{\cos y} = -\tan y \tag{57}$$

$$\cot(-y) = \frac{\cos(-y)}{\sin(-y)} = \frac{\cos y}{-\sin y} = -\cot y \tag{58}$$

$$\sec(-y) = \frac{1}{\cos(-y)} = \frac{1}{\cos y} = \sec y \tag{59}$$

$$\text{cosec}(-y) = \frac{1}{\sin(-y)} = \frac{1}{-\sin y} = -\text{cosec}\, y \tag{60}$$

or, *the sin., tan., cot., and cosec. of the negative of an angle are the negative of the sin., tan., cot., and cosec. of the angle itself; and the cos. and sec. are the same as those of the angle itself.*

38. In (37) and (39) let $x = 90°$; we find after reduction

$$\sin(90° - y) = \cos y \qquad \cos(90° - y) = \sin y$$

which agree with the definition of cosine, but give no new relations. But in (36) and (38) let $x = 90°$, we find

$$\sin(90° + y) = \cos y, \qquad \cos(90° + y) = -\sin y \tag{61}$$

whence by (9) and (2),

$$\tan(90° + y) = -\cot y \qquad \cot(90° + y) = -\tan y \tag{62}$$

$$\sec(90° + y) = -\text{cosec}\, y \qquad \text{cosec}(90° + y) = \sec y \tag{63}$$

or, *the sin. and cosec. of an angle are equal to the cos. and sec. of the excess of the angle above 90°; and the cos., tan., cot., and sec. are equal to the negatives of the sin., cot., tan., and cosec. of the excess of the angle above 90°.*

39. In (37) and (39) let $x = 180°$; we find

$$\sin(180° - y) = \sin y \qquad \cos(180° - y) = -\cos y \quad (64)$$
$$\tan(180° - y) = -\tan y \qquad \cot(180° - y) = -\cot y \quad (65)$$
$$\sec(180° - y) = -\sec y \qquad \text{cosec}(180° - y) = \text{cosec}\, y \quad (66)$$

or, *the sin. and cosec. of the supplement of an angle are the same as those of the angle itself; and the cos., tan., cot., and sec. are the negative of those of the angle itself.*

40. If y is acute (that is, less than 90°), all its trig. functions are positive; and since its supplement $180° - y$ is obtuse (that is, greater than 90°), it follows from the preceding article, that *the sin. and cosec. of an obtuse angle are positive, while its cos., tan.. cot., and sec. are negative.*

41. In (36) and (38) let $x = 180°$; we find

$$\sin(180° + y) = -\sin y \qquad \cos(180° + y) = -\cos y \quad (67)$$
$$\tan(180° + y) = \tan y \qquad \cot(180° + y) = \cot y \quad (68)$$
$$\sec(180° + y) = -\sec y \qquad \text{cosec}(180° + y) = -\text{cosec}\, y \quad (69)$$

by means of which, if y is acute, we obtain the values of the sines, &c. of angles between 180° and 270°.

42. In (37) and (39) let $x = 270°$; we find

$$\sin(270° - y) = -\cos y \qquad \cos(270° - y) = -\sin y \quad (70)$$
$$\tan(270° - y) = \cot y \qquad \cot(270° - y) = \tan y \quad (71)$$
$$\sec(270° - y) = -\text{cosec}\, y \qquad \text{cosec}(270° - y) = -\sec y \quad (72)$$

43. In (36) and (38) let $x = 270°$; we find

$$\sin(270° + y) = -\cos y \qquad \cos(270° + y) = \sin y \quad (73)$$
$$\tan(270° + y) = -\cot y \qquad \cot(270° + y) = -\tan y \quad (74)$$
$$\sec(270° + y) = \text{cosec}\, y \qquad \text{cosec}(270° + y) = -\sec y \quad (75)$$

44. In (37) and (39) let $x = 360°$; we find

$$\sin(360° - y) = -\sin y \qquad \cos(360° - y) = \cos y \quad (76)$$
$$\tan(360° - y) = -\tan y \qquad \cot(360° - y) = -\cot y \quad (77)$$
$$\sec(360° - y) = \sec y \qquad \text{cosec}(360° - y) = -\text{cosec}\, y \quad (78)$$

or the functions of $360° - y$ are the same as those of $-y$ (Art. 37).

45. In (36) and (38) let $x = 360°$; we find

$$\sin(360° + y) = \sin y \qquad \cos(360° + y) = \cos y \quad (79)$$

or, *the functions of an angle which exceeds* 360° *are the same as those of the excess above* 360°.

It follows that the functions of $720° + y$ are the same as those of $360° + y$, and therefore the same as those of y; and in like manner for an angle which exceeds any multiple of $360°$.

46. Since $y - 90°$ is the negative of $90° - y$, we obtain from Art. 37,

$$\left.\begin{aligned} \sin(y - 90°) &= -\sin(90° - y) = -\cos y \\ \cos(y - 90°) &= \cos(90° - y) = \sin y \end{aligned}\right\} \quad (80)$$

whence also tan., &c.; and in the same manner we may find the functions of $y - 180°$, $y - 270°$, $y - 360°$, &c.

47. We shall now give the geometrical interpretation of the preceding results.

Fig. 13.

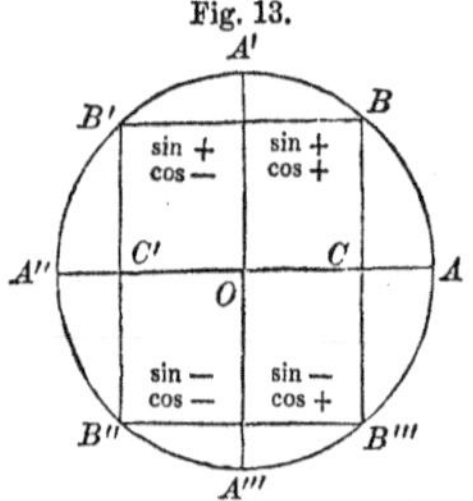

In Fig. 13, let the radius revolve from the position OA to OA', OA'', &c., as in Art. 32, thus describing a continuously increasing angular magnitude; or, which is equivalent, let the arc commencing at A increase continuously to AB, AA', AB', &c. Then the changes in the values of the several trigonometric lines may be traced as follows.

1st. *The sine* being, by Art. 21, the perpendicular from one extremity of the arc upon the diameter drawn through the other extremity, we shall have $\sin AB = BC$, $\sin AB' = B'C'$, $\sin AA''B'' = B''C'$, $\sin AA''B''' = B'''C$, and if we make

$$AB = A''B' = A''B'' = AB''' = y$$

we have

$$\begin{aligned} \sin y &= BC \\ \sin(180° - y) &= B'C' \\ \sin(180° + y) &= B''C' \\ \sin(360° - y) &= B'''C \end{aligned}$$

The lines BC, $B'C'$, $B''C'$, $B'''C$, however, represent only the *numerical* values of the sines, and are here equal. But the results above obtained from our formulæ enable us to distinguish between them by means of their algebraic signs. Thus, by (64), (67), (76),

$$\begin{aligned} \sin(180° - y) &= \sin y \\ \sin(180° + y) &= -\sin y \\ \sin(360° - y) &= -\sin y \end{aligned}$$

so that the sines from 0° to 180° are positive, while those from 180° to 360° are negative; or the sines which are *above* the diameter AA'' are positive, while those which are *below* this diameter are negative; or still more generally, the sines that have *opposite directions*, with reference to the fixed diameter from which they are measured, have *opposite signs.*

2d. *The cosine* being, by Art. 21, the distance from the center to the foot of the sine, we have

$$\cos y = O\,C$$
$$\cos(180° - y) = O\,C'$$
$$\cos(180° + y) = O\,C'$$
$$\cos(360° - y) = O\,C$$

but by (64), (67), (76),

$$\cos(180° - y) = -\cos y$$
$$\cos(180° + y) = -\cos y$$
$$\cos(360° - y) = \cos y$$

so that the cosines on the *right* of the diameter $A'A'''$ are positive, while those on the *left* of this diameter are negative; or rather the cosines that have *opposite directions*, with reference to the diameter from which they are measured, have *opposite signs.*

We have here only exhibited a well-known principle in the application of analysis to geometry, viz.: *that all lines measured in opposite directions from a fixed line have opposite signs.*

To interpret the results (56), it is only necessary to observe that a negative arc will be one reckoned from A towards B''', or in the opposite direction to that of the positive arc, so that

$$\sin A\,B''' = \sin(-y) = B'''\,C = -B\,C = -\sin y$$
$$\cos A\,B''' = \cos(-y) = O\,C = \cos y$$

as in (56).

The same principle applies to the tangents, but it will be simpler in practice to obtain their signs (as also those of the secants), analytically, from those of the sine and cosine, as has been already shown. It will be sufficient to bear in mind the following table, which is also expressed by Fig. 13.

	1st QUAD.	2d QUAD.	3d QUAD.	4th QUAD.
SINE	+	+	−	−
COSINE	+	−	−	+

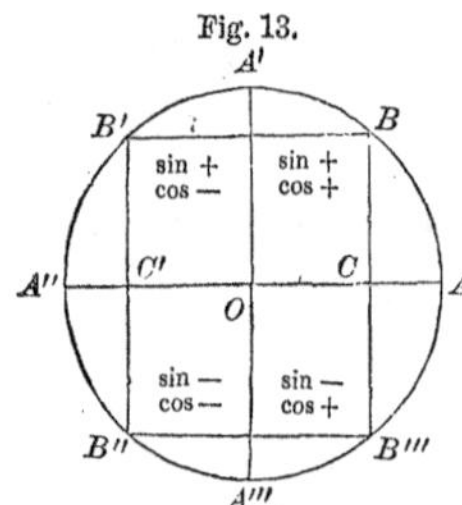

48. The particular values of the sine and cosine at A, A', A'', &c., or sin. and cos. of $0°$, $90°$, $180°$, &c., may also be found by Fig. 13, upon the same principles; but this we leave to the student.

49. General Remark.—In the demonstration of the fundamental formulæ for $\sin(x \pm y)$, and $\cos(x \pm y)$, Art. 31, the angles x, y and $x \pm y$ were all taken less than $90°$ and positive. In this chapter these formulæ have been applied to angles of any magnitude, and the resulting functions have been shown to take opposite signs when the lines representing them take opposite directions. It follows that, in deducing trigonometric formulæ from geometrical figures, we need not embarrass our demonstrations with the consideration of the various cases of the problem, or of the various values of the angles of the figure. The formula deduced from any supposed position of the lines of the figure will be of general application, provided in the practical application of this formula to the particular cases, we observe those values and signs of the trigonometric functions which have now been determined.

50. The results of this chapter may be expressed by a few general formulæ. From (79) it appears that all the trigonometric functions return to the same values after one or more complete revolutions of $360°$. If we represent the semi-circumference, or two right angles, by π (Art. 11), and let $n =$ any whole number or zero, we shall have

$$\sin 4n\frac{\pi}{2} = 0 \qquad \cos 4n\frac{\pi}{2} = 1 \tag{81}$$

$$\sin(4n+1)\frac{\pi}{2} = 1 \qquad \cos(4n+1)\frac{\pi}{2} = 0 \tag{82}$$

$$\sin(4n+2)\frac{\pi}{2} = 0 \qquad \cos(4n+2)\frac{\pi}{2} = -1 \tag{83}$$

$$\sin(4n+3)\frac{\pi}{2} = -1 \qquad \cos(4n+3)\frac{\pi}{2} = 0 \tag{84}$$

whence

$$\tan 4n\frac{\pi}{2} = 0 \qquad \tan(4n+1)\frac{\pi}{2} = \infty$$

$$\tan(4n+2)\frac{\pi}{2} = 0 \qquad \tan(4n+3)\frac{\pi}{2} = \infty$$

or the tan. of the even multiples of $\frac{\pi}{2} = 0$, and of the odd multiples $= \infty$, so that we may write more simply

$$\tan 2n\frac{\pi}{2} = 0 \qquad \tan(2n+1)\frac{\pi}{2} = \infty \tag{85}$$

In these formulæ we have only to give n one of the values 0, 1, 2, 3, 4, &c., to obtain the functions of any given multiple of the right angle. Thus, we find

$\sin 450^\circ = \sin 5\,\frac{\pi}{2} = \sin (4+1)\,\frac{\pi}{2} = 1$ by making $n = 1$, in (82).

Since the subtraction of $8\,n\,\frac{\pi}{2}$ from the arc will not change the functions, the above formulæ are also true when n is a negative whole number.

51. In a similar manner we obtain

$$\sin\left[4\,n\,\frac{\pi}{2}+y\right]=\sin y \qquad \cos\left[4\,n\,\frac{\pi}{2}+y\right]=\cos y \qquad (86)$$

$$\sin\left[(4\,n+1)\,\frac{\pi}{2}+y\right]=\cos y \qquad \cos\left[(4\,n+1)\,\frac{\pi}{2}+y\right]=-\sin y \quad (87)$$

$$\sin\left[(4\,n+2)\,\frac{\pi}{2}+y\right]=-\sin y \qquad \cos\left[(4\,n+2)\,\frac{\pi}{2}+y\right]=-\cos y \quad (88)$$

$$\sin\left[(4\,n+3)\,\frac{\pi}{2}+y\right]=-\cos y \qquad \cos\left[(4\,n+3)\,\frac{\pi}{2}+y\right]=\sin y \qquad (89)$$

$$\tan\left[2\,n\,\frac{\pi}{2}+y\right]=\tan y \qquad \tan\left[(2\,n+1)\,\frac{\pi}{2}+y\right]=-\cot y \quad (90)$$

in which n may be any whole number, positive or negative, and y any angle, positive or negative.

52. A still more concise form may be given to the formulæ of the two preceding articles, as follows: n being, as before, any whole number, positive or negative.

$$\sin 2\,n\,\frac{\pi}{2}=0 \qquad \cos 2\,n\,\frac{\pi}{2}=(-1)^n \qquad (91)$$

$$\sin (2\,n+1)\,\frac{\pi}{2}=(-1)^n \qquad \cos (2\,n+1)\,\frac{\pi}{2}=0 \qquad (92)$$

$$\sin\left[2\,n\,\frac{\pi}{2}+y\right]=(-1)^n\sin y \qquad \cos\left[2\,n\,\frac{\pi}{2}+y\right]=(-1)^n\cos y \quad (93)$$

$$\sin\left[(2n+1)\,\frac{\pi}{2}+y\right]=(-1)^n\cos y \qquad \cos\left[(2n+1)\,\frac{\pi}{2}+y\right]=-(-1)^n\sin y \quad (94)$$

and from these (85) and (90) may be directly deduced.

53. We have seen that an angle being given, there is but one corresponding sine. On the other hand, a sine being given, there is an indefinite number of angles corresponding; for if a denote the given sine, and y any corresponding angle, then a is also the sine of all the angles

$$\pi - y, \qquad 2\,\pi + y, \qquad 3\,\pi - y, \text{ \&c.}$$
$$-\pi - y, \qquad -2\,\pi + y, \qquad -3\,\pi - y, \text{ \&c.}$$

or in general

$$a = \sin y = \sin\left[n\,\pi + (-1)^n\,y\right] \qquad (95)$$

In like manner if a is a given cosine, and y any corresponding angle,

$$a = \cos y = \cos (2 n \pi \pm y) \tag{96}$$

and if a is a given tangent corresponding to the angle y,

$$a = \tan y = \tan (n \pi + y) \tag{97}$$

SINE AND TANGENT OF A SMALL ANGLE OR ARC.

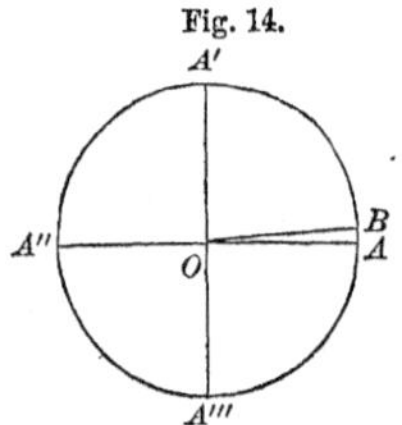

Fig. 14.

54. When the angle $AOB = x$, Fig. 14, is very small, the sine and tangent are *very nearly* equal to the arc AB, which measures the angle, the radius being unity; and the cosine and secant are nearly equal to $OA = 1$ (Art. 21). Therefore, to find the sine or tangent of a very small angle *approximately*, we have only to find the length of the arc by Art. 9; thus

$$\sin 1'' = \text{arc } 1'' = 0{\cdot}000004848137$$

$$\log. \sin 1'' = 4{\cdot}6855749$$

and x being a small angle, or arc, expressed in seconds,

$$\sin x = \tan x = x \sin 1'' \tag{98}$$

If x is expressed in minutes,

$$\sin x = \tan x = x \sin 1' \tag{99}$$

If x expresses the length of the arc, the radius being unity,

$$\sin x = \tan x = x \tag{100}$$

The employment of these approximate values must be governed by the degree of accuracy required in a particular application. It is found, for example, that they are sufficiently accurate when the nearest second only is required in our results, provided the angle does not much exceed 1°.

55. If x and y are any two small angles, it follows from the preceding article that

$$\sin x : \sin y = x \sin 1'' : y \sin 1'' = x : y$$

that is, *the sines (or tangents) of small angles are proportional to the angles themselves.* The application of this theorem, however, like that of the preceding, must depend upon the accuracy required in the problem in which it is employed.*

* For a full discussion of the limits under which this theorem may be employed, see a paper, by the author of this work, in the Astronomical Journal, (Cambridge, Mass.) Vol. i. p. 84.

CHAPTER IV.

GENERAL FORMULÆ.

56. WE have already obtained four fundamental equations, (36), (37), (38), and (39), involving two angles, x and y. From these we shall now deduce a number of formulæ, either required in the subsequent parts of this work, or of general utility in the applications of trigonometry.

57. The sum and difference of the equations (36) and (37) are

$$\sin(x+y)+\sin(x-y)=2\sin x\cos y \tag{101}$$

$$\sin(x+y)-\sin(x-y)=2\cos x\sin y \tag{102}$$

and the sum and difference of (38) and (39) are

$$\cos(x+y)+\cos(x-y)=2\cos x\cos y \tag{103}$$

$$\cos(x+y)-\cos(x-y)=-2\sin x\sin y \tag{104}$$

58. If we put

$$x+y=x'$$

$$x-y=y'$$

whence

$$2x=x'+y', \qquad x=\tfrac{1}{2}(x'+y')$$

$$2y=x'-y', \qquad y=\tfrac{1}{2}(x'-y')$$

equation (101) will become

$$\sin x'+\sin y'=2\sin\tfrac{1}{2}(x'+y')\cos\tfrac{1}{2}(x'-y')$$

and (102), (103), and (104) admit of a similar transformation. But since x' and y' admit of all varieties of value, we may omit the accents and apply the formulæ to any two angles x and y; we have thus

$$\sin x+\sin y=2\sin\tfrac{1}{2}(x+y)\cos\tfrac{1}{2}(x-y) \tag{105}$$

$$\sin x-\sin y=2\cos\tfrac{1}{2}(x+y)\sin\tfrac{1}{2}(x-y) \tag{106}$$

$$\cos x+\cos y=2\cos\tfrac{1}{2}(x+y)\cos\tfrac{1}{2}(x-y) \tag{107}$$

$$\cos x-\cos y=-2\sin\tfrac{1}{2}(x+y)\sin\tfrac{1}{2}(x-y) \tag{108}$$

Each of these equations may be enunciated as a theorem; thus

(105) expresses that "the sum of the sines of any two angles is equal to twice the sine of half the sum of the angles multiplied by the cosine of half their difference."

These formulæ are of frequent use (especially in computations performed by logarithms), in transforming a sum or difference into a product.

59. Dividing (105) by (106), we have by (14) and (15)

$$\frac{\sin x + \sin y}{\sin x - \sin y} = \tan \tfrac{1}{2}(x + y) \cot \tfrac{1}{2}(x - y)$$

or by (2)

$$\frac{\sin x + \sin y}{\sin x - \sin y} = \frac{\tan \frac{1}{2}(x + y)}{\tan \frac{1}{2}(x - y)} \tag{109}$$

and from (107) and (108) we find in the same manner

$$\frac{\cos x - \cos y}{\cos x + \cos y} = -\tan \tfrac{1}{2}(x + y) \tan \tfrac{1}{2}(x - y) \tag{110}$$

We find also

$$\frac{\sin x + \sin y}{\cos x + \cos y} = \tan \tfrac{1}{2}(x + y) \tag{111}$$

$$\frac{\sin x - \sin y}{\cos x + \cos y} = \tan \tfrac{1}{2}(x - y) \tag{112}$$

$$\frac{\sin x + \sin y}{\cos x - \cos y} = -\cot \tfrac{1}{2}(x - y) \tag{113}$$

$$\frac{\sin x - \sin y}{\cos x - \cos y} = -\cot \tfrac{1}{2}(x + y) \tag{114}$$

60. Divide the equations (36), (37), (38) and (39) by $\cos x \cos y$; then by (14) we have

$$\frac{\sin (x + y)}{\cos x \cos y} = \tan x + \tan y \tag{115}$$

$$\frac{\sin (x - y)}{\cos x \cos y} = \tan x - \tan y \tag{116}$$

$$\frac{\cos (x + y)}{\cos x \cos y} = 1 - \tan x \tan y \tag{117}$$

$$\frac{\cos (x - y)}{\cos x \cos y} = 1 + \tan x \tan y \tag{118}$$

61. Divide (36), (37), (38) and (39) by $\sin x \sin y$; and by $\sin x$ $\cos y$; then

$$\frac{\sin(x \pm y)}{\sin x \sin y} = \cot y \pm \cot x \tag{119}$$

$$\frac{\cos(x \pm y)}{\sin x \sin y} = \cot x \cot y \mp 1 \tag{120}$$

$$\frac{\sin(x \pm y)}{\sin x \cos y} = 1 \pm \cot x \tan y \tag{121}$$

$$\frac{\cos(x \pm y)}{\sin x \cos y} = \cot x \mp \tan y \tag{122}$$

62. Divide (115) by (117), and (116) by (118); then by (14)

$$\tan(x + y) = \frac{\tan x + \tan y}{1 - \tan x \tan y} \tag{123}$$

$$\tan(x - y) = \frac{\tan x - \tan y}{1 + \tan x \tan y} \tag{124}$$

by which, when the tangents of two angles are given, we may compute the tangent of their sum or difference. To find the cotangent of the sum or difference when the cotangents of the angles are given, divide (120) by (119),

$$\cot(x \pm y) = \frac{\cot y \cot x \mp 1}{\cot y \pm \cot x} \tag{125}$$

63. Dividing (115) by (116), and (117) by (118), (or from the equations of Art. 61), we have

$$\frac{\sin(x + y)}{\sin(x - y)} = \frac{\tan x + \tan y}{\tan x - \tan y} = \frac{\cot y + \cot x}{\cot y - \cot x} \tag{126}$$

$$\frac{\cos(x + y)}{\cos(x - y)} = \frac{1 - \tan x \tan y}{1 + \tan x \tan y} = \frac{\cot x \cot y - 1}{\cot x \cot y + 1} \tag{127}$$

64. Formulæ for secants are obtained from those for cosines by means of (2); thus we find

$$\sec(x \pm y) = \frac{1}{\cos x \cos y \mp \sin x \sin y}$$

and multiplying numerator and denominator by $\sec x \sec y$,

$$\sec(x \pm y) = \frac{\sec x \sec y}{1 \mp \tan x \tan y} \tag{128}$$

5

Also since

$$\sec x \pm \sec y = \frac{1}{\cos x} \pm \frac{1}{\cos y} = \frac{\cos y \pm \cos x}{\cos x \cos y}$$

we find by (107) and (108)

$$\sec x + \sec y = \frac{2 \cos \frac{1}{2} (x + y) \cos \frac{1}{2} (x - y)}{\cos x \cos y} \tag{129}$$

$$\sec x - \sec y = \frac{2 \sin \frac{1}{2} (x + y) \sin \frac{1}{2} (x - y)}{\cos x \cos y} \tag{130}$$

In the same manner from (105) and (106)

$$\operatorname{cosec} x + \operatorname{cosec} y = \frac{2 \sin \frac{1}{2} (x + y) \cos \frac{1}{2} (x - y)}{\sin x \sin y} \tag{131}$$

$$\operatorname{cosec} x - \operatorname{cosec} y = -\frac{2 \cos \frac{1}{2} (x + y) \sin \frac{1}{2} (x - y)}{\sin x \sin y} \tag{132}$$

These formulæ, although generally omitted in treatises on trigonometry, will be found useful in a subsequent part of this work.

65. The product of (36) and (37), and of (38) and (39), are

$$\sin (x + y) \sin (x - y) = \sin^2 x \cos^2 y - \cos^2 x \sin^2 y$$

$$\cos (x + y) \cos (x - y) = \cos^2 x \cos^2 y - \sin^2 x \sin^2 y$$

By (13) we have $\cos^2 x = 1 - \sin^2 x$ and $\cos^2 y = 1 - \sin^2 y$, which substituted in the preceding equations, give

$$\sin (x + y) \sin (x - y) = \sin^2 x - \sin^2 y = \cos^2 y - \cos^2 x \tag{133}$$

$$\cos (x + y) \cos (x - y) = \cos^2 x - \sin^2 y = \cos^2 y - \sin^2 x \tag{134}$$

66. In (36), (38) and (123), let $y = x$, we find

$$\sin 2x = 2 \sin x \cos x \tag{135}$$

$$\cos 2x = \cos^2 x - \sin^2 x \tag{136}$$

$$\tan 2x = \frac{2 \tan x}{1 - \tan^2 x} \tag{137}$$

by which the functions of the double angle may be found from those of the simple angle.

67. To find the functions of the half angle from those of the whole angle, we have, from (13) and (136),

$$\cos^2 x + \sin^2 x = 1$$

$$\cos^2 x - \sin^2 x = \cos 2x$$

the sum and difference of which are

$$2 \cos^2 x = 1 + \cos 2x$$

$$2 \sin^2 x = 1 - \cos 2x$$

As these express the relations of an angle $2x$ and its half x, their meaning will not be changed by writing x and $\frac{1}{2}x$ instead of $2x$ and x; whence

$$2\cos^2 \tfrac{1}{2}x = 1 + \cos x \tag{138}$$

$$2\sin^2 \tfrac{1}{2}x = 1 - \cos x \tag{139}$$

the quotient of which is

$$\tan^2 \tfrac{1}{2}x = \frac{1 - \cos x}{1 + \cos x} \tag{140}$$

68. The following may be proposed as exercises.

$$\sin x = \frac{2\tan\frac{1}{2}x}{1 + \tan^2\frac{1}{2}x} = \frac{2}{\cot\frac{1}{2}x + \tan\frac{1}{2}x} \tag{141}$$

$$\tan x = \frac{2\tan\frac{1}{2}x}{1 - \tan^2\frac{1}{2}x} = \frac{2}{\cot\frac{1}{2}x - \tan\frac{1}{2}x} \tag{142}$$

$$\tan^2\tfrac{1}{2}x + 2\cot x\tan\tfrac{1}{2}x - 1 = 0 \tag{143}$$

$$\tan^2\tfrac{1}{2}x - 2\operatorname{cosec} x\tan\tfrac{1}{2}x + 1 = 0 \tag{144}$$

$$\cos x = \frac{1 - \tan^2\frac{1}{2}x}{1 + \tan^2\frac{1}{2}x} \tag{145}$$

$$\tan\tfrac{1}{2}x = \operatorname{cosec} x - \cot x = \frac{1 - \cos x}{\sin x} \tag{146}$$

$$\cot\tfrac{1}{2}x = \operatorname{cosec} x + \cot x = \frac{1 + \cos x}{\sin x} \tag{147}$$

$$\tan\tfrac{1}{2}x = \frac{1 + \sin x - \cos x}{1 + \sin x + \cos x} \tag{148}$$

69. Several useful formulæ result from the preceding, by introducing 45° or 30°. If $x = 45°$ in (36), (37), (38), and (39), we have, by (33),

$$\sin(45° \pm y) = \cos(45° \mp y) = \frac{\cos y \pm \sin y}{\sqrt{2}} \tag{149}$$

whence

$$\tan(45° \pm y) = \cot(45° \mp y) = \frac{\cos y \pm \sin y}{\cos y \mp \sin y} \tag{150}$$

in which either the upper signs must be taken throughout, or the lower signs throughout.

If we divide the numerator and denominator of (150) by $\cos y$ or $\sin y$,

$$\tan(45° \pm y) = \frac{1 \pm \tan y}{1 \mp \tan y} = \frac{\cot y \pm 1}{\cot y \mp 1} \tag{151}$$

From this, by (57),

$$\tan(y - 45^\circ) = \frac{\tan y - 1}{\tan y + 1} \qquad (152)$$

70. Again, let $x = 90^\circ \pm y$ in (138), (139) (140), and (146),

$$\sin(45^\circ \pm \tfrac{1}{2} y) = \cos(45^\circ \mp \tfrac{1}{2} y) = \sqrt{\left(\frac{1 \pm \sin y}{2}\right)} \qquad (153)$$

$$\tan(45^\circ \pm \tfrac{1}{2} y) = \sqrt{\left(\frac{1 \pm \sin y}{1 \mp \sin y}\right)} \qquad (154)$$

$$\tan(45^\circ \pm \tfrac{1}{2} y) = \frac{1 \pm \sin y}{\cos y} = \frac{\cos y}{1 \mp \sin y} \qquad (155)$$

From the last we find

$$\tan(45^\circ + \tfrac{1}{2} y) + \tan(45^\circ - \tfrac{1}{2} y) = \frac{2}{\cos y} = 2 \sec y \qquad (156)$$

$$\tan(45^\circ + \tfrac{1}{2} y) - \tan(45^\circ - \tfrac{1}{2} y) = \frac{2 \sin y}{\cos y} = 2 \tan y \qquad (157)$$

the quotient of which is

$$\frac{\tan(45^\circ + \tfrac{1}{2} y) - \tan(45^\circ - \tfrac{1}{2} y)}{\tan(45^\circ + \tfrac{1}{2} y) + \tan(45^\circ - \tfrac{1}{2} y)} = \sin y \qquad (158)$$

71. In (101), (102), (103) and (104), let $x = 30^\circ$; then by (27) and (28)

$$\sin(30^\circ + y) + \sin(30^\circ - y) = \cos y \qquad (159)$$

$$\sin(30^\circ + y) - \sin(30^\circ - y) = \sin y \sqrt{3} \qquad (160)$$

$$\cos(30^\circ + y) + \cos(30^\circ - y) = \cos y \sqrt{3} \qquad (161)$$

$$\cos(30^\circ + y) - \cos(30^\circ - y) = -\sin y \qquad (162)$$

and in a similar manner we may introduce 60°; but it is unnecessary to extend these substitutions, as they involve no difficulty, and can be made as occasion demands.

FORMULÆ FOR MULTIPLE ANGLES.

72. From (101) and (102) we have

$$\sin(y + x) = 2 \sin y \cos x - \sin(y - x)$$

$$\sin(y + x) = 2 \cos y \sin x + \sin(y - x)$$

in which let $y = (m - 1)x$; then

$$\sin mx = 2 \sin(m - 1)x \cos x - \sin(m - 2)x \qquad (163)$$

$$\sin mx = 2 \cos(m - 1)x \sin x + \sin(m - 2)x \qquad (164)$$

which are the general formulæ for computing the sine of any multiple mx, from the lower multiples $(m - 1)x$ and $(m - 2)x$, and the simple angle x.

If we make m successively 1, 2, 3, 4, &c., these formulæ give

$$\begin{aligned}
\sin x &= \sin x &&= \sin x \\
\sin 2x &= 2 \sin x \cos x &&= 2 \cos x \sin x \\
\sin 3x &= 2 \sin 2x \cos x - \sin x &&= 2 \cos 2x \sin x + \sin x \\
\sin 4x &= 2 \sin 3x \cos x - \sin 2x &&= 2 \cos 3x \sin x + \sin 2x \\
&\&c. && \&c.
\end{aligned}$$

73. From (103) and (104)

$$\cos (y + x) = 2 \cos y \cos x - \cos (y - x)$$
$$\cos (y + x) = -2 \sin y \sin x + \cos (y - x)$$

which, if we put $y = (m - 1) x$, become

$$\cos mx = 2 \cos (m - 1) x \cos x - \cos (m - 2) x \quad (165)$$
$$\cos mx = -2 \sin (m - 1) x \sin x + \cos (m - 2) x \quad (166)$$

If m is taken successively equal to 1, 2, 3, 4, &c.

$$\begin{aligned}
\cos x &= \cos x &&= \cos x \\
\cos 2x &= 2 \cos x \cos x - 1 &&= -2 \sin x \sin x + 1 \\
\cos 3x &= 2 \cos 2x \cos x - \cos x &&= -2 \sin 2x \sin x + \cos x. \\
\cos 4x &= 2 \cos 3x \cos x - \cos 2x &&= -2 \sin 3x \sin x + \cos 2x \\
&\&c. && \&c.
\end{aligned}$$

74. In (123) let $y = (m - 1) x$; then

$$\tan mx = \frac{\tan (m - 1) x + \tan x}{1 - \tan (m - 1) x \tan x} \quad (167)$$

whence

$$\tan x = \tan x \qquad \tan 3x = \frac{\tan 2x + \tan x}{1 - \tan 2x \tan x}$$

$$\tan 2x = \frac{2 \tan x}{1 - \tan^2 x} \qquad \tan 4x = \frac{\tan 3x + \tan x}{1 - \tan 3x \tan x}$$

&c.

75. If in the expression for $\sin 3x$, Art. 72, we substitute the value of $\sin 2x$, we find

$$\sin 3x = 4 \sin x \cos^2 x - \sin x$$

by which we find the sine of the multiple directly from the functions of the simple angle. If this be substituted in the expression for $\sin 4x$, the latter will also be expressed in terms of the simple angle. By these successive substitutions we easily obtain the following tables:

$$\begin{aligned}
\sin x &= \sin x \\
\sin 2x &= 2 \sin x \cos x \\
\sin 3x &= 4 \sin x \cos^2 x - \sin x \\
\sin 4x &= 8 \sin x \cos^3 x - 4 \sin x \cos x \\
&\&c.
\end{aligned}$$

76.

$$\begin{aligned}
\cos x &= \cos x \\
\cos 2x &= 2 \cos^2 x - 1 \\
\cos 3x &= 4 \cos^3 x - 3 \cos x \\
\cos 4x &= 8 \cos^4 x - 8 \cos^2 x + 1 \\
&\&c.
\end{aligned}$$

77. If in these equations we substitute for $\cos^2 x = 1 - \sin^2 x$ they become

$$\begin{aligned} \sin\ x &= \sin x \\ \sin 2x &= 2 \sin x \sqrt{(1 - \sin^2 x)} \\ \sin 3x &= 3 \sin x - 4 \sin^3 x \\ \sin 4x &= (4 \sin x - 8 \sin^3 x) \sqrt{(1 - \sin^2 x)} \\ &\text{\&c.} \end{aligned}$$

78.

$$\begin{aligned} \cos\ x &= \sqrt{(1 - \sin^2 x)} \\ \cos 2x &= 1 - 2 \sin^2 x \\ \cos 3x &= (1 - 4 \sin^2 x) \sqrt{(1 - \sin^2 x)} \\ \cos 4x &= 1 - 8 \sin^2 x + 8 \sin^4 x \\ &\text{\&c.} \end{aligned}$$

From the preceding tables it appears that the cosine of the multiple angle may always be expressed rationally in terms of the cosine of the simple angle; but that the sine of only the odd multiples and the cosine of only the even multiples can be expressed rationally in terms of the sine of the simple angle.

79. By successive substitutions we find from the formulæ of Art. 74.

$$\begin{aligned} \tan\ x &= \tan x \\ \tan 2x &= \frac{2 \tan x}{1 - \tan^2 x} \\ \tan 3x &= \frac{3 \tan x - \tan^3 x}{1 - 3 \tan^2 x} \\ \tan 4x &= \frac{4 \tan x - 4 \tan^3 x}{1 - 6 \tan^2 x + \tan^4 x} \\ &\text{\&c.} \end{aligned}$$

80. The preceding results are but particular applications of general formulæ to be given hereafter, (Chapter XV.) They are introduced here for the convenience of reference in elementary applications. The powers of the sine or cosine of the simple angle may also be expressed in the multiples of the angle: but they are most readily obtained from the general formulæ of Chapter XV.

Relations of Three Angles.

81. Let x, y, and z be any three angles; we have, by (36) and (38),

$$\begin{aligned} \sin(x + y + z) &= \sin(x + y) \cos z + \cos(x + y) \sin z \\ &= \sin x \cos y \cos z + \cos x \sin y \cos z \\ &\quad + \cos x \cos y \sin z - \sin x \sin y \sin z \qquad (168) \end{aligned}$$

$$\begin{aligned} \cos(x + y + z) &= \cos(x + y) \cos z - \sin(x + y) \sin z \\ &= \cos x \cos y \cos z - \sin x \sin y \cos z \\ &\quad - \sin x \cos y \sin z - \cos x \sin y \sin z \qquad (169) \end{aligned}$$

and in the same way we may develop the sines and cosines of $x + y - z$, $x - y + z$, &c.; but we may find these directly from (168) and (169) by changing the sign of z, y, &c., and observing (56).

The quotient of (168) divided by (169) gives, after dividing the numerator and denominator by $\cos x \cos y \cos z$,

$$\tan(x+y+z) = \frac{\tan x + \tan y + \tan z - \tan x \tan y \tan z}{1 - \tan x \tan y - \tan x \tan z - \tan y \tan z} \qquad (170)$$

82. Let x, y and z be any three angles, and from the equations

$$\sin(x-z) = \sin x \cos z - \cos x \sin z$$
$$\sin(y-z) = \sin y \cos z - \cos y \sin z$$

let $\cos z$ be eliminated; we find

$$\sin y \sin(x-z) - \sin x \sin(y-z) = \sin z(\sin x \cos y - \cos x \sin y)$$
$$= \sin z \sin(x-y)$$

If $\sin z$ is eliminated, we find

$$\cos y \sin(x-z) - \cos x \sin(y-z) = \cos z \sin(x-y)$$

These equations may be more elegantly expressed, as follows:

$$\sin x \sin(y-z) + \sin y \sin(z-x) + \sin z \sin(x-y) = 0 \qquad (171)$$
$$\cos x \sin(y-z) + \cos y \sin(z-x) + \cos z \sin(x-y) = 0 \qquad (172)$$

A number of similar relations may be deduced from these by substituting $90° \pm x$, &c., for x, &c.

83. Let

$$v = \tfrac{1}{2}(x+y+z)$$

we have by (104)

$$2 \sin v \sin(v-x) = \cos x - \cos(2v-x) = \cos x - \cos(y+z)$$
$$2 \sin(v-y) \sin(v-z) = \cos(y-z) - \cos(2v-y-z)$$
$$= \cos(y-z) - \cos x$$

the product of which is

$$4 \sin v \sin(v-x) \sin(v-y) \sin(v-z) = \cos x\,[\cos(y-z) + \cos(y+z)]$$
$$- \cos^2 x - \cos(y+z)\cos(y-z)$$

Reducing the second member by (103) and (134);

$$4 \sin v \sin(v-x) \sin(v-y) \sin(v-z) = 2 \cos x \cos y \cos z - \cos^2 x$$
$$- \cos^2 y - \cos^2 z + 1 \qquad (173)$$

In the same manner we find

$$4 \cos v \cos(v-x) \cos(v-y) \cos(v-z) = 2 \cos x \cos y \cos z + \cos^2 x$$
$$+ \cos^2 y + \cos^2 z - 1 \qquad (174)$$

84. The following may be proposed as exercises.

$$\sin x + \sin y + \sin z - \sin(x+y+z) = 4 \sin \tfrac{1}{2}(x+y) \sin \tfrac{1}{2}(x+z) \sin \tfrac{1}{2}(y+z) \qquad (175)$$

$$\cos x + \cos y + \cos z + \cos(x+y+z) = 4 \cos \tfrac{1}{2}(x+y) \cos \tfrac{1}{2}(x+z) \cos \tfrac{1}{2}(y+z) \qquad (176)$$

$$\tan x + \tan y + \tan z - \tan x \tan y \tan z = \frac{\sin(x+y+z)}{\cos x \cos y \cos z} \qquad (177)$$

$$\cot x + \cot y + \cot z - \cot x \cot y \cot z = -\frac{\cos(x+y+z)}{\sin x \sin y \sin z} \qquad (178)$$

$$\left.\begin{aligned} &4\,[\sin(x+y+z)+2\sin x\sin y\sin z]^2 = 4\,[\sin(x+y)\cos z+\cos(x-y)\sin z]^2 \\ &= [1-\cos(2\,x+2\,y)]\,(1+\cos 2\,z)+[1+\cos(2\,x+2\,y)]\,(1-\cos 2z) \\ &\qquad\qquad +2\,(\sin 2\,x+\sin 2\,y)\sin 2\,z \\ &= 2\,(1+\sin 2\,x\sin 2\,y+\sin 2\,x\sin 2\,z+\sin 2\,y\sin 2\,z-\cos 2\,x\cos 2\,y\cos 2\,z) \end{aligned}\right\} \quad (179)$$

85. Let the sum of three angles x, y and z be π, or a multiple of π, that is, an *even* multiple of $\frac{\pi}{2}$ a condition which is expressed by the equation

$$x+y+z=2\,n\,.\,\frac{\pi}{2} \qquad (180)$$

then, $\tan(x+y+z)=0$, and the first member of (170) being thus reduced to zero, the numerator of the second number must be zero, or

$$\tan x+\tan y+\tan z=\tan x\tan y\tan z \qquad (181)$$

an equation, it must be remembered, that is true only under the condition (180). Since x, y and z may be selected in an infinite variety of ways so as to satisfy (180), it follows from (181) that there is an infinite number of solutions of the problem, "to find three numbers whose sum is equal to their product."

Let the sum of three angles x, y and z be $\frac{\pi}{2}$ or an *odd* multiple of $\frac{\pi}{2}$; that is, Let

$$x+y+z=(2\,n+1)\,\frac{\pi}{2} \qquad (182)$$

then, $\tan(x+y+z)=\infty$, and the denominator of (170) must be zero, or

$$\tan x\tan y+\tan x\tan z+\tan y\tan z=1$$

which, divided by $\tan x\tan y\tan z$, gives

$$\cot x+\cot y+\cot z=\cot x\cot y\cot z \qquad (183)$$

a relation that holds only under the condition (182).

86. Let

$$x+y+z=n\,\pi=2\,n\,\frac{\pi}{2} \qquad (184)$$

We have by (93) and (91)

$$\begin{aligned} \cos(x+y-z) &= \cos(n\,\pi-2\,z) = (-1)^n\cos 2\,z \\ \cos(x-y+z) &= \cos(n\,\pi-2\,y) = (-1)^n\cos 2\,y \\ \cos(y+z-x) &= \cos(n\,\pi-2\,x) = (-1)^n\cos 2\,x \\ \cos(y+z+x) &= \cos n\,\pi = (-1)^n \end{aligned}$$

the sums of the first two and of the second two are by (103)

$$\begin{aligned} 2\cos x\cos(y-z) &= (-1)^n(\cos 2\,z+\cos 2\,y) \\ 2\cos x\cos(y+z) &= (-1)^n(\cos 2\,x+1) \end{aligned}$$

and the sum and difference of these equations are

$$\begin{aligned} 4\cos x\cos y\cos z &= (-1)^n(\cos 2\,z+\cos 2\,y+\cos 2\,x+1) \\ 4\cos x\sin y\sin z &= (-1)^n(\cos 2\,z+\cos 2\,y-\cos 2\,x-1) \end{aligned}$$

or

$$\pm 4\cos x\cos y\cos z = \cos 2\,x+\cos 2\,y+\cos 2\,z+1 \qquad (185)$$

$$\pm 4\cos x\sin y\sin z = -\cos 2\,x+\cos 2\,y+\cos 2\,z-1 \qquad (186)$$

the upper sign being taken when n in (184) is even, the lower when n is odd.

In the same manner we obtain

$$\mp 4 \sin x \sin y \sin z = \sin 2x + \sin 2y + \sin 2z \quad (187)$$

$$\mp 4 \sin x \cos y \cos z = -\sin 2x + \sin 2y + \sin 2z \quad (188)$$

the signs being taken as above.

Again, let

$$x + y + z = (2n + 1)\frac{\pi}{2} \quad (189)$$

we shall find by the same process

$$\pm 4 \sin x \sin y \sin z = \cos 2x + \cos 2y + \cos 2z - 1 \quad (190)$$

$$\pm 4 \sin x \cos y \cos z = -\cos 2x + \cos 2y + \cos 2z + 1 \quad (191)$$

$$\pm 4 \cos x \cos y \cos z = \sin 2x + \sin 2y + \sin 2z \quad (192)$$

$$\pm 4 \cos x \sin y \sin z = -\sin 2x + \sin 2y + \sin 2z \quad (193)$$

$+$ or $-$ according as n in (189) is even or odd.

Inverse Trigonometric Functions.

87. If

$$y = \sin x$$

y is an *explicit* function of x, and, since x and y are mutually dependent, x is an *implicit* function of y; but to express x in the form of an explicit function of y, we write*

$$x = \sin^{-1} y$$

which is read, "x equal to the angle (or arc) whose sine is y," and x is called the *inverse function* of y, or of sine x.

In like manner $\tan^{-1} y$ is "the angle or arc whose tangent is y," &c.

88. Many of the formulæ already given may be conveniently expressed with the aid of this notation. Thus, by (16),

$$x = \sec^{-1} \sqrt{(1 + \tan^2 x)}$$

or if we put $y = \tan x$

$$\tan^{-1} y = \sec^{-1} \sqrt{(1 + y^2)}$$

* This notation was suggested by the use of the negative exponents in algebra. If we have $y = nx$, we also have $x = n^{-1} y$, where y is a function of x, and x is the corresponding inverse function of y. The latter equation might be read "x is a quantity which multiplied by n gives y." It may be necessary to caution the beginner against the error of supposing that $\sin^{-1} y$ is equivalent to $\frac{1}{\sin y}$.

For a general view of the nature of inverse functions, see Peirce's Diff. Calc. Arts. 13, et seq.

And in the same way the formulæ of Art. 28 give

$$\sin^{-1} y = \operatorname{cosec}^{-1} \frac{1}{y} = \cos^{-1} \surd (1 - y^2) = \tan^{-1} \frac{y}{\surd (1 - y^2)}$$

$$\cos^{-1} y = \sec^{-1} \frac{1}{y} = \sin^{-1} \surd (1 - y^2) = \tan^{-1} \frac{\surd (1 - y^2)}{y}$$

$$\tan^{-1} y = \cot^{-1} \frac{1}{y} = \sin^{-1} \frac{y}{\surd (1 + y^2)} = \cos^{-1} \frac{1}{\surd (1 + y^2)}$$

Formulæ (123) and (124) may be written

$$x \pm y = \tan^{-1} \frac{\tan x \pm \tan y}{1 \mp \tan x \tan y}$$

or putting $t = \tan x$, $t' = \tan y$,

$$\tan^{-1} t \pm \tan^{-1} t' = \tan^{-1} \frac{t \pm t'}{1 - t t'} \qquad (194)$$

Also the formulæ of Arts. 67 and 68 give

$$\cos^{-1} y = 2 \sin^{-1} \sqrt{\left(\frac{1 - y}{2}\right)} = 2 \cos^{-1} \sqrt{\left(\frac{1 + y}{2}\right)} = 2 \tan^{-1} \sqrt{\left(\frac{1 - y}{1 + y}\right)}$$

$$2 \tan^{-1} y = \sin^{-1} \frac{2 y}{1 + y^2} = \tan^{-1} \frac{2 y}{1 - y^2} = \cos^{-1} \left(\frac{1 - y^2}{1 + y^2}\right)$$

89. We may also employ the notation $\sin^{-1} (\cos x)$ or "the arc whose sine is equal to the cosine of x," i. e. "the complement of x"; and $\sin (\cos^{-1} y)$ or "the sine of the arc whose cosine is y," &c. We shall have accordingly

$$\sin (\sin^{-1} y) = y \qquad \tan (\tan^{-1} y) = y \text{ \&c.}$$

$$\sin^{-1} (\sin x) = x \qquad \tan^{-1} (\tan x) = x \text{ \&c.}$$

But it must be observed that since the same sine or tangent corresponds to an infinite number of angles, (Art. 53,) these last equations should be written

$$\sin^{-1} (\sin x) = n \pi + (-1)^n x, \qquad \tan^{-1} (\tan x) = n \pi + x$$

which are equivalent to (95) and (97).

CHAPTER V.

TRIGONOMETRIC TABLES.

90. Before proceeding to the numerical computation of triangles and to other applications of the preceding formulæ, the student should make himself acquainted with the arrangement of, and the mode of consulting, the trigonometric tables. We shall here speak of those points only that are common to all tables, but it will be necessary to consult also the explanations that are always prefixed to a table in order to understand any peculiarity that may attach to it. We suppose also that he is acquainted with the nature and use of the common tables of logarithms of numbers.

There are two principal trigonometric tables;* the first, called the *Table of Natural Sines, &c.*, contains simply the numerical values of the sines, tangents, &c. for each given value of the angle; the

* The most convenient seven-figure tables yet published in this country are *Stanley's*, already mentioned, p. 12. Attached to these are also five-figure tables, and a table of anti-logarithms.

Computers, engaged in extensive and varied calculations, generally provide themselves not only with tables of seven figures, but also with those of six, of five, and even of four figures—the selection and use of a particular table in any case being determined by the degree of precision sought for in the results. We might, indeed, employ seven-figure, or even ten-figure tables in all cases, and reject the final figures of our results, when a lower degree of approximation is thought sufficient; but it is clearly a loss of time and labor to employ other figures besides those which are necessary in arriving at the proposed degree of precision.

The best six-figure tables are to be found in Bremiker's *Nova Tabula Berolinensis*, (Berlin, 1852,) which are distinguished for simplicity of arrangement, as well as accuracy.

Bowditch's five-figure tables, in his Epitome of Navigation, are valuable on account of their undoubted accuracy.

Four-figure tables are be found in various collections, as for instance, in Schumacher's *Hülfstafeln*, (edited by *Warnstorff.*)

Of the foreign seven-figure tables we may cite, Taylor's, Hutton's, Babbage's, Shortrede's, in England; Callet's, Bagay's, Borda's, in France. Taylor's, Shortrede's, and Bagay's give the log. functions to every second of the quadrant; Borda's give the functions corresponding to the centesimal division of angles, (Art. 6.)

For computations requiring more than seven figures recourse must be had to the ten-figure tables of Vlacq, *Thesaurus Logarithmorum Completus*, edited by Vega, (Leipzig, 1794).

second, called the *Table of Logarithmic Sines, &c.*, contains the logarithms of the numbers in the first table. As the greater part of the computations of trigonometry are carried on by logarithms, the latter table is by far the most useful.

Table of Natural Sines, &c.

91. The arrangement of this table will be understood from a simple inspection. It contains the sines, &c. of angles between zero and 90°, generally for every minute, and the functions of angles consisting of a number of degrees, minutes, and seconds, have to be found by interpolations similar in their nature to those that are required in using tables of logarithms of numbers. This interpolation is based upon the supposition that the differences of the sines, &c., are proportional to the differences of the angles; and this proportion, though theoretically inexact, gives, in general, a sufficient approximation, provided the differences of the angles of the table are sufficiently small. When the greatest accuracy is desired, the tables should give the angles to every second, or at least to every 10″, and the sines, &c., should be given to at least seven decimal places.

92. As every angle between 45° and 90° is the complement of another between 45° and 0°, every sine of an angle less than 45° is the cosine of another greater than 45°; every tangent is a cotangent, &c.; hence the angles at the top of the tables generally extend only to 45°, and the same functions answer for the remaining 45°, by giving them at the bottom of the table the names of the complemental functions.

93. As the sines, &c., pass through all their possible *numerical* values, while the angle varies from 0° to 90°, the tables are not extended beyond 90°; but we easily deduce the functions of all other angles by the principles of Chap. III.

For the functions of an angle between 90° and 180°, we may take the *same* functions of its supplement, observing to prefix the proper algebraic sign, Art. 39. Thus, from Hutton's Tables we find

$$\begin{aligned}
\sin 140^\circ\, 16' &= \sin 39^\circ\, 44' = 0{\cdot}6392153 \\
\cos 140^\circ\, 16' &= -\cos 39^\circ\, 44' = -0{\cdot}7690278 \\
\tan 140^\circ\, 16' &= -\tan 39^\circ\, 44' = -0{\cdot}8311992 \\
\cot 140^\circ\, 16' &= -\cot 39^\circ\, 44' = -1{\cdot}2030810 \\
\sec 140^\circ\, 16' &= -\sec 39^\circ\, 44' = -1{\cdot}3003431 \\
\operatorname{cosec} 140^\circ\, 16' &= \operatorname{cosec} 39^\circ\, 44' = 1{\cdot}5644181
\end{aligned}$$

remembering that in the 2d quadrant all the functions are negative except the sine, and its reciprocal, the cosecant.

Or, we may (Art. 38) deduct 90° from the given angle and take from the table the *complemental* functions of the remainder, prefixing the signs as before; thus

$$\sin 140^\circ\ 16' = \quad \cos 50^\circ\ 16' = \text{\&c.}$$
$$\cos 140^\circ\ 16' = -\sin 50^\circ\ 16' = \text{\&c.}$$

which is the better practical method, as the subtraction of 90° may be performed mentally.

94. For angles between 180° and 270°, we deduct 180° and take the *same* functions of the remainder, prefixing the signs that belong to the 3d quadrant, Art. 41; thus

$$\sin 220^\circ\ 26' = -\sin 40^\circ\ 26'$$
$$\cos 220^\circ\ 26' = -\cos 40^\circ\ 26'$$
$$\tan 220^\circ\ 26' = +\tan 40^\circ\ 26'\ \text{\&c.}$$

95. For angles between 270° and 360°, we may deduct 270° and take the *complemental* functions of the remainder, prefixing the signs that belong to the 4th quadrant, Art. 43; thus

$$\sin 331^\circ\ 27' = -\cos 61^\circ\ 27'$$
$$\cos 331^\circ\ 27' = +\sin 61^\circ\ 27'$$
$$\tan 331^\circ\ 27' = -\cot 61^\circ\ 27'\ \text{\&c.}$$

Or we may take the *same* functions of the difference between the angle and 360°, Art. 44, observing the signs.

96. Above 360° we deduct 360°, and take the same functions with their signs, Art. 45; and if the angle exceeds 720°, 1080°, &c., we deduct 720°, 1080°, &c. ; thus

$$\cos\ 840^\circ\ 45' = \cos 120^\circ\ 45^\circ = -\sin 30^\circ\ 45'$$
$$\tan 1372^\circ\ 13' = \tan 292^\circ\ 13' = -\cot 22^\circ\ 13'$$

Table of Logarithmic Sines, &c.

97. In this table we find the logarithms of the numbers in the Table of Natural Sines arranged in precisely the same manner; it will therefore require but little additional explanation.

As the sines and cosines are all less than unity (being by their definitions proper fractions), their logarithms properly have negative indices; but these are avoided in the usual manner by increasing the index by 10, so that we find the index 9 instead of -1, 8 instead of -2, &c. The tangents under 45° being also less than unity, the indices of their logs. are also increased by 10.

In some tables, to preserve uniformity, the indices of the logs. of all the functions are increased by 10, so that the log. secants and cosecants are always greater than 10. In using these tables, we have the general rule that for each log. function added in forming a sum we must deduct 10 from that sum.

98. Since

$$\sec A = \frac{1}{\cos A}$$

we have $$\log \sec A = -\log \cos A$$

or since the tabular log. cos. is increased by 10,

$$\log \sec A = 10 - \log \cos A$$

that is, *the log. secant is the arithmetical complement of the tabular log. cosine.* For a like reason log. cosec. is the ar. co. of the log. sin.; and log. cot. is the ar. co. of the log. tan.

Also since

$$\tan A = \frac{\sin A}{\cos A}$$

$$\log \tan A = \log \sin A - \log \cos A$$

by which property, together with the preceding, we may obtain by subtraction only, the log. tan. cot. sec. and cosec. from a table containing only the log. sin. and cos.

99. When the natural sines, &c. are negative, we shall in this work indicate it by prefixing the negative sign also to their logarithms;* thus we shall write

$$\cos 140^\circ\ 16' = -0{\cdot}7690278$$

and $$\log \cos 140^\circ\ 16' = -9{\cdot}8859420$$

---

* Strictly speaking, *negative* numbers have no logarithms, but in practice, the multiplication, division, &c. of numbers is performed without reference to their signs, i. e. as if they were all positive, and the sign of the result is then deduced from the signs of the factors according to the rules of algebra. We employ logarithms simply to effect the first of these operations, i. e. the multiplication, division, &c. of the numbers considered as positive; and to facilitate the second operation, or the determination of the sign, we prefix to the logs. the signs which are prefixed to the numbers to which they belong.

As the logs. of all the trig. functions are positive (being rendered so by the addition of 10 where necessary), it will easily be remembered that the sign in the latter case is not that of the logarithm, but of the number to which it belongs.

Elementary Method of Constructing the Trigonometric Table.

100. By dividing $\pi = 3{\cdot}1415926$ by the number of seconds in 180°, we found (Art. 9) the length of the arc 1″, and (Art. 54), the sine of 1″, which is sensibly equal to the arc. In the same manner we find, by dividing by 10800,

$$\sin 1' = 0{\cdot}0002908882$$

and by (7)

$$\cos 1' = \sqrt{(1 - \sin^2 1')} = \sqrt{[(1 + \sin 1')(1 - \sin 1')]}$$
$$= \sqrt{(1{\cdot}0002908882 \times {\cdot}9997091118)}$$

or performing the arithmetical operations

$$\cos 1' = 0{\cdot}9999999577$$

Then by (101) and (103)

$$\sin(x + y) = 2 \sin x \cos y - \sin(x - y)$$
$$\cos(x + y) = 2 \cos x \cos y - \cos(x - y)$$

in which we can suppose y to be constantly equal to 1′ and x to become successively 1′, 2′, 3′, &c. Thus, first substituting $y = 1'$,

$$\sin(x + 1') = 2 \sin x \cos 1' - \sin(x - 1')$$
$$\cos(x + 1') = 2 \cos x \cos 1' - \cos(x - 1')$$

then if $x = 1', 2', 3'$, &c., we find for the sines

$$\sin 2' = 2 \cos 1' \sin 1' - \sin 0' = 0{\cdot}0005817764$$
$$\sin 3' = 2 \cos 1' \sin 2' - \sin 1' = 0{\cdot}0008726646$$
$$\sin 4' = 2 \cos 1' \sin 3' - \sin 2' = 0{\cdot}0011635526$$
$$\sin 5' = 2 \cos 1' \sin 4' - \sin 3' = 0{\cdot}0014544407$$
&c. &c.

and for the cosines

$$\cos 2' = 2 \cos 1' \cos 1' - \cos 0' = 0{\cdot}9999998308$$
$$\cos 3' = 2 \cos 1' \cos 2' - \cos 1' = 0{\cdot}9999996193$$
$$\cos 4' = 2 \cos 1' \cos 3' - \cos 2' = 0{\cdot}9999993232$$
$$\cos 5' = 2 \cos 1' \cos 4' - \cos 3' = 0{\cdot}9999989425$$
&c. &c.

The whole difficulty in this operation consists in the multiplication of each successive sine or cosine by $2 \cos 1' = 1{\cdot}9999999154$; but this multiplication is much shortened by observing that

$$2 \cos 1' = 1{\cdot}9999999154 = 2 - {\cdot}0000000846$$

so that if we put

$$m = {\cdot}0000000846$$

we have $2 \cos 1' = 2 - m$ and therefore

$$\sin 2' = 2 \sin 1' - \sin 0' - m \sin 1'$$
$$\sin 3' = 2 \sin 2' - \sin 1' - m \sin 2'$$
$$\sin 4' = 2 \sin 3' - \sin 2' - m \sin 3'$$
&c.

$$\cos 2' = 2 \cos 1' - \cos 0' - m \cos 1'$$
$$\cos 3' = 2 \cos 2' - \cos 1' - m \cos 2'$$
$$\cos 4' = 2 \cos 3' - \cos 2' - m \cos 3'$$
&c.

which are computed with great facility.

101. It is not necessary, however, to continue this process beyond 30°; for by (159) and (162) we have

$$\sin (30° + y) = \cos y - \sin (30° - y)$$
$$\cos (30° + y) = \cos (30° - y) - \sin y$$

so that the table is continued above 30° by the simple subtraction of the sines and cosines under 30° previously found. Thus, making y successively 1′, 2′, 3′, &c.

$$\sin 30° 1' = \cos 1' - \sin 29° 59'$$
$$\sin 30° 2' = \cos 2' - \sin 29° 58'$$
$$\sin 30° 3' = \cos 3' - \sin 29° 57'$$
&c.

$$\cos 30° 1' = \cos 29° 59' - \sin 1'$$
$$\cos 30° 2' = \cos 29° 58' - \sin 2'$$
$$\cos 30° 3' = \cos 29° 57' - \sin 3'$$
&c.

This last process requires to be continued only to 45° since the sines and cosines of the angles above 45° will be respectively the cosines and sines of their complements below 45°.

102. The tangents and cotangents may be found from the sines and cosines by the formulæ

$$\tan x = \frac{\sin x}{\cos x} \qquad \cot x = \frac{\cos x}{\sin x}$$

and the secants and cosecants by the formulæ

$$\sec x = \frac{1}{\cos x} \qquad \operatorname{cosec} x = \frac{1}{\sin x}$$

103. In so extended a computation as the construction of the entire table, it is necessary to verify the accuracy of the work from time to time, by separate and independent calculations. By means of (138) and (139) we can find from the cosine of an angle the sine and cosine of its half; hence from the cos. $45° = \sqrt{\frac{1}{2}}$ we can find sin. and cos. of 22° 30′, and from these the sin. and cos. of 11° 15′ by the same formulæ; and from cos. $30° = \frac{1}{2}\sqrt{3}$ we can find sin. and cos. of 15°, 7° 30′, and 3° 45′. If these agree with those found by the first process, the whole work may be considered as correct.

104. There are various other angles whose functions can be expressed under finite forms more or less simple, and may therefore be employed for the purpose of verification.

Let $x = 18°$; then $3x + 2x = 90°$ and $\cos 3x = \sin 2x$, whence, by Art. 76,

$$\begin{aligned} 4\cos^3 x - 3\cos x &= \cos 3x = 2\sin x\cos x \\ 4\cos^2 x - 3 &= 2\sin x \\ 4(1 - \sin^2 x) - 3 &= 2\sin x \\ \sin^2 x + \tfrac{1}{2}\sin x &= \tfrac{1}{4} \end{aligned}$$

which equation of the 2d degree being resolved, gives $\sin x =$

$$\sin 18° = \cos 72° = \frac{\sqrt{5} - 1}{4}$$

whence
$$\cos 18° = \sin 72° = \frac{\sqrt{(10 + 2\sqrt{5})}}{4}$$

From these by (138) and (139), we find the sine and cosine of 9° and 36°; then by (37) and (39) those of $36° - 30° = 6°$, whence those of 3°; after which it will be easy to form a table of the exact values of the sines and cosines for every 3° of the quadrant.* These expressions, however, are not of much use, directly, in the construction of tables, as we have much better methods; but they lead to a *formula of verification* which is of some importance. We find

$$\cos 36° = \frac{\sqrt{5} + 1}{4}$$

* A table of this kind is given by Cagnoli in his *Trigonometrie.*

And by (102)

$$\sin(72^\circ + y) - \sin(72^\circ - y) = 2\cos 72^\circ \sin y = \frac{\sqrt{5} - 1}{2}\sin y$$

$$\sin(36^\circ + y) - \sin(36^\circ - y) = 2\cos 36^\circ \sin y = \frac{\sqrt{5} + 1}{2}\sin y$$

the difference of these equations gives

$$\sin(36^\circ + y) - \sin(36^\circ - y) = \sin(72^\circ + y) - \sin(72^\circ - y) + \sin y$$

which is *Euler's formula of verification.* By giving y any value at pleasure, the correctness of five sines of the tables is examined. By substituting $90^\circ - y$ for y in this formula it is easily reduced to the following

$$\sin(90^\circ - y) + \sin(18^\circ + y) + \sin(18^\circ - y) = \sin(54^\circ + y) + \sin(54^\circ - y)$$

which is known as Legendre's formula, though not essentially different from Euler's.

105. The method that has here been given for computing the trigonometric table, though simple in principle is nevertheless sufficiently operose. The method by infinite series, to be given hereafter, will be found to be much more rapid and simple in practice.

CHAPTER VI.

SOLUTION OF PLANE RIGHT TRIANGLES.

106. In order to solve a plane right triangle, two parts in addition to the right angle must be given, one of which must be a side. The solution is effected directly by means of our definitions of sine, &c., which are expressed by the equations (1). As three of the six functions are only the reciprocals of the other three, we shall base the solutions upon the following three; (Fig. 15):

Fig. 15.

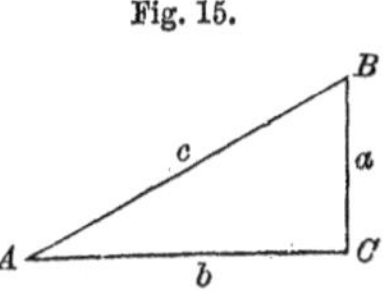

$$\sin A = \frac{a}{c} \qquad \cos A = \frac{b}{c} \qquad \tan A = \frac{a}{b}$$

Since each of these equations expresses a relation between three parts—an angle and two sides—it follows that in order to apply them, or in order to solve the triangle *trigonometrically*, there must be given two of these parts; and that of the three parts considered, one must be an angle while the other two are sides. Thus, if an angle and side are given, the third part sought must be a side; but if two sides are given, the third part sought must be an angle.

In every instance the choice of the proper equation will be determined by the precept,—*employ that trigonometric function of the angle which is equal to the ratio of the two sides considered.*

107. Case. I. *Given the hypotenuse and one angle*, or c and A.

To find a. We consider a, c and A; and since the ratio of a and c is given by the sine, we have

$$\sin A = \frac{a}{c} \quad \text{whence} \quad a = c \sin A \tag{195}$$

To find b. Considering b, c and A, we have the ratio of b and c expressed by the cosine, or

$$\cos A = \frac{b}{c} \quad \text{whence} \quad b = c \cos A \tag{196}$$

To find B. We have $B = 90° - A$.

The required quantities a and b being equal to the product of two factors, the computation is conveniently performed by the addition of the logarithms of these factors.

EXAMPLES.

1. Given $c = 672{\cdot}3412$, $A = 35°\ 16'\ 25''$; to find the other parts.

By (195)		By (196)	
$c = 672{\cdot}3412$	log 2·8275897		log 2·8275897
$A = 35°\ 16'\ 25''$	log sin 9·7615382		log cos 9·9119049
$a = 388{\cdot}2647$	log* 2·5891279	$b = 548{\cdot}9018$	log* 2·7394946

Ans. $a = 388{\cdot}2647$
$b = 548{\cdot}9018$
$B = 54°\ 43'\ 35''$

2. Given $c = 42567{\cdot}2$, $B = 87°\ 49'\ 10''$; find the other parts.

Ans. $a = 1619{\cdot}626$
$b = 42536{\cdot}37$
$A = 2°\ 10'\ 50''$

108. CASE II. *Given the hypotenuse and one side*, or c and b. To solve this case trigonometrically, we must first find an angle.

To find A. We have

$$\cos A = \frac{b}{c} \tag{197}$$

To find a. We have, by the preceding case,

$$\sin A = \frac{a}{c} \qquad a = c \sin A \tag{198}$$

But a may be found by geometry from the equation

$$a^2 + b^2 = c^2 \quad \text{whence} \quad a^2 = c^2 - b^2$$

$$a = \sqrt{(c^2 - b^2)} = \sqrt{[(c + b)(c - b)]} \tag{199}$$

EXAMPLES.

1. Given $c = 672{\cdot}3412$, $b = 548{\cdot}9018$; find A and a.

By (197)		By (198)	
$b = 548{\cdot}9018$	log 2·7394946		
$c = 672{\cdot}3412$	log 2·8275897		log 2·8275897
$A = 35°\ 16'\ 25''$	log cos 9·9119049		log sin 9·7615382
		$a = 388{\cdot}2647$	log 2·5891279

* *Ten* is rejected from each of these indices because the logarithms of the sine and cosine in the table are ten too great. Art. 97.

By (199)

$$
\begin{array}{rll}
c = & 672{\cdot}3412 & \\
b = & 548{\cdot}9018 & \\
c + b = & 1221{\cdot}2430 & \log 3{\cdot}0868021 \\
c - b = & 123{\cdot}4394 & \log 2{\cdot}0914538 \\
 & & 2)5{\cdot}1782559 \\
a = & 388{\cdot}2647 & \log 2{\cdot}5891279
\end{array}
$$

Ans. $A = 35° 16'25''$
$B = 54° 43'35''$
$a = 388{\cdot}2647$

2. Given $c = {\cdot}092357$, $b = {\cdot}058018$; find a.
Ans. $a = {\cdot}071859$

109. Case III. *Given an angle and its adjacent side*, or A and b. *To find a.* We have

$$\tan A = \frac{a}{b} \qquad \text{whence} \qquad a = b \tan A. \tag{200}$$

To find c. We have

$$\cos A = \frac{b}{c} \text{ whence, by (2), } \quad c = \frac{b}{\cos A} = b \sec A \tag{201}$$

or directly from the secant

$$\sec A = \frac{c}{b} \qquad \text{whence} \qquad c = b \sec A$$

Examples.

1. Given $A = 88° \ 59'$, $b = 2{\cdot}234875$; find the other parts.
Ans. $a = 125{\cdot}9365$
$c = 125{\cdot}9563$
$B = 1° \ 1'$

2. Given $B = 60°$, $a = 10$; find c. (See Art. 29).
Ans. $c = 20$.

110. Case IV. *Given an angle and its opposite side*, or A and a. *To find c.* We have

$$\sin A = \frac{a}{c} \qquad\qquad c = \frac{a}{\sin A} = a \operatorname{cosec} A \tag{202}$$

To find b. We have

$$\tan A = \frac{a}{b} \qquad\qquad b = \frac{a}{\tan A} = a \cot A \tag{203}$$

Examples.

1. Given $A = 25° 18' 48''$, $a = \cdot 085623$; find b.
Ans. $b = \cdot 1810278$

2. Given $B = 39° 17' 5''$, $b = \cdot 01$; find c.
Ans. $c = \cdot 0157934$

111. Case V. *Given the two sides*, or a and b.
To find A and B. We have

$$\tan A = \cot B = \frac{a}{b} \tag{204}$$

To find c. We have

$$\sin A = \frac{a}{c} \qquad c = \frac{a}{\sin A} = a \operatorname{cosec} A \tag{205}$$

We may also find c directly by geometry, from

$$c^2 = a^2 + b^2 \quad \text{whence} \quad c = \sqrt{(a^2 + b^2)}$$

but this is not readily computed by logarithms.

Examples.

1. Given $a = 30$, $b = 40$; find c. *Ans.* $c = 50$

2. Given $a = 8\cdot 678912$, $b = 2\cdot 463878$; find A and c.
Ans. $A = 74° 9' 4''\cdot 1$
$c = 9\cdot 021875$

Additional Formulæ for Right Triangles.

112. By inspecting the tables it will be seen that when the angles are very small, the cosines differ very little from each other; consequently a small angle cannot be found with very great accuracy from its cosine. For a similar reason an angle that is nearly 90° cannot be accurately computed from its sine. It is therefore desirable, when a required angle is small, to find it by its sine, and when near 90° by its cosine, or in either case by its tangent or cotangent; and for this purpose special formulæ are sometimes necessary. We shall deduce several such formulæ, from which one adapted to a particular case may be selected.

113. From (197) we find, by (139)

$$1 - \cos A = 2 \sin^2 \tfrac{1}{2} A = 1 - \frac{b}{c} = \frac{c - b}{c}$$

$$\sin \tfrac{1}{2} A = \sqrt{\left(\frac{c - b}{2c}\right)} \tag{206}$$

which may be used instead of (197), when A is small, that is when b is nearly equal to c. It gives also

$$c - b = 2c \sin^2 \tfrac{1}{2} A \tag{207}$$

by which $c - b$ may be accurately found when A is small.

Also from (197), by (140)

$$\tan \tfrac{1}{2} A = \sqrt{\left(\frac{1 - \cos A}{1 + \cos A}\right)} = \sqrt{\left(\frac{c - b}{c + b}\right)} = \frac{c - b}{a} \tag{208}$$

114. From the equation

$$\sin A = \frac{a}{c}$$

we deduce by (153) and (154)

$$\sin (45° \pm \tfrac{1}{2} A) = \sqrt{\left(\frac{c \pm a}{2\,c}\right)} \tag{209}$$

$$\cos (45° \pm \tfrac{1}{2} A) = \sqrt{\left(\frac{c \mp a}{2\,c}\right)} \tag{210}$$

$$\tan (45° \pm \tfrac{1}{2} A) = \sqrt{\left(\frac{c \pm a}{c \mp a}\right)} \tag{211}$$

and from $\tan A = \dfrac{a}{b}$ we find by (151),

$$\tan (45° \pm A) = \frac{b \pm a}{b \mp a} \tag{212}$$

115. By (136) we have

$$\cos 2\,A = \cos^2 A - \sin^2 A = \frac{b^2 - a^2}{c^2}$$

which, since $2\,A = A + 90° - B = 90° - (B - A)$, gives

$$\sin (B - A) = \frac{(b + a)\,(b - a)}{c^2} \tag{213}$$

By (135)

$$\cos (B - A) = \sin 2\,A = 2 \sin A \cos A = \frac{2\,a\,b}{c^2} \tag{214}$$

and from (213) and (214)

$$\tan (B - A) = \frac{(b + a)\,(b - a)}{2\,a\,b} \tag{215}$$

by which $B - A$ is found with great accuracy when b and a are nearly equal.

Example. Given $c = 4602{\cdot}836$, $b = 4602{\cdot}21059$ to find A.

By (206).

$c - b = 0{\cdot}62541$	log 9·7961648
$2\,c = 9205{\cdot}672$	log 3·9640555
	2)5·8321093
$\tfrac{1}{2} A = 28'\ 20''{\cdot}18$	log sin $\tfrac{1}{2} A$ = 7·9160547
$A = 56'\ 40''{\cdot}36$	

The ordinary process gives log cos $A = 9{\cdot}9999410$, whence $A = 56'\ 40''$. These results are obtained by Stanley's Tables, in which the log. sines, &c., are given for every 10″ for the first 15°. A greater discrepancy between the two results would be found by tables in which the functions were given only for each minute.

A slight error remains in the value of $\tfrac{1}{2} A = 28'\ 20''{\cdot}18$, on account of the large differences of the log. sines in this part of the table, or rather on account of the rapid change of these differences. We avoid the use of these large differences, and gain somewhat in accuracy, by employing the approximate value of sin. $\frac{1}{2} A$ given by (98), whence

$$\sin \tfrac{1}{2} A = \tfrac{1}{2} A \sin 1'', \qquad \tfrac{1}{2} A = \frac{\sin \frac{1}{2} A}{\sin 1''}$$

Thus we have found above

$$\begin{aligned} \log \sin \tfrac{1}{2} A &= 7{\cdot}9160547 \\ \text{Art. 54, } \log \sin 1'' &= 4{\cdot}6855749 \\ \log \tfrac{1}{2} A &= 3{\cdot}2304798 \end{aligned}$$

$$\tfrac{1}{2} A = 1700''{\cdot}12 = 28'\ 20''{\cdot}12$$

But to obtain $\frac{1}{2} A$ with the utmost precision, recourse must be had to the following process, which is constantly employed in observatories, and wherever small angles are to be computed with extreme accuracy. Special tables are prepared containing for every minute from 0° to 2° the logarithms of

$$\frac{\sin x}{x} = h \quad \text{and} \quad \frac{\tan x}{x} = k$$

which do not vary rapidly, and may therefore be taken with accuracy from the tables. Then we have

$$\sin x = x \,.\, \frac{\sin x}{x} = x \,.\, h \qquad x = \frac{\sin x}{h}$$

$$\tan x = x \,.\, \frac{\tan x}{x} = x \,.\, k \qquad x = \frac{\tan x}{k}$$

A table of this kind will be found on page 156 of Stanley's Tables, where the notation used is

$$q = \log \sin x, \qquad n = \log x$$

and therefore in the column marked $q - n$ we find the $\log \frac{\sin x}{x}$. Thus in the above example we have found

$$\log \sin \tfrac{1}{2} A = q \quad = 7{\cdot}9160547$$

and from the table

$$q - n = 4{\cdot}6855700$$

$$\tfrac{1}{2} A = 1700''{\cdot}14 = 28'\ 20''{\cdot}14 \qquad \log \tfrac{1}{2} A = \quad n = 3{\cdot}2304847$$

which is the true value of $\frac{1}{2} A$ within 0″·01.

Stanley's Table contains also the values of

$$\log \frac{\tan x}{x} = q - n \qquad (q = \log \tan x, \quad n = \log x)$$

$$\log \frac{x}{\sin x} = q + n \qquad (q = \log \operatorname{cosec} x, \quad n = \log x)$$

$$\log \frac{x}{\tan x} = q + n \qquad (q = \log \cot x, \quad n = \log x)$$

the use of which may easily be inferred from the example just given.

CHAPTER VII.

FORMULÆ FOR THE SOLUTION OF PLANE OBLIQUE TRIANGLES.

116. As every oblique triangle may be resolved into two right triangles by a perpendicular from one of the angles upon the opposite side, we are enabled to deduce all the formulæ for their solution from those of the preceding chapter.

117. *The sides of a plane triangle are proportional to the sines of their opposite angles.*

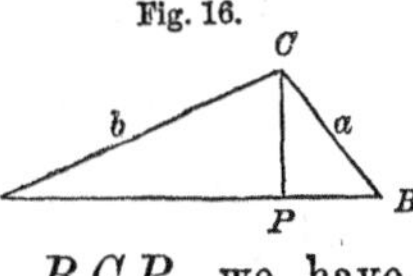

Fig. 16.

Denote the angles of the triangle ABC, Fig. 16, by A, B and C, and the sides opposite these angles respectively by a, b and c. From C draw CP perp. to AB and put $CP = p$. Then in the right triangles ACP, BCP, we have, by (195)

$$p = b \sin A, \qquad p = a \sin B$$

whence

$$b \sin A = a \sin B$$

which, converted into a proportion, gives

$$a : b = \sin A : \sin B \tag{216}$$

and in the same way we may prove that

$$a : c = \sin A : \sin C$$

$$b : c = \sin B : \sin C$$

and these three proportions may be written as one, thus:

$$a : b : c = \sin A : \sin B : \sin C \tag{217}$$

or thus,

$$\frac{a}{\sin A} = \frac{b}{\sin B} = \frac{c}{\sin C} \tag{218}$$

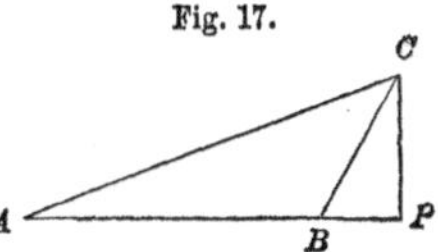

Fig. 17.

When the perpendicular falls without the triangle, Fig. 17, the angle CBP is the supplement of B, but by Art. 39, it has the same sine, so that the triangle CBP gives

$$p = a \sin CBP = a \sin B$$

the same as was found from Fig. 16. The proposition is therefore general in its application.*

118. *The sum of any two sides of a plane triangle is to their difference as the tangent of half the sum of the opposite angles is to the tangent of half their difference.*

For, by the preceding article,

$$a : b = \sin A : \sin B$$

whence, by composition and division,

$$a + b : a - b = \sin A + \sin B : \sin A - \sin B$$

But from (109) if $x = A$, $y = B$ we obtain the proportion

$$\sin A + \sin B : \sin A - \sin B = \tan \tfrac{1}{2}(A + B) : \tan \tfrac{1}{2}(A - B)$$

which, compared with the above, gives

$$a + b : a - b = \tan \tfrac{1}{2}(A + B) : \tan \tfrac{1}{2}(A - B) \qquad (219)$$

This may also be written

$$\frac{a + b}{a - b} = \frac{\tan \frac{1}{2}(A + B)}{\tan \frac{1}{2}(A - B)} \qquad (220)$$

and we may infer the same relation between b, c, B, C and a, c, A, C.

119. *The square of any side of a triangle is equal to the sum of the squares of the other two sides diminished by twice the rectangle of these sides multiplied by the cosine of their included angle.*

Fig. 16.

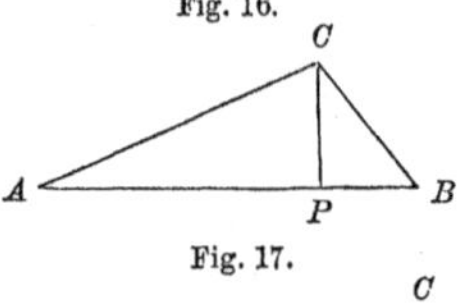

Fig. 17.

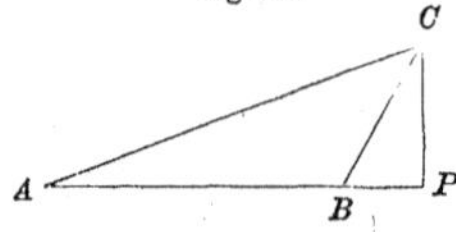

In the triangle ABC, Figs. 16 and 17,* we have either

$$BP = c - AP \quad \text{or } BP = AP - c$$

but in both cases

$$BP^2 = AP^2 + c^2 - 2c \times AP$$

Adding CP^2 to both members, we find

$$a^2 = b^2 + c^2 - 2c \times AP$$

But the triangle ACP gives by (196)

$$AP = b \cos A$$

---

* The consideration of Fig. 17 was not strictly necessary according to the principle stated in Art. 49. It may, however, be useful for the student to verify that principle when convenient.

which substituted in the preceding equation gives

$$a^2 = b^2 + c^2 - 2\,bc\cos A \tag{221}$$

as was to be proved. We have in the same way

$$b^2 = a^2 + c^2 - 2\,ac\cos B \tag{222}$$

$$c^2 = a^2 + b^2 - 2\,ab\cos C \tag{223}$$

120. The same result is obtained from the following equations (which are evident from Fig. 16, where $c = AP + PB$)

$$\left.\begin{aligned} a &= b\cos C + c\cos B \\ b &= c\cos A + a\cos C \\ c &= a\cos B + b\cos A \end{aligned}\right\} \tag{224}$$

From the first of these

$$c\cos B = a - b\cos C$$

whence $\quad c^2\cos^2 B = a^2 - 2\,ab\cos C + b^2\cos^2 C$

and from (218), $\quad c^2\sin^2 B = \qquad\qquad b^2\sin^2 C$

the sum of which two equations is, by (13),

$$c^2 = a^2 - 2\,ab\cos C + b^2$$

121. From (221) we find

$$\cos A = \frac{b^2 + c^2 - a^2}{2\,bc} \tag{225}$$

by which an angle is found when the three sides are given; but to adapt it for convenient computation by logarithms, the following transformations are necessary:

Subtract both members from unity; then

$$1 - \cos A = \frac{2\,bc - b^2 - c^2 + a^2}{2\,bc} = \frac{a^2 - (b-c)^2}{2\,bc}$$

But, by (139), making $x = A$, we have

$$1 - \cos A = 2\sin^2 \tfrac{1}{2} A$$

Also, the numerator of the second member being the difference of two squares may be resolved into two factors, viz.: the sum and the difference of a and $b - c$; thus,

$$a^2 - (b-c)^2 = [a - (b-c)] \times [a + (b-c)] = (a - b + c)(a + b - c)$$

Substituting these values in the above equation and dividing by 2

$$\sin^2 \tfrac{1}{2} A = \frac{(a - b + c)(a + b - c)}{4\,bc} \tag{226}$$

This may be simplified by representing the half sum of the three sides by s, or by putting

$$a + b + c = 2s$$

whence

$$a - b + c = a + b + c - 2b = 2s - 2b = 2(s - b)$$
$$a + b - c = a + b + c - 2c = 2s - 2c = 2(s - c)$$

which substituted in (226) give

$$\sin^2 \tfrac{1}{2} A = \frac{(s - b)(s - c)}{bc} \tag{227}$$

In the same manner we should find from (222) and (223)

$$\sin^2 \tfrac{1}{2} B = \frac{(s - a)(s - c)}{ac}, \qquad \sin^2 \tfrac{1}{2} C = \frac{(s - a)(s - b)}{ab} \tag{228}$$

122. Add both members of (225) to unity; then

$$1 + \cos A = \frac{2\,bc + b^2 + c^2 - a^2}{2\,bc} = \frac{(b + c)^2 - a^2}{2\,bc}$$

But by (138)

$$1 + \cos A = 2 \cos^2 \tfrac{1}{2} A$$

Also, $\qquad (b + c)^2 - a^2 = (b + c + a)(b + c - a)$

therefore

$$\cos^2 \tfrac{1}{2} A = \frac{(b + c + a)(b + c - a)}{4\,bc}$$

Substituting s in the numerator as in the preceding article

$$\cos^2 \tfrac{1}{2} A = \frac{s(s - a)}{bc} \tag{229}$$

and therefore also

$$\cos^2 \tfrac{1}{2} B = \frac{s(s - b)}{ac}, \qquad \cos^2 \tfrac{1}{2} C = \frac{s(s - c)}{ab} \tag{230}$$

123. Dividing (227) by (229), we have, by (14)

$$\tan^2 \tfrac{1}{2} A = \frac{(s - b)(s - c)}{s(s - a)} \tag{231}$$

and in the same manner

$$\tan^2 \tfrac{1}{2} B = \frac{(s - a)(s - c)}{s(s - b)}, \qquad \tan^2 \tfrac{1}{2} C = \frac{(s - a)(s - b)}{s(s - c)} \tag{232}$$

124. The preceding formulæ are sufficient for the resolution of all the usual cases of plane triangles; but there are others that are occasionally useful. From (218) we find, by (105), (106) and (135),

$$\frac{a+b}{c}=\frac{\sin A+\sin B}{\sin C}=\frac{\sin\frac{1}{2}(A+B)\cos\frac{1}{2}(A-B)}{\sin\frac{1}{2}C\cos\frac{1}{2}C}$$

$$\frac{a-b}{c}=\frac{\sin A-\sin B}{\sin C}=\frac{\cos\frac{1}{2}(A+B)\sin\frac{1}{2}(A-B)}{\sin\frac{1}{2}C\cos\frac{1}{2}C}$$

But since $A+B+C=180°$,

$$A+B=180°-C, \qquad \tfrac{1}{2}(A+B)=90°-\tfrac{1}{2}C$$

$$\sin\tfrac{1}{2}(A+B)=\cos\tfrac{1}{2}C, \qquad \cos\tfrac{1}{2}(A+B)=\sin\tfrac{1}{2}C$$

by means of which we find

$$\frac{a+b}{c}=\frac{\cos\frac{1}{2}(A-B)}{\sin\frac{1}{2}C}=\frac{\cos\frac{1}{2}(A-B)}{\cos\frac{1}{2}(A+B)} \tag{233}$$

$$\frac{a-b}{c}=\frac{\sin\frac{1}{2}(A-B)}{\cos\frac{1}{2}C}=\frac{\sin\frac{1}{2}(A-B)}{\sin\frac{1}{2}(A+B)} \tag{234}$$

The quotient of (233) divided by (234) gives (220).

125. Adding unity to both members of (233), or subtracting it, we have, by (103) and (104)

$$\frac{a+b+c}{c}=\frac{\cos\frac{1}{2}(A+B)+\cos\frac{1}{2}(A-B)}{\cos\frac{1}{2}(A+B)}=\frac{2\cos\frac{1}{2}A\cos\frac{1}{2}B}{\sin\frac{1}{2}C}$$

$$\frac{a+b-c}{c}=\frac{\cos\frac{1}{2}(A-B)-\cos\frac{1}{2}(A+B)}{\cos\frac{1}{2}(A+B)}=\frac{2\sin\frac{1}{2}A\sin\frac{1}{2}B}{\sin\frac{1}{2}C}$$

Similarly from (234) we find

$$\frac{c+a-b}{c}=\frac{\sin\frac{1}{2}(A+B)+\sin\frac{1}{2}(A-B)}{\sin\frac{1}{2}(A+B)}=\frac{2\sin\frac{1}{2}A\cos\frac{1}{2}B}{\cos\frac{1}{2}C}$$

$$\frac{c-a+b}{c}=\frac{\sin\frac{1}{2}(A+B)-\sin\frac{1}{2}(A-B)}{\sin\frac{1}{2}(A+B)}=\frac{2\cos\frac{1}{2}A\sin\frac{1}{2}B}{\cos\frac{1}{2}C}.$$

Substituting $s=\frac{1}{2}(a+b+c)$, these equations become

$$\frac{s}{c}=\frac{\cos\frac{1}{2}A\cos\frac{1}{2}B}{\sin\frac{1}{2}C} \tag{235}$$

$$\frac{s-c}{c}=\frac{\sin\frac{1}{2}A\sin\frac{1}{2}B}{\sin\frac{1}{2}C} \tag{236}$$

$$\frac{s-b}{c}=\frac{\sin\frac{1}{2}A\cos\frac{1}{2}B}{\cos\frac{1}{2}C} \tag{237}$$

$$\frac{s-a}{c}=\frac{\cos\frac{1}{2}A\sin\frac{1}{2}B}{\cos\frac{1}{2}C} \tag{238}$$

From these equations we can deduce immediately (227) &c.; for example, exchanging c for b in (236), we have

$$\frac{s-b}{b} = \frac{\sin \frac{1}{2} A \sin \frac{1}{2} C}{\sin \frac{1}{2} B}$$

which, multiplied by (236) gives (227).

126. Four times the product of (227) and (229) is by (135)

$$\sin^2 A = \frac{4}{b^2 c^2} \cdot s(s-a)(s-b)(s-c)$$

whence

$$\sin A = \frac{2}{bc} \sqrt{[s(s-a)(s-b)(s-c)]} \qquad (239)$$

Exchanging A for B and C successively, this gives also

$$\sin B = \frac{2}{ac} \sqrt{[s(s-a)(s-b)(s-c)]} \qquad (240)$$

$$\sin C = \frac{2}{ab} \sqrt{[s(s-a)(s-b)(s-c)]} \qquad (241)$$

In these equations put*

$$K = \sqrt{[s(s-a)(s-b)(s-c)]} \qquad (242)$$

then

$$\sin A = \frac{2K}{bc} \qquad \sin B = \frac{2K}{ac} \qquad \sin C = \frac{2K}{ab} \qquad (243)$$

The quotient of the first of these divided by the second is

$$\frac{\sin A}{\sin B} = \frac{ac}{bc} = \frac{a}{b}$$

which brings us back to the theorem of Art. 117.

127. The sum of A, B and C being 180°, and the sum of $\frac{1}{2} A$, $\frac{1}{2} B$ and $\frac{1}{2} C$ being 90°, we have, by Arts. 85 and 86, the following relations among the angles of a plane triangle.

$$\tan A + \tan B + \tan C = \tan A \tan B \tan C$$
$$\cot \tfrac{1}{2} A + \cot \tfrac{1}{2} B + \cot \tfrac{1}{2} C = \cot \tfrac{1}{2} A \cot \tfrac{1}{2} B \cot \tfrac{1}{2} C$$
$$\sin A + \sin B + \sin C = 4 \cos \tfrac{1}{2} A \cos \tfrac{1}{2} B \cos \tfrac{1}{2} C$$
$$\sin A + \sin B - \sin C = 4 \sin \tfrac{1}{2} A \sin \tfrac{1}{2} B \cos \tfrac{1}{2} C$$
$$\cos A + \cos B + \cos C - 1 = 4 \sin \tfrac{1}{2} A \sin \tfrac{1}{2} B \sin \tfrac{1}{2} C$$
$$\cos A + \cos B - \cos C + 1 = 4 \cos \tfrac{1}{2} A \cos \tfrac{1}{2} B \sin \tfrac{1}{2} C$$

in the last of which we may interchange A, B and C. These relations may be substituted in the equations of Art. 125.

---

* K is the area of the triangle. See Art. 148.

128. The following equations are added as exercises.

$$\frac{\tan \frac{1}{2} A}{\tan \frac{1}{2} B} = \frac{s - b}{s - a}, \qquad \tan \tfrac{1}{2} A \tan \tfrac{1}{2} B = \frac{s - c}{s}$$

$$\frac{\sin (A - B)}{\sin (A + B)} = \frac{(a + b)(a - b)}{c^2}$$

$$\tan \tfrac{1}{2} A \tan \tfrac{1}{2} B \cot \tfrac{1}{2} C = \frac{(s - c)^2}{K}$$

$$\cot \tfrac{1}{2} A + \cot \tfrac{1}{2} B + \cot \tfrac{1}{2} C = \frac{s^2}{K}$$

$$\sin \tfrac{1}{2} A \sin \tfrac{1}{2} B \sin \tfrac{1}{2} C = \frac{K^2}{abcs}$$

$$\cos \tfrac{1}{2} A \cos \tfrac{1}{2} B \cos \tfrac{1}{2} C = \frac{Ks}{abc}$$

$$\tan \tfrac{1}{2} A \tan \tfrac{1}{2} B \tan \tfrac{1}{2} C = \frac{K}{s^2}$$

$$K^2 = s(s - a)(s - b)(s - c)$$

$$= \frac{1}{16}\left[4 b^2 c^2 - (a^2 - b^2 - c^2)^2\right]$$

$$= \frac{1}{16}\left(2 a^2 b^2 + 2 a^2 c^2 + 2 b^2 c^2 - a^4 - b^4 - c^4\right)$$

CHAPTER VIII.

SOLUTION OF PLANE OBLIQUE TRIANGLES.

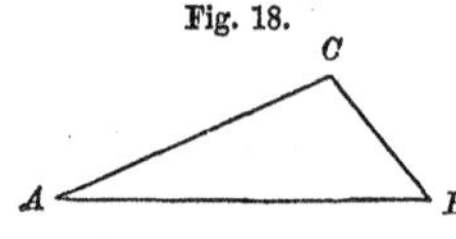

Fig. 18.

129. Case I. *Given two angles and one side*, or A, B and a. Fig. 18.

To find the third angle. We have

$$C = 180° - (A + B)$$

To find b and c. We apply the theorem of Art. 117, and state the proportions thus: the sine of the angle opposite the given side is to the sine of the angle opposite the required side, as the given side is to the required side. Thus we have

$$\sin A : \sin B = a : b$$

whence $$b = \frac{a \sin B}{\sin A} = a \sin B \operatorname{cosec} A \tag{244}$$

and $$\sin A : \sin C = a : c$$

whence $$c = \frac{a \sin C}{\sin A} = a \sin C \operatorname{cosec} A \tag{245}$$

Examples.

1. Given $A = 50° \ 38' \ 52''$, $B = 60° \ 7' \ 25''$ and $a = 412{\cdot}6708$, to find C, b and c.

$$A + B = 110° \ 46' \ 17''$$
$$C = 69° \ 13' \ 43''$$

	By (244).	By (245).
$A = 50° \ 38' \ 52''$	log cosec 0·1116730	log cosec 0·1116730
$B = 60° \ \ 7' \ 25''$	log sin 9·9380702	
$C = 69° \ 13' \ 43''$		log sin 9·9708129
$a = 412{\cdot}6708$	log 2·6156037	log 2·6156037
	log b 2·6653469	log c 2·6980896
	$b = 462{\cdot}7505$	$c = 498{\cdot}9875$

2. Given $A = 100° 16' 35''$, $B = 25° 16' 13''$, and $b = 29{\cdot}167$, find a and c.

Ans. $a = 67{\cdot}22857$
$c = 55{\cdot}59178$

130. Case I. Given A, B and a. *Second solution.* We find $C = 180° - (A + B)$; then, by (233) and (234)

$$b + c = a \cdot \frac{\cos \frac{1}{2}(B - C)}{\cos \frac{1}{2}(B + C)} = a \cdot \frac{\cos \frac{1}{2}(B - C)}{\sin \frac{1}{2} A} \tag{246}$$

$$b - c = a \cdot \frac{\sin \frac{1}{2}(B - C)}{\sin \frac{1}{2}(B + C)} = a \cdot \frac{\sin \frac{1}{2}(B - C)}{\cos \frac{1}{2} A} \tag{247}$$

which give the sum and difference of the required sides; adding half the difference to half the sum, we find the greater side, and subtracting half the difference from half the sum, we find the less side.

131. Case I. Given A, B and a. *Third Solution.* When A and B are nearly equal, and great accuracy is desired, we may compute the difference between a and b; for we have, from (244),

$$a - b = a - \frac{a \sin B}{\sin A} = a \cdot \frac{\sin A - \sin B}{\sin A}$$

or

$$a - b = \frac{2\, a \cos \frac{1}{2}(A + B) \sin \frac{1}{2}(A - B)}{\sin A} \tag{248}$$

Example. Given $A = 35° 40' 12''{\cdot}3$, $B = 35° 37' 48''{\cdot}6$, and $a = 26246{\cdot}948$.

$A = 35° 40' 12''{\cdot}3$	log cosec 0·2342442	0·23424
$B = 35° 37' 48''{\cdot}6$	log 2 0·3010300	0·30103
$\frac{1}{2}(A + B) = 35° 39' 0''{\cdot}5$	log cos 9·9098720	9·90987
$\frac{1}{2}(A - B) = 0° 1' 11''{\cdot}9$	log sin 6·5423038	6·54230
$a = 26246{\cdot}948$	log 4·4190788	4·41908
$a - b = 25{\cdot}499$	log 1·4065288	1·40652
$b = 26221{\cdot}449$		

One of the advantages of this process is, that $a - b$ may be found with sufficient accuracy with five-figure tables, as in the second column of logarithms above. If a had been given to ten figures instead of eight, we should still have been able with the seven-figure logs. to find $a - b$ to seven figures, and therefore b to ten figures, which could not be done by the ordinary methods without ten-figure tables.

132. Case II. *Given two sides and an angle opposite one of them*, or a, b and A.

To find B. To find the angle opposite the other given side, we apply Art. 117, and state the proportion thus: the side opposite the given angle is to the side opposite the required angle as the sine of the given angle is to the sine of the required angle. Thus, with the present data, we have

$$a : b = \sin A : \sin B \quad \text{whence} \quad \sin B = \frac{b \sin A}{a} \tag{249}$$

To find C. We have $C = 180° - (A + B)$

To find c. Having found C, we now have the data of Case I.: therefore, by (245)

$$c = a \sin C \operatorname{cosec} A \tag{250}$$

133. It is shown in geometry that when two sides and an angle opposite one of them are given, there may be constructed two triangles, as in Fig. 19, whenever the given angle is acute and the given side opposite to it is less than the other given side. In one of them, the required angle B is acute, and in the other it is obtuse, and the two values are supplements of each other; for

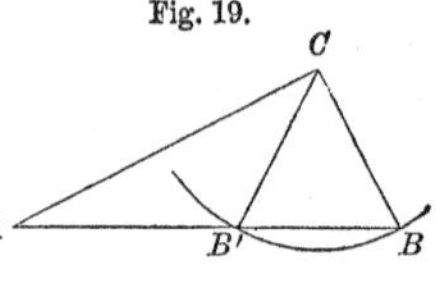

$$B = BB'C = 180° - AB'C$$

These two values of B are given in the trigonometric solution by the consideration that $\sin B$ found by (249) is at once the sine of an acute angle, and the sine of its supplement, Art. 39.

In general, when an angle is determined only by its sine, it admits of two values, supplements of each other, unless the conditions of the problem are such as to exclude one of these values. In the present case, the obtuse value of B is excluded when a is greater than b, and there is but one triangle whether A is acute or obtuse, as in Fig. 20.

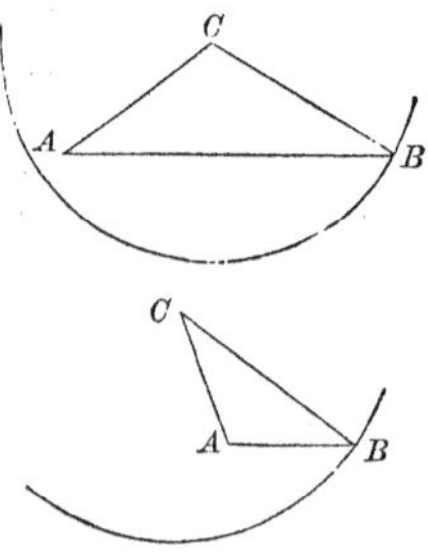

134. If the given parts were such that

$$a = b \sin A$$

a would be equal to the perpendicular from C upon the side c, and we should have but one solution, namely, a right triangle, B and its supplement both being 90°.

135. If the given parts were such that

$$a < b \sin A$$

a would be less than the perpendicular from C and the problem would be impossible. It would also be impossible if $a < b$ while $A > 90°$.

136. When there are two solutions, represent the two values of B by B' and B'', then the two values of C will be

$$C' = 180° - (A + B') = 180° - B' - A = B'' - A \tag{251}$$

$$C'' = 180° - (A + B'') = 180° - B'' - A = B' - A \tag{252}$$

and the two values of c will be

$$c' = a \sin C' \operatorname{cosec} A \qquad c'' = a \sin C'' \operatorname{cosec} A \qquad (253)$$

EXAMPLES.

1. Given $a = 31{\cdot}23879$, $b = 49{\cdot}00117$ and $A = 32° 18'$; find B, C and c.

$a = 31{\cdot}23879$	ar. co. log 8·5053058		
$b = 49{\cdot}00117$	log 1·6902064		
$A = 32° 18'$	log sin 9·7278277		
$B' = 56° 56' 56''{\cdot}3$	log sin 9·9233399		
$B'' = 123° 3' 3''{\cdot}7$			
$C' = 90° 45' 3''{\cdot}7$	log sin 9·9999627		
$C'' = 24° 38' 56''{\cdot}3$			log sin 9·6201962
	log cosec A 0·2721723		0·2721723
	log a 1·4946942		1·4946942
	log c' 1·7668292		log c'' 1·3870627
	$c' = 58{\cdot}45601$		$c'' = 24{\cdot}38163$

Ans. $B = 56° 56' 56''{\cdot}3$, $C = 90° 45' 3''{\cdot}7$, $c = 58{\cdot}45601$ or $B = 123° 3' 3''{\cdot}7$, $C = 24° 38' 56''{\cdot}3$, $c = 24{\cdot}38163$

2. Given $a = {\cdot}051234$, $b = {\cdot}042356$, $A = 55°$; find B, C and c.

Ans. $B = 42° 37' 32''{\cdot}7$
$C = 82° 22' 27''{\cdot}3$
$c = {\cdot}06199202$

3. Given $a = {\cdot}042356$, $b = {\cdot}051234$, $A = 55°$; find B, C and c.

Ans. $B = 82° 14' 35''{\cdot}7$, $C = 42° 45' 24''{\cdot}3$, $c = {\cdot}03510331$ or $B = 97° 45' 24''{\cdot}3$, $C = 27° 14' 35''{\cdot}7$, $c = {\cdot}02366993$

4. Given $a = 40$, $b = 50$, $A = 53° 7' 48''{\cdot}4$; find B.

Ans. $B = 90°$.

5. Given $a = 40$, $b = 50$, $A = 60°$; solve the triangle.

Ans. Impossible.

6. Given $b = 40$, $c = 50$, $B = 100°$; solve the triangle.

Ans. Impossible.

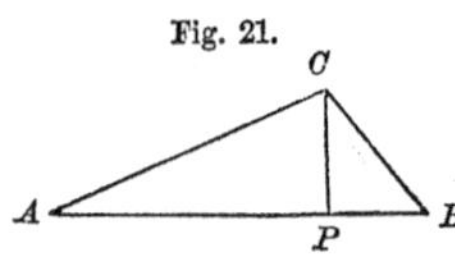

Fig. 21.

137. CASE II. Given a, b and A. *Second Solution.* We may solve, separately, the two right triangles APC, BPC, Fig. 21, which is a convenient method when there are two solutions. We first find B by (249); then we have

$$AP = b \cos A, \qquad BP = a \cos B$$

and

$$c = AP + BP$$

The cosine of the obtuse value of B is negative, (Art. 39), so that BP is then negative, and we have the two values of c from the formula

$$c = AP \pm BP$$

There will be but one solution, if $BP > AP$, for c cannot be negative.

138. CASE III. *Given two sides and the included angle*, or a, b and C. *To find A and B.* We have first

$$A + B = 180° - C$$

$$\tfrac{1}{2}(A + B) = 90° - \tfrac{1}{2} C$$

from which we next find the half difference of A and B by the theorem of Art. 118, which gives

$$a + b : a - b = \tan \tfrac{1}{2}(A + B) : \tan \tfrac{1}{2}(A - B)$$

$$\tan \tfrac{1}{2}(A - B) = \frac{a - b}{a + b} \tan \tfrac{1}{2}(A + B) = \frac{a - b}{a + b} \cot \tfrac{1}{2} C \qquad (254)$$

The half difference added to the half sum gives the greater angle, (opposite to the greater given side), and the half difference subtracted from the half sum gives the less angle.

To find c. We have the data of Case I., and therefore

$$c = a \sin C \operatorname{cosec} A = b \sin C \operatorname{cosec} B \qquad (255)$$

EXAMPLES.

1. Given $a = \cdot 062387$, $b = \cdot 023475$, and $C = 110°32'$; find A, B and c.

$$A + B = 180° - C = 69° 28'$$

$a + b =$	$\cdot 085862$	ar. co. log	$1\cdot 0661990$
$a - b =$	$\cdot 038912$	log	$8\cdot 5900836$
$\tfrac{1}{2}(A + B) =$	$34° 44'$	log tan	9.8409174
$\tfrac{1}{2}(A - B) =$	$17° 26' 33''$	log tan	$9\cdot 4972000$
$A =$	$52° 10' 33''$		
$B =$	$17° 17' 27''$	log cosec	0.5269189
$C =$	$110° 32'$	log sin	9.9714931
$b =$	$\cdot 023475$	log	$8\cdot 3706056$
$c =$	$\cdot 0739635$	log	$8\cdot 8690176$

Ans. $A = 52° 10' 33''$
$B = 17° 17' 27''$
$c = \cdot 0739635$

2. Given $a = 31{\cdot}0005$, $b = 15{\cdot}1101$, $C = 10° 15'$; find A and B.

Ans. $A = 160° 17' 13''{\cdot}7$
$B = 9° 27' 46''{\cdot}3$

3. Given $a = 2{\cdot}463878$, $b = 9{\cdot}021875$ and $C = 74° 9' 4''{\cdot}2$; find A and B.

Ans. $A = 15° 50' 55''{\cdot}8$
$B = 90° \ 0' \ 0''$

4. Given $b = 15{\cdot}1101$, $c = 31{\cdot}0005$, $A = 10° 15'$; find B and C.

Ans. $B = 9° 27' 46''{\cdot}3$
$C = 160° 17' 13''{\cdot}7$

139. Having found A and B as above, the most convenient mode of finding c is by (233) or (234), which give

$$c = (a + b) \frac{\cos \frac{1}{2} (A + B)}{\cos \frac{1}{2} (A - B)} \tag{256}$$

$$c = (a - b) \frac{\sin \frac{1}{2} (A + B)}{\sin \frac{1}{2} (A - B)} \tag{257}$$

for we have, from the process of finding A and B, the log. of $a + b$, or of $a - b$, and the values of $\frac{1}{2} (A + B)$ and $\frac{1}{2} (A - B)$, so that we have only two new logs. to find, which are taken out at the same opening of the tables with the tangents of $\frac{1}{2} (A + B)$ and $\frac{1}{2} (A - B)$.

140. Case III. Given a, b and C. *Second Solution.* When a and b are given by their logarithms, which occurs when they are deduced by a logarithmic process from other data (as, for example, in the computation of a series of triangles in a survey), we proceed as follows. Let x be an *auxiliary* angle, such that

$$\tan x = \frac{a}{b} \tag{258}$$

an assumption always admissible, since a tangent may have any value from 0 to ∞.

We deduce

$$\frac{\tan x - 1}{\tan x + 1} = \frac{a - b}{a + b}$$

or by (152)
$$\tan (x - 45°) = \frac{a - b}{a + b}$$

which substituted in (254) gives

$$\tan \tfrac{1}{2} (A - B) = \tan (x - 45°) \tan \tfrac{1}{2} (A + B) \tag{259}$$

We find x from (258) and employ its value in (259). As this method does not require the preparation of $a + b$ and $a - b$, it is quite as short in practice as (254).

Example.

Given $\log a = 8{\cdot}7950941$, $\log b = 8{\cdot}3706056$, and $C = 110^\circ\ 32'$. (Same as Ex. 1. Art. 138.)

	$\log a = 8{\cdot}7950941$
	$\log b = 8{\cdot}3706056$
$x = 69^\circ\ 22'\ 46''{\cdot}8$	$\log\tan 0{\cdot}4244885$
$x - 45^\circ = 24^\circ\ 22'\ 46''{\cdot}8$	$\log\tan 9{\cdot}6562825$
$\frac{1}{2}(A + B) = 34^\circ\ 44'$	$\log\tan 9{\cdot}8409174$
$\frac{1}{2}(A - B) = 17^\circ\ 26'\ 32''{\cdot}9$	$\log\tan 9{\cdot}4971999$

141. Case III. Given a, b, and C. *Third Solution.* To express A or B directly in terms of the data, we have, from (218) and (224)

$$c \sin A = a \sin C$$
$$c \cos A = b - a \cos C$$

the quotient of which is

$$\tan A = \frac{a \sin C}{b - a \cos C} \tag{260}$$

and in the same manner

$$\tan B = \frac{b \sin C}{a - b \cos C} \tag{261}$$

142. Case III. Given a, b and C. *Fourth Solution.* To find c directly from the data, we have, by (223)

$$c^2 = a^2 + b^2 - 2\, ab \cos C$$

which, however, is not adapted for logarithmic computation. It may be adapted as follows. Substitute by (139)

$$\cos C = 1 - 2 \sin^2 \tfrac{1}{2} C$$

then
$$c^2 = a^2 + b^2 - 2\, ab + 4\, ab \sin^2 \tfrac{1}{2} C$$
$$= (a - b)^2 + 4\, ab \sin^2 \tfrac{1}{2} C$$

$$c = (a - b) \sqrt{\left(1 + \frac{4\, ab \sin^2 \frac{1}{2} C}{(a - b)^2}\right)} \tag{262}$$

Let x be an auxiliary angle, such that

$$\tan^2 x = \frac{4\, ab \sin^2 \frac{1}{2} C}{(a - b)^2}$$

or

$$\tan x = \frac{2 \sin \frac{1}{2} C}{a - b} \sqrt{ab} \tag{263}$$

then the radical in the above equation becomes $\sqrt{(1 + \tan^2 x)} = \sec x$; therefore,

$$c = (a - b) \sec x \tag{264}$$

143. We may also adapt (223) for logarithmic computation by means of (138), which gives

$$\cos C = -1 + 2 \cos^2 \tfrac{1}{2} C$$

whence

$$c^2 = a^2 + b^2 + 2\, ab - 4\, ab \cos^2 \tfrac{1}{2} C$$

$$c = (a + b) \sqrt{\left(1 - \frac{4\, ab \cos^2 \frac{1}{2} C}{(a + b)^2}\right)} \tag{265}$$

Let

$$\sin x = \frac{2 \cos \frac{1}{2} C}{a + b} \sqrt{ab} \tag{266}$$

then the radical becomes $\sqrt{(1 - \sin^2 x)} = \cos x$; therefore,

$$c = (a + b) \cos x \tag{267}$$

144. It is to be observed, that the supposition (263) is always possible, since a tangent may have any value between 0 and ∞, and therefore an angle x may always be found having any given number as its tangent. As the greatest value of a sine is unity, it is not so obvious that the supposition (266) is always possible; but whatever the values of a and b

$$(a - b)^2 > 0$$

therefore

$$(a + b)^2 > 4\, ab$$

whence

$$\frac{2\sqrt{ab}}{a + b} < 1$$

therefore the second member of (266) is always < 1.

Example.

Given $a = \cdot062387$, $b = \cdot023475$, $C = 110° 32'$; (same as Ex. 1, Art. 138)

By (266) and (267).

$a = \cdot062387$	log 8·7950941	
$b = \cdot023475$	log 8·3706056	
	2)7·1656997	
	log $\sqrt{ab}$ = 8·5828499	
$a + b = \cdot085862$	ar. co. log 1·0661990	log 8·9338010
$\frac{1}{2} C = 55° 16'$	l. cos 9·7556902	
	log 2 0·3010300	
	l. sin x 9·7057691	l. cos x 9·9352161
	c = ·07396344	log 8·8690171

145. CASE IV. *Given the three sides*, or a, b and c. *To find A, B and C.* We have from (227) and (228),

$$\left.\begin{aligned}\sin\tfrac{1}{2}A &= \sqrt{\left(\frac{(s-b)(s-c)}{bc}\right)}\\ \sin\tfrac{1}{2}B &= \sqrt{\left(\frac{(s-a)(s-c)}{ac}\right)}\\ \sin\tfrac{1}{2}C &= \sqrt{\left(\frac{(s-a)(s-b)}{ab}\right)}\end{aligned}\right\} \quad (268)$$

or by (229) and (230)

$$\left.\begin{aligned}\cos\tfrac{1}{2}A &= \sqrt{\left(\frac{s(s-a)}{bc}\right)}\\ \cos\tfrac{1}{2}B &= \sqrt{\left(\frac{s(s-b)}{ac}\right)}\\ \cos\tfrac{1}{2}C &= \sqrt{\left(\frac{s(s-c)}{ab}\right)}\end{aligned}\right\} \quad (269)$$

or by (231) and (232)

$$\left.\begin{aligned}\tan\tfrac{1}{2}A &= \sqrt{\left(\frac{(s-b)(s-c)}{s(s-a)}\right)}\\ \tan\tfrac{1}{2}B &= \sqrt{\left(\frac{(s-a)(s-c)}{s(s-b)}\right)}\\ \tan\tfrac{1}{2}C &= \sqrt{\left(\frac{(s-a)(s-b)}{s(s-c)}\right)}\end{aligned}\right\} \quad (270)$$

In these formulæ $s = \frac{1}{2}(a + b + c)$. Either of these three methods may, in general, be employed, but (268) is to be preferred when the half angle is less than 45°, and (269) when the half angle is more than 45°.* When all the angles are required, (270) will be the simplest, as it requires but four different logs. to be taken from the tables. It is accurate for all values of the angle.

---

* See Art. 112.

EXAMPLES.

1. Given $a = 10$, $b = 12$, $c = 14$; find the angles.

By (268).

$a = 10$		ar co l 9·0000000	ar co l 9·0000000
$b = 12$	ar co l 8·9208188		ar co l 8·9208188
$c = 14$	ar co l 8·8538720	ar co l 8·8538720	
$2s = 36$			
$s = 18$			
$s - a = 8$		log 0·9030900	log 0·9030900
$s - b = 6$	log 0·7781513		log 0·7781513
$s - c = 4$	log 0·6020600	log 0·6020600	
	2)9·1549021	2)9·3590220	2)9·6020601
log sines	$\frac{1}{2}A$ 9·5774510	$\frac{1}{2}B$ 9·6795110	$\frac{1}{2}C$ 9·8010300

$\frac{1}{2}A = 22^\circ\, 12'\, 27''\cdot 6$ $\quad \frac{1}{2}B = 28^\circ\, 33'\, 39''\cdot 0$ $\quad \frac{1}{2}C = 39^\circ\, 13'\, 53''\cdot 5$

$A = 44^\circ\, 24'\, 55''\cdot 2$ $\quad B = 57^\circ\, 7'\, 18''\cdot 0$ $\quad C = 78^\circ\, 27'\, 47''\cdot 0$

Verification. $\quad A + B + C = 180^\circ$

2. Given $a = \cdot 8706$, $b = \cdot 0916$, $c = \cdot 7902$; find the angles.

Ans. $A = 149^\circ\, 49'\, 0''.4$
$B = 3^\circ\, 1'\, 56''.2$
$C = 27^\circ\, 9'\, 3''.4$

3. Given $a = \cdot 5123864$, $b = \cdot 3538971$, $c = \cdot 3090507$; find C.

Ans. $C = 36^\circ\, 18'\, 10''\cdot 2$

146. The computation by (270), when all the angles are required, will be much facilitated by the introduction of an auxiliary quantity*

$$r = \sqrt{\left(\frac{(s-a)(s-b)(s-c)}{s}\right)} \qquad (271)$$

from which we find by (270)

$$\tan \tfrac{1}{2} A = \frac{r}{s-a}, \qquad \tan \tfrac{1}{2} B = \frac{r}{s-b}, \qquad \tan \tfrac{1}{2} C = \frac{r}{s-c} \qquad (272)$$

* This quantity r is the radius of the inscribed circle. See (289).

EXAMPLE. Given $a = 6053$, $b = 4082$, $c = 7068$. We find

$s = 8601{\cdot}5$	ar. co.	log 6·0654258
$s - a = 2548{\cdot}5$		log 3·4062846
$s - b = 4519{\cdot}5$		log 3·6550904
$s - c = 1533{\cdot}5$		log 3·1856838
		2)6·3124846
	$\log r$ =	3.1562423
$\frac{1}{2} A = 29° \; 20' \; 54''{\cdot}47$	$\log \tan \frac{1}{2} A = \log \frac{r}{s-a}$ =	9·7499577
$\frac{1}{2} B = 17° \; 35' \; 31''{\cdot}70$	$\log \tan \frac{1}{2} B = \log \frac{r}{s-b}$ =	9·5011519
$\frac{1}{2} C = 43° \; 3' \; 33''{\cdot}83$	$\log \tan \frac{1}{2} C = \log \frac{r}{s-c}$ =	9·9705585

Verification. $90° \; 0' \; 0''{\cdot}00$

Fig. 22.

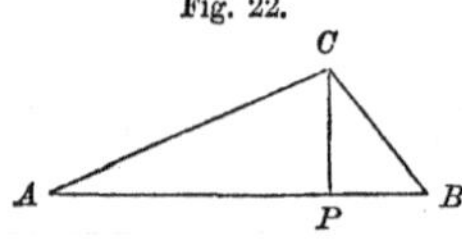

147. The case where the three sides are given is sometimes solved as follows. From C, Fig. 22, draw CP perp. to c. Then

$$AC^2 = AP^2 + CP^2, \qquad BC^2 = BP^2 + CP^2$$

the difference of which is

$$AC^2 - BC^2 = AP^2 - BP^2$$

or
$$(AC + BC)(AC - BC) = (AP + BP)(AP - BP)$$

and if $AP - BP = d$, this equation gives

$$d = \frac{(b + a)(b - a)}{c} \tag{273}$$

Then, since $AP + BP = c$, and $AP - BP = d$, we have

$$AP = \tfrac{1}{2}(c + d), \qquad BP = \tfrac{1}{2}(c - d) \tag{274}$$

and in the right triangles ACP, BCP

$$\cos A = \frac{AP}{b}, \qquad \cos B = \frac{BP}{a} \tag{275}$$

so that (273), (274) and (275) solve the problem. When $d > c$, BP is negative, $\cos B$ is negative, and B is an obtuse angle, (Art. 39).

AREA OF A PLANE TRIANGLE.

148. Representing the area by K, and the perpendicular CP, Fig. 22, by p, we have, by geometry,

$$K = \tfrac{1}{2} cp \tag{276}$$

In the triangle ACP, we have $p = b \sin A$, whence

$$K = \tfrac{1}{2} bc \sin A \tag{277}$$

by which the area is computed from two sides and the included angle.

Substituting in (277) the value of $\sin A$ from (239),

$$K = \sqrt{[s(s-a)(s-b)(s-c)]} \tag{278}$$

by which the area is computed from the three sides.

CHAPTER IX.

MISCELLANEOUS PROBLEMS RELATING TO PLANE TRIANGLES.

149. *In a given plane triangle, to find the perpendicular from one of the angles upon the opposite side.*

Let p be the perpendicular from C upon c. We have

$$p = b \sin A \tag{279}$$

or by (239) and (278),

$$p = \frac{2}{c} \sqrt{[s(s-a)(s-b)(s-c)]} = \frac{2K}{c} \tag{280}$$

the expression for p in terms of the three sides, where $s = \frac{1}{2}(a + b + c)$ and K is the area of the triangle.

If we substitute in (279) $\sin A = \frac{a}{c} \sin C$, it becomes

$$p = \frac{ab}{c} \sin C \tag{281}$$

or, if we substitute the value of $b = c\,\frac{\sin B}{\sin C}$

$$p = c\,\frac{\sin A \sin B}{\sin C} = c\,\frac{\sin A \sin B}{\sin (A+B)} \tag{282}$$

When the triangle is right-angled at C, (282) becomes

$$p = c \sin A \cos A = \frac{c}{2} \sin 2A$$

the expression for the perpendicular upon the hypotenuse.

150. If p', p'', p''', denote the perpendiculars upon the sides a, b, c respectively, we have from (280)

$$\frac{1}{p'} + \frac{1}{p''} + \frac{1}{p'''} = \frac{a+b+c}{2K} = \frac{s}{K} \tag{283}$$

151. *To find the radius of the circle circumscribed about a plane triangle.*

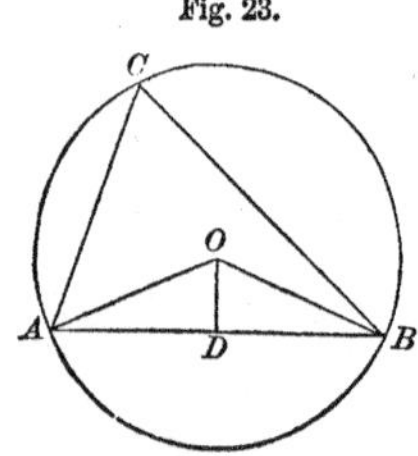

Fig. 23.

The center O of the circle, Fig. 23, lies in the perpendicular erected from the middle point of one of the sides, as AB. Let the radius $= R$. We have, by geometry,

$$AOD = \tfrac{1}{2} AOB = C$$

and in the triangle AOD,

$$\sin AOD = \sin C = \frac{AD}{AO} = \frac{c}{2R}$$

whence

$$R = \frac{c}{2 \sin C} \tag{284}$$

Substituting the value of $\sin C$ from (241),

$$R = \frac{abc}{4\sqrt{[s(s-a)(s-b)(s-c)]}} = \frac{abc}{4K} \tag{285}$$

From (229) and (230), we easily find

$$\cos \tfrac{1}{2} A \cos \tfrac{1}{2} B \cos \tfrac{1}{2} C = \frac{Ks}{abc}$$

which combined with (285) gives

$$R = \frac{s}{4 \cos \frac{1}{2} A \cos \frac{1}{2} B \cos \frac{1}{2} C} \tag{286}$$

152. *To find the radius of the circle inscribed in a plane triangle.*

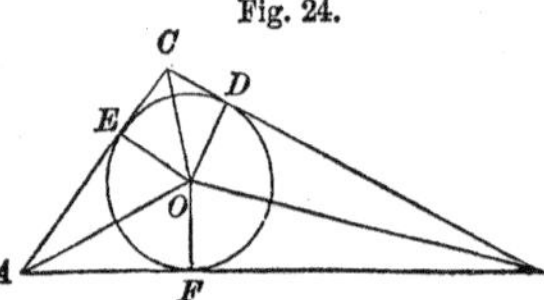

Fig. 24.

The required center O, Fig. 24, is in the intersection of the three lines bisecting the angles, and each of the perpendiculars OD, OE, OF, is equal to the required radius $= r$. The value of OF in terms of $AB = c$, $OAB = \frac{1}{2} A$, and $OBA = \frac{1}{2} B$ is by (282)

$$r = c \, . \, \frac{\sin \frac{1}{2} A \sin \frac{1}{2} B}{\sin \frac{1}{2}(A + B)} = c \, . \, \frac{\sin \frac{1}{2} A \sin \frac{1}{2} B}{\cos \frac{1}{2} C} \tag{287}$$

This is reduced by means of (236) to

$$r = (s - c) \tan \tfrac{1}{2} C \tag{288}$$

Substituting the value of $\tan \frac{1}{2} C$,

$$r = \sqrt{\left(\frac{(s-a)(s-b)(s-c)}{s}\right)} = \frac{K}{s} \tag{289}$$

This is reduced by means of (231) and (232) to

$$r = s \tan \tfrac{1}{2} A \tan \tfrac{1}{2} B \tan \tfrac{1}{2} C \tag{290}$$

153. Besides the inscribed circle, strictly so called, there are three other circles that touch the three sides, (or sides produced), and are exterior to the triangle, as in Fig. 25. These have been named *escribed* circles. Their centers are found

Fig. 25.

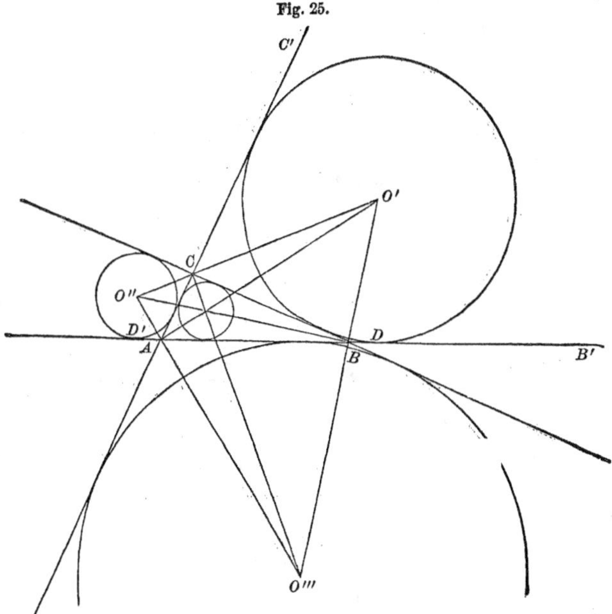

geometrically, by bisecting the exterior angles BCC', CBB', &c. Designate the centers of the circles lying within the angles A, B, and C respectively, by O', O'', and O''', and their radii by r', r'', r'''. We find the perpendiculars from O', &c., upon BC, &c. by (282) to be

$$\left.\begin{aligned} r' &= a \, . \, \frac{\cos\frac{1}{2}B\cos\frac{1}{2}C}{\sin\frac{1}{2}(B+C)} = a \, . \, \frac{\cos\frac{1}{2}B\cos\frac{1}{2}C}{\cos\frac{1}{2}A} \\ r'' &= b \, . \, \frac{\cos\frac{1}{2}A\cos\frac{1}{2}C}{\sin\frac{1}{2}(A+C)} = b \, . \, \frac{\cos\frac{1}{2}A\cos\frac{1}{2}C}{\cos\frac{1}{2}B} \\ r''' &= c \, . \, \frac{\cos\frac{1}{2}A\cos\frac{1}{2}B}{\sin\frac{1}{2}(A+B)} = c \, . \, \frac{\cos\frac{1}{2}A\cos\frac{1}{2}B}{\cos\frac{1}{2}C} \end{aligned}\right\} \qquad (291)$$

By means of (285) we reduce those values to

$$r' = s\tan\tfrac{1}{2}A, \qquad r'' = s\tan\tfrac{1}{2}B, \qquad r''' = s\tan\tfrac{1}{2}C \qquad (292)$$

Substituting the values of $\tan\frac{1}{2}A$, &c.

$$\left.\begin{aligned} r' &= \sqrt{\left(\frac{s(s-b)(s-c)}{s-a}\right)} = \frac{K}{s-a} \\ r'' &= \sqrt{\left(\frac{s(s-a)(s-c)}{s-b}\right)} = \frac{K}{s-b} \\ r''' &= \sqrt{\left(\frac{s(s-a)(s-b)}{s-c}\right)} = \frac{K}{s-c} \end{aligned}\right\} \qquad (293)$$

Also, by means of (236) applied successively to a, b and c, we may reduce (291) to the following:

$$\left.\begin{aligned} r' &= (s - a) \tan \tfrac{1}{2} A \cot \tfrac{1}{2} B \cot \tfrac{1}{2} C \\ r'' &= (s - b) \cot \tfrac{1}{2} A \tan \tfrac{1}{2} B \cot \tfrac{1}{2} C \\ r''' &= (s - c) \cot \tfrac{1}{2} A \cot \tfrac{1}{2} B \tan \tfrac{1}{2} C \end{aligned}\right\} \quad (294)$$

154. *Relations between the radii of the circumscribed, inscribed, and three escribed circles of the preceding article, and the three perpendiculars from the angles upon the opposite sides.*

The four equations of (289) and (292) give

$$r\, r'\, r''\, r''' = \frac{K^4}{s\,(s - a)\,(s - b)\,(s - c)} = \frac{K^4}{K^2} = K^2 \quad (295)$$

Dividing this successively by r^2, r'^2, &c.

$$\left.\begin{aligned} \frac{r'\, r''\, r'''}{r} &= s^2 & \qquad \frac{r\, r''\, r'''}{r'} &= (s - a)^2 \\ \frac{r\, r'\, r'''}{r''} &= (s - b)^2 & \qquad \frac{r\, r'\, r''}{r'''} &= (s - c)^2 \end{aligned}\right\} \quad (296)$$

Again, we have, (Art. 127),

$$\tan \tfrac{1}{2} A \tan \tfrac{1}{2} B + \tan \tfrac{1}{2} A \tan \tfrac{1}{2} C + \tan \tfrac{1}{2} B \tan \tfrac{1}{2} C = 1$$

and substituting in this the value of the tangents from (292)

$$r'\, r'' + r'\, r''' + r''\, r''' = s^2 = \frac{r'\, r''\, r'''}{r}$$

$$\frac{1}{r'} + \frac{1}{r''} + \frac{1}{r'''} = \frac{1}{r} \quad (297)$$

From (292) we find

$$\frac{\tan \frac{1}{2} A}{\tan \frac{1}{2} B} = \frac{r'}{r''} \qquad r'' \tan \tfrac{1}{2} A = r' \tan \tfrac{1}{2} B$$

from which it follows that in Fig. 25, the distances $A\,D$ and $B\,D'$, are equal (D, D' being the points of contact of the circles O', O'' with $A\,B$ produced), and therefore $B\,D = A\,D'$. Other curious geometrical properties may be traced with the aid of our equations.

From (284),

$$R = \frac{a}{4 \sin \frac{1}{2} A \cos \frac{1}{2} A} = \frac{b}{4 \sin \frac{1}{2} B \cos \frac{1}{2} B} = \frac{c}{4 \sin \frac{1}{2} C \cos \frac{1}{2} C}$$

which combined with (287) and (291) give, by Art. 127,

$$\left.\begin{aligned} \frac{r}{R} &= 4 \sin \tfrac{1}{2} A \sin \tfrac{1}{2} B \sin \tfrac{1}{2} C = \cos A + \cos B + \cos C - 1 \\ \frac{r'}{R} &= 4 \sin \tfrac{1}{2} A \cos \tfrac{1}{2} B \cos \tfrac{1}{2} C = - \cos A + \cos B + \cos C + 1 \\ \frac{r''}{R} &= 4 \cos \tfrac{1}{2} A \sin \tfrac{1}{2} B \cos \tfrac{1}{2} C = \cos A - \cos B + \cos C + 1 \\ \frac{r'''}{R} &= 4 \cos \tfrac{1}{2} A \cos \tfrac{1}{2} B \sin \tfrac{1}{2} C = \cos A + \cos B - \cos C + 1 \end{aligned}\right\} \quad (298)$$

Changing the signs of the first of these equations, the sum of the four is

$$\frac{r' + r'' + r''' - r}{R} = 4, \qquad R = \tfrac{1}{4}\,(r' + r'' + r''' - r) \tag{299}$$

Finally, if p', p'', p''' denote the three perpendiculars from the angles upon the sides a, b, c respectively, we have by (283), (289) and (297) the following relation:

$$\frac{1}{p'} + \frac{1}{p''} + \frac{1}{p'''} = \frac{1}{r'} + \frac{1}{r''} + \frac{1}{r'''} = \frac{1}{r} \tag{299*}$$

155. *To find the distance between the centers of the circumscribed and inscribed circles.**

Fig. 26.

Let P, Fig. 26, be the center of the circumscribed, and O, that of the inscribed circle. Put $P\,O = D$. By Arts. 151 and 152,

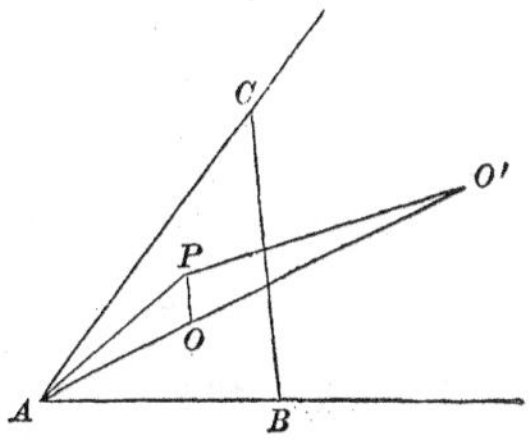

$$P\,A\,B = 90° - C, \qquad O\,A\,B = \tfrac{1}{2}\,A$$

whence

$$P\,A\,O = 90° - C - \tfrac{1}{2}\,A = \tfrac{1}{2}\,(B - C)$$

and by (221)

$$P\,O^2 = P\,A^2 + O\,A^2 - 2\,P\,A\,.\,O\,A \cos P\,A\,O$$

or

$$D^2 = R^2 + \frac{r^2}{\sin^2 \frac{1}{2} A} - \frac{2\,Rr \cos \frac{1}{2}\,(B - C)}{\sin \frac{1}{2} A}$$

By (298)

$$\frac{r^2}{\sin^2 \frac{1}{2} A} = \frac{4\,Rr \sin \frac{1}{2}\,B \sin \frac{1}{2}\,C}{\sin \frac{1}{2} A}$$

therefore

$$D^2 = R^2 - \frac{2\,Rr \cos \frac{1}{2}\,(B + C)}{\sin \frac{1}{2} A}$$

or

$$D^2 = R^2 - 2\,Rr \tag{300}$$

156. Let $P\,O' = D'$, Fig. 26, O' being the center of the escribed circle lying within the angle A. If r' = radius of this circle, we have, as in the preceding article

$$D'^2 = R^2 + \frac{r'^2}{\sin^2 \frac{1}{2} A} - \frac{2\,Rr' \cos \frac{1}{2}\,(B - C)}{\sin \frac{1}{2} A}$$

$$\frac{r'^2}{\sin^2 \frac{1}{2} A} = \frac{4\,Rr' \cos \frac{1}{2}\,B \cos \frac{1}{2}\,C}{\sin \frac{1}{2} A}$$

$$\left.\begin{aligned} D'^2 &= R^2 + 2\,Rr' \\ D''^2 &= R^2 + 2\,Rr'' \\ D'''^2 &= R^2 + 2\,Rr''' \end{aligned}\right\} \tag{301}$$

---

* Hymers' Trig. Appendix, Art. 58.

the expressions for the distances of the centers of the three escribed circles from that of the circumscribed circle.

The sum of (300) and (301) gives by (299)

$$D^2 + D'^2 + D''^2 + D'''^2 = 12\,R^2 \tag{302}$$

157. *Given two sides of a plane triangle and the difference of their opposite angles,* (or a, b, and $A - B$), *to solve the triangle.*

We have $\frac{1}{2}(A + B)$ directly from (220), which also solves the case where two angles and the sum or difference of two sides are given.

158. *Given the angles and the sum of the sides,* (or A, B, C, and $a + b + c = 2\,s$). By (235)

$$c = s \cdot \frac{\sin \frac{1}{2} C}{\cos \frac{1}{2} A \cos \frac{1}{2} B}$$

and a and b are found by similar formulæ.

159. *Given one angle, the opposite side, and the sum of the squares of the other two sides,* (or C, c, and $a^2 + b^2 = e^2$).

In the identical equations

$$(a + b)^2 = e^2 + 2\,ab, \qquad (a - b)^2 = e^2 - 2\,ab$$

substitute the value of $2\,ab$ given by (223), namely,

$$2\,ab = \frac{e^2 - c^2}{\cos C}$$

we find $$(a + b)^2 = e^2 + \frac{e^2 - c^2}{\cos C}, \qquad (a - b)^2 = e^2 - \frac{e^2 - c^2}{\cos C}$$

which determine $a + b$, and $a - b$, and therefore a and b.

To compute these equations by logarithms, let

$$g^2 = \frac{e^2 - c^2}{\cos C} = \frac{(e + c)\,(e - c)}{\cos C} \tag{303}$$

then $$(a + b)^2 = e^2 + g^2, \qquad (a - b)^2 = e^2 - g^2$$

that is $a + b$ is the hypotenuse of a right triangle whose sides are e and g; and $a - b$ is one side of a right triangle whose other side is g, and whose hypotenuse is e. Let the angle opposite g be denoted by x in the first triangle and by x' in the second, then by the formulæ of right triangles

$$\left.\begin{aligned} \tan x &= \frac{g}{e} \qquad & a + b &= e \sec x \\ \sin x' &= \frac{g}{e} \qquad & a - b &= e \cos x' \end{aligned}\right\} \tag{304}$$

so that the problem is solved by logarithms by finding $\log g$ from (303) and employing its value in (304).

The above may serve as an example of a *geometrical* method of introducing the auxiliary quantities, which is occasionally useful. The analytical process in the present instance is similar to that of Art. 143; thus

$$a + b = e\sqrt{\left(1 + \frac{g^2}{e^2}\right)} \qquad a - b = e\sqrt{\left(1 - \frac{g^2}{e^2}\right)}$$

therefore if $\tan x = \frac{g}{e}$ we have $\sqrt{\left(1+\frac{g^2}{e^2}\right)} = \sec x$

and if $\sin x' = \frac{g}{e}$ we have $\sqrt{\left(1-\frac{g^2}{e^2}\right)} = \cos x'$

whence the same formulæ as before.

160. *Given an angle, its opposite side, and the difference of the squares of the other two sides*, (or C, c, and $a^2 - b^2 = f^2$).

We have by multiplying (233) by (234)

$$\frac{\sin(A-B)}{\sin C} = \frac{a^2-b^2}{c^2} = \frac{f^2}{c^2}$$

$$\sin(A-B) = \frac{f^2}{c^2}\sin C$$

whence $A - B$, and since $A + B = 180° - C$, the angles are determined. There will be two solutions given by $\sin(A - B)$ except where the obtuse value of $A - B$ is greater than $A + B$.

161. *Given the three perpendiculars from the three angles upon the opposite sides.*

Denote the perps. upon a, b and c respectively by a', b' and c', and let

$$a'' = \frac{1}{a'}, \quad b'' = \frac{1}{b'}, \quad c'' = \frac{1}{c'}$$

If $k = 2$ area of the triangle

$$aa' = bb' = cc' = k$$

and therefore

$$a = a''k, \quad b = b''k, \quad c = c''k$$

Substituting these values of a, b and c in (225), (227,) &c.

$$\cos A = \frac{b''^2 + c''^2 - a''^2}{2\,b''c''}, \quad \sin^2 \tfrac{1}{2} A = \frac{(s''-b'')(s''-c'')}{b''c''}, \text{ \&c.}$$

in which $2\,s'' = a'' + b'' + c''$.

162. *Given the radii of the circumscribed and inscribed circles, and the perpendicular from one of the angles upon the opposite side, to solve the triangle.*

Let c be the side to which the perpendicular (p) is drawn. We have found for R, r and p the expressions

$$R = \frac{c}{2\sin C} = \frac{c}{2\sin(A+B)} = \frac{c}{4\sin\frac{1}{2}(A+B)\cos\frac{1}{2}(A+B)}$$

$$r = c\,.\,\frac{\sin\frac{1}{2}A\sin\frac{1}{2}B}{\sin\frac{1}{2}(A+B)}$$

$$p = c\,.\,\frac{\sin A\sin B}{\sin(A+B)}$$

Eliminating c, we have

$$\frac{p}{2R} = \sin A \sin B \qquad (m)$$

$$\frac{r}{R} = 4 \cos \tfrac{1}{2}(A + B) \sin \tfrac{1}{2} A \sin \tfrac{1}{2} B \qquad (n)$$

from which two equations A and B are to be found. Developing $\cos \frac{1}{2}(A + B)$, (n) becomes

$$\frac{r}{R} = 4 \sin \tfrac{1}{2} A \cos \tfrac{1}{2} A \sin \tfrac{1}{2} B \cos \tfrac{1}{2} B - 4 \sin^2 \tfrac{1}{2} A \sin^2 \tfrac{1}{2} B$$

$$= \sin A \sin B - 4 \sin^2 \tfrac{1}{2} A \sin^2 \tfrac{1}{2} B$$

which subtracted from (m) gives

$$\frac{p - 2r}{2R} = 4 \sin^2 \tfrac{1}{2} A \sin^2 \tfrac{1}{2} B \qquad (o)$$

Dividing the square of (m) by (o), we find

$$\frac{p^2}{2R(p - 2r)} = 4 \cos^2 \tfrac{1}{2} A \cos^2 \tfrac{1}{2} B$$

whence

$$\sin \tfrac{1}{2} A \sin \tfrac{1}{2} B = \sqrt{\left(\frac{p - 2r}{8R}\right)} = \frac{p - 2r}{2\sqrt{[2R(p - 2r)]}}$$

$$\cos \tfrac{1}{2} A \cos \tfrac{1}{2} B = \frac{p}{2\sqrt{[2R(p - 2r)]}}$$

The difference and sum of these two equations give

$$\left.\begin{aligned} \cos \tfrac{1}{2}(A + B) &= \frac{r}{\sqrt{[2R(p - 2r)]}} \\ \cos \tfrac{1}{2}(A - B) &= \frac{p - r}{\sqrt{[2R(p - 2r)]}} \end{aligned}\right\} \qquad (305)$$

which determine $\frac{1}{2}(A + B)$ and $\frac{1}{2}(A - B)$ and therefore A and B. The sides are then found by the formula

$$c = 2R \sin C$$

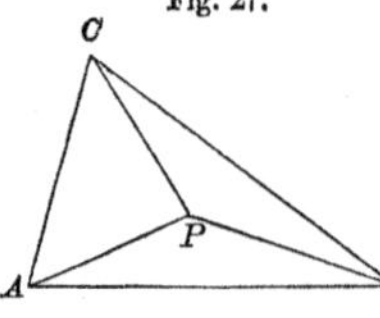

Fig. 27.

163. *In a given plane triangle ABC, Fig. 27, to find a point P such that the three lines drawn from this point to the angles A, B and C shall make given angles with each other.*

Let the given angles be $BPC = \alpha$ and $APC = \beta$ and the required angles $PAC = x$ $PBC = y$

The sum of the angles of the quadrilateral $ACBP$ is

$$x + y + \alpha + \beta + C = 360°$$

whence

$$\tfrac{1}{2}(x + y) = 180° - \tfrac{1}{2}(\alpha + \beta + C) \qquad (306)$$

In the triangles APC, BPC, we have

$$PC = \frac{b \sin x}{\sin \beta} = \frac{a \sin y}{\sin \alpha} \qquad \frac{\sin x}{\sin y} = \frac{a}{b} \cdot \frac{\sin \beta}{\sin \alpha} = m$$

from which

$$\frac{\sin x + \sin y}{\sin x - \sin y} = \frac{\tan \frac{1}{2}(x+y)}{\tan \frac{1}{2}(x-y)} = \frac{m+1}{m-1}$$

$$\tan \tfrac{1}{2}(x-y) = \frac{m-1}{m+1} \tan \tfrac{1}{2}(x+y)$$

To compute this equation by logarithms, let

$$\left.\begin{array}{l} \tan \gamma = m = \dfrac{a \sin \beta}{b \sin \alpha} \\ \text{then by (152),} \quad \tan \frac{1}{2}(x-y) = \tan(\gamma - 45°) \tan \frac{1}{2}(x+y) \end{array}\right\} \quad (307)$$

so that the angles x and y are found by (306) and (307).

164. The following problems are proposed as exercises.

In a plane triangle ABC —

1. Given c, the perp. upon $c = p$ and $a + b = m$.

$$\sin^2 x = \frac{4p^2}{(m+c)(m-c)} \qquad a - b = c \cos x$$

$$\tan \tfrac{1}{2} C = \frac{2pc}{(m+c)(m-c)}$$

2. Given c, the perp. upon $c = p$, and $a - b = n$.

$$\tan^2 x = \frac{4p^2}{(c+n)(c-n)} \qquad a + b = c \sec x$$

$$\tan \tfrac{1}{2} C = \frac{(c+n)(c-n)}{2pc}$$

3. Given C, c, and $ab = q^2$.

$$\tan x = \frac{2q}{c} \cos \tfrac{1}{2} C \qquad a + b = c \sec x$$

$$\sin x' = \frac{2q}{c} \sin \tfrac{1}{2} C \qquad a - b = c \cos x'$$

4. Given C, the perp. from $C = p$, and $a + b = m$.

$$\tan x = \frac{m}{p} \tan \tfrac{1}{2} C \qquad c = m \tan \tfrac{1}{2} x$$

5. Given C, the perp. from $C = p$, and $a - b = n$.

$$\tan x = \frac{n}{p} \cot \tfrac{1}{2} C \qquad c = n \cot \tfrac{1}{2} x$$

6. Given c, C, and $a + b = m$.

$$\cos \tfrac{1}{2}(A - B) = \frac{m}{c} \sin \tfrac{1}{2} C \qquad a - b = c\,\frac{\sin \frac{1}{2}(A-B)}{\cos \frac{1}{2} C}$$

7. Given c, A, and $a + b = m$.

$$\tan \tfrac{1}{2} B = \frac{m - c}{m + c} \cot \tfrac{1}{2} A$$

8. Given $a + b = m$, the perp. upon $c = p$, and the difference of the segments of $c = d$.

$$c = m \sqrt{\left(1 - \frac{4 p^2}{m^2 - d^2}\right)}$$

$$\tan \tfrac{1}{2} (A + B) = \sqrt{\left[\frac{4 m^2 p^2}{(m^2 - d^2)(m^2 - d^2 - 4 p^2)}\right]}$$

$$\tan \tfrac{1}{2} (A - B) = \frac{2 pd}{m^2 - d^2}$$

or with an auxiliary angle

$$\sin^2 x = \frac{4 p^2}{(m + d)(m - d)} \qquad c = m \cos x$$

$$\tan \tfrac{1}{2} (A + B) = \frac{m}{2 p} \sin x \tan x \qquad \tan \tfrac{1}{2} (A - B) = \frac{d}{2 p} \sin^2 x$$

9. Given the perimeter $= 2 s$, C, and the perp. from $C = p$.

$$\tan^2 x = \frac{p}{2 s} \cot \tfrac{1}{2} C \qquad c = s \cos^2 x$$

10. Given c, $a + b = m$ and the radius of the inscribed circle $= r$.

$$\sin x = \frac{2 r}{c} \sqrt{\left(\frac{m + c}{m - c}\right)} \qquad a - b = c \cos x$$

$$\tan \tfrac{1}{2} C = \frac{2 r}{m - c}$$

11. Given c, $a - b = n$, and the radius of the inscribed circle $= r$.

$$\tan x = \frac{(c + n)(c - n)}{4 r^2} \qquad a + b = c \cot (x - 45^\circ)$$

12. Given the radii r', r'', r''', of the three escribed circles. (Arts. 153, 154).

$$\tan^2 \tfrac{1}{2} A = \frac{r'^2}{r' r'' + r' r''' + r'' r'''}$$

165. *Given the sides of a quadrilateral inscribed in a circle, to find its angles and area.*

Fig. 28.

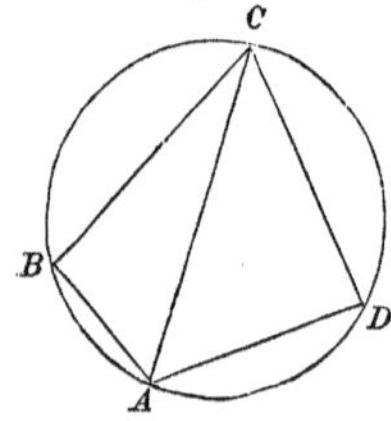

In Fig. 28, let $AB = a$, $BC = b$, $CD = c$, $DA = d$. Let $2 s = a + b + c + d$ and K = area of $ABCD$; then from the triangles ABC, ADC, observing that $B = 180^\circ - D$ we find

$$\sin^2 \tfrac{1}{2} B = \frac{(s - a)(s - b)}{ab + cd}, \qquad \cos^2 \tfrac{1}{2} B = \frac{(s - c)(s - d)}{ab + cd}$$

$$\tan^2 \tfrac{1}{2} B = \frac{(s - a)(s - b)}{(s - c)(s - d)}$$

$$K = \sqrt{[(s - a)(s - b)(s - c)(s - d)]} \qquad (308)$$

CHAPTER X.

SOLUTION OF CERTAIN TRIGONOMETRIC EQUATIONS AND OF NUMERICAL EQUATIONS OF THE SECOND AND THIRD DEGREES.

166. THE solution of a problem in which the unknown quantity is an angle, often depends upon that of one or more equations, involving different functions of the angle, which cannot be reduced by merely algebraic transformations. We shall select a few simple examples of such equations from among those that most frequently occur in astronomy.

167. *To find z from the equation*

$$\sin(\alpha + z) = m \sin z \tag{309}$$

in which α and m are given. We have, by (119),

$$\sin(\alpha + z) = \sin\alpha \sin z\,(\cot z + \cot\alpha)$$

which becomes identical with (309) by taking

$$\sin\alpha\,(\cot z + \cot\alpha) = m$$

whence the required solution

$$\cot z = \frac{m}{\sin\alpha} - \cot\alpha \tag{310}$$

If the proposed equation were

$$\sin(\alpha - z) = m \sin z \tag{311}$$

we should find

$$\cot z = \frac{m}{\sin\alpha} + \cot\alpha \tag{312}$$

Unless z is limited by the nature of the problem in which these equations are employed, there will be an indefinite number of solutions; for all the angles z, $z + 180°$, $z + 360°$, $z + 540°$, &c., in general all the angles $z + n\pi$ have the same cotangent. [See (68), (79).] In most cases, however, we consider only the first two of these solutions, taking the values of z always less than 360°.

Similar remarks apply in all cases where an angle is determined by a *single* trigonometric function; but if the problem is such as to give the values of *two* functions of the required angle, as the sine and cosine, the solution is entirely determinate under 360°, since there cannot be two different angles less than 360° that have the same sine and cosine.

168. The solution of the preceding article requires the use of a table of natural cotangents; to obtain a formula adapted for logarithmic computation entirely, we deduce from (309) the following

$$\frac{\sin(\alpha + z) + \sin z}{\sin(\alpha + z) - \sin z} = \frac{m+1}{m-1}$$

But by (109), if $x = \alpha + z$, $y = z$, we have

$$\frac{\sin(\alpha + z) + \sin z}{\sin(\alpha + z) - \sin z} = \frac{\tan(z + \frac{1}{2}\alpha)}{\tan\frac{1}{2}\alpha}$$

which substituted above, gives

$$\tan(z + \tfrac{1}{2}\alpha) = \frac{m+1}{m-1}\tan\tfrac{1}{2}\alpha$$

which determines $z + \frac{1}{2}\alpha$, whence z is found by deducting $\frac{1}{2}\alpha$.

The computation of this equation is facilitated in most cases by introducing an *auxiliary* angle, such that

$$\tan\phi = m$$

an assumption always admissible, since while the angle varies from 0 to 90° the tangent varies from 0 to ∞, so that an angle ϕ may always be found having any given number as its tangent.

We have then by (152),

$$\frac{m+1}{m-1} = \frac{\tan\phi + 1}{\tan\phi - 1} = \cot(\phi - 45°)$$

and the preceding solution becomes

$$\tan\phi = m, \qquad \tan(z + \tfrac{1}{2}\alpha) = \cot(\phi - 45°)\tan\tfrac{1}{2}\alpha \tag{313}$$

The logarithmic solution of (311) is found in the same manner to be

$$\tan\phi = m, \qquad \tan(z - \tfrac{1}{2}\alpha) = \cot(\phi + 45°)\tan\tfrac{1}{2}\alpha \tag{314}$$

169. *To find z from the equation*

$$\tan(\alpha + z) = m \tan z \tag{315}$$

We deduce

$$\frac{\tan(\alpha + z) + \tan z}{\tan(\alpha + z) - \tan z} = \frac{m + 1}{m - 1}$$

so that by (126) and (152) the solution is

$$\tan \phi = m, \qquad \sin(\alpha + 2z) = \cot(\phi - 45°) \sin \alpha \tag{316}$$

170. *To find z from the equation*

$$\tan(\alpha + z) \tan z = m \tag{317}$$

We deduce

$$\frac{1 + \tan(\alpha + z) \tan z}{1 - \tan(\alpha + z) \tan z} = \frac{1 + m}{1 - m}$$

so that by (127) and (151) the solution is

$$\tan \phi = m, \qquad \cos(\alpha + 2z) = \tan(45° - \phi) \cos \alpha \tag{318}$$

171. *To find z from the equation*

$$\sin(\alpha \pm z) \sin z = m \tag{319}$$

By (108) we find

$$\cos \alpha - \cos(\alpha \pm 2z) = \pm 2 \sin(\alpha \pm z) \sin z = \pm 2m$$

whence

$$\cos(\alpha \pm 2z) = \cos \alpha \mp 2m \tag{319*}$$

which determines $\alpha \pm 2z$, and hence $2z$.

From (319*) we have four values of $\alpha \pm 2z$ between 0° and 720°; therefore, four values of $2z$ between the same limits, and four values of z between 0° and 360°.

In general, we shall have four solutions under 360° in all cases where the *double angle* is determined by a *single function.*

The logarithmic solution of (319) varies with the signs of m and z. Thus, if the equation is

$$\sin(\alpha + z) \sin z = m$$

m being essentially positive, we have by (133)

$$\cos^2 \tfrac{1}{2}\alpha - \cos^2(z + \tfrac{1}{2}\alpha) = \sin(\alpha + z) \sin z = m$$

$$\cos^2(z + \tfrac{1}{2}\alpha) = \cos^2 \tfrac{1}{2}\alpha - m$$

and by (133) again this is solved by

$$\cos^2 \phi = m, \qquad \cos^2(z + \tfrac{1}{2}\alpha) = \sin(\phi + \tfrac{1}{2}\alpha) \sin(\phi - \tfrac{1}{2}\alpha)$$

and the other cases are solved by similar methods.

172. The preceding examples will suffice to indicate the method to be followed with all the equations of the following table. The solutions of the equations involving cosines may be obtained from those involving sines, by exchanging z for $90° \pm z$, or α for $90° \pm \alpha$.

Logarithmic solutions of the first four will be obtained by imitating the process of Art. 171.

EQUATIONS.	SOLUTIONS.
1. $\sin(\alpha \pm z) \sin z = m$	$\cos(\alpha \pm 2z) = \cos\alpha \mp 2m$
2. $\cos(\alpha \pm z) \cos z = m$	$\cos(\alpha \pm 2z) = 2m - \cos\alpha$
3. $\sin(\alpha \pm z) \cos z = m$	$\sin(\alpha \pm 2z) = 2m - \sin\alpha$
4. $\cos(\alpha \pm z) \sin z = m$	$\sin(\alpha \pm 2z) = \sin\alpha \pm 2m$
5. $\sin(\alpha \pm z) = m \sin z$	$\tan\phi = m,\quad \tan(z \pm \frac{1}{2}\alpha) = \cot(\phi \mp 45°) \tan\frac{1}{2}\alpha$
6. $\cos(\alpha \pm z) = m \cos z$	$\tan\phi = m,\quad \tan(\frac{1}{2}\alpha \pm z) = \tan(45° - \phi) \cot\frac{1}{2}\alpha$
7. $\sin(\alpha \pm z) = m \cos z$	$\tan\phi = m,$ $\tan(45° - \frac{1}{2}\alpha \mp z) = \tan(45° - \phi) \tan(45° + \frac{1}{2}\alpha)$
8. $\cos(\alpha \pm z) = m \sin z$	$\tan\phi = m,$ $\tan(45° - \frac{1}{2}\alpha \mp z) = \tan(45° \mp \phi) \tan(\frac{1}{2}\alpha - 45°)$
9. $\tan(\alpha \pm z) \tan z = m$	$\tan\phi = m,\quad \cos(\alpha \pm 2z) = \tan(45° \mp \phi) \cos\alpha$
10. $\tan(\alpha \pm z) = m \tan z$	$\tan\phi = m,\quad \sin(2z \pm \alpha) = \cot(\phi \mp 45°) \sin\alpha$

In the numerical solutions the signs of the angles and their functions must be carefully observed. The signs of the functions should be prefixed to their logarithms, according to Art. 99.

The auxiliary angle ϕ may be taken numerically less than 90° in all cases, but positive or negative according to the sign of its tangent. It can easily be shown that we shall thus obtain the same values of z as by taking ϕ in the 2d quadrant when its tangent is negative, or in the 3d quadrant when its tangent is positive.

EXAMPLE.

Find z from (317) when $\alpha = 65°$ and $m = 1{\cdot}5196154$. By (318)

$\log\tan\phi = \log m$*	$+\ 0{\cdot}1817337$			
ϕ	$56°\,39'\,9''$			
$45° - \phi$	$-\ 11°\,39'\,9''$			
$\log\tan(45° - \phi)$	$-\ 9{\cdot}3143426$			
$\log\cos\alpha$	$+\ 9{\cdot}6259483$			
$\log\tan(45° - \phi)\cos\alpha = \log\cos(\alpha + 2z)$	$-\ 8{\cdot}9402909$			
$\alpha + 2z$	95°	or 265°	or 455°	or 625°
α	65°			
$2z$	30°	or 200°	or 390°	or 560°
z	15°	or 100°	or 195°	or 280°

* It must be remembered that in this employment of the signs + and —, these signs belong to the natural numbers; and when the logs. are *added* or *subtracted*, the sign of the result is to be determined according to the rules of *multiplication* and *division* in algebra.

173. *To find z from the equation*

$$\sin(\alpha + z) = m \sin(\beta + z).$$

Put $z' = \beta + z$, $\alpha' = \alpha - \beta$, then this equation becomes

$$\sin(\alpha' + z') = m \sin z'$$

which is of the form (309) and may be solved by (309*) or (311); then $z = z' - \beta$.

In the same manner equations of this form, involving cosines or tangents, may be reduced to those of the preceding table.

174. *To find k and z from the equations*

$$\left.\begin{aligned} k \sin z &= m \\ k \cos z &= n \end{aligned}\right\} \quad (320)$$

We have, by division,

$$\tan z = \frac{m}{n}$$

which gives two values of z, one less, the other greater than 180°; whence, also, two values of k from either of the equations

$$k = \frac{m}{\sin z} = \frac{n}{\cos z}$$

The solution becomes entirely determinate (z not exceeding 360°) as follows:

1st. *When the sign of k is given.* For if k is positive, $\sin z$ has the sign of m, and $\cos z$ the sign of n, and z must be taken in the quadrant denoted by these signs. If k is negative, the signs of $\sin z$ and $\cos z$ are the opposite to those of m and n, and z must be taken accordingly.

2d. *When z is restricted by either the condition* $z < 180°$, or $z > 180°$. For under either of these conditions the tangent gives but one solution. If $z < 180°$, k has the sign of m; and if $z > 180°$ k has the opposite sign to that of m.

3d. *When z is restricted to acute values, positive or negative.* For under this condition a positive tangent will give z between 0° and + 90°; and a negative tangent, between 0° and − 90°; and k will always have the sign of n.

It follows that m and n being any given numbers whatever, we may always satisfy the conditions expressed by (320), 1st, by a positive number k and an angle z between 0° and 360°; 2d, by a number k (unrestricted as to sign) and an angle $z < 180°$; 3d, by a number k (unrestricted as to sign) and an angle $z > 180°$; and 4th, by a number k, and an angle z in the 1st or 4th quadrant.

Example.

To find k and z from (320), (k being a positive number), when $m = -0{\cdot}3076258$, $n = +0{\cdot}4278735$.

	$k \sin z$	$-\ 0{\cdot}3076258$
	$k \cos z$	$+\ 0{\cdot}4278735$
(a)	$\log k \sin z$	$-\ 9{\cdot}4880228$
(b)	$\log k \cos z$	$+\ 9{\cdot}6313147$
(a) − (b)	$\log \tan z$	$-\ 9{\cdot}8567081$
	z	$324^\circ\ 17'\ 6''{\cdot}6$
(c)	$\log \sin z$	$-\ 9{\cdot}7662280$
(a) − (c)	$\log k$	$+\ 9{\cdot}7217948$
	k	$+\ 0{\cdot}5269808$

Upon this problem and the deductions we have made from it, rests the method of introducing the auxiliary angles required in solving many of the formulæ of spherical trigonometry. It is applicable to any equation that can be reduced to the form of that solved in the following article.

175. *To solve the equation*

$$m \cos z + n \sin z = q \tag{321}$$

m, n and q being given.

The first member will be reduced to the form $k \sin (\phi + z)$ by assuming k and ϕ such that

$$k \sin \phi = m, \qquad k \cos \phi = n \tag{322}$$

whence

$$k \sin \phi \cos z + k \cos \phi \sin z = q$$

$$\sin (\phi + z) = \frac{q}{k} \tag{323}$$

Therefore, if k and ϕ be found from (322) by the preceding article, (k being limited to positive values), we can then find by (323) the value $\phi + z$ and therefore of z. There will be two solutions from the two values of $\phi + z$ given by (323).

If we restrict ϕ to values less than 180°, (as we may do according to the last article), we may find it by the equation

$$\left.\begin{array}{ll} & \tan \phi = \dfrac{m}{n} \\ \text{and then} & \sin(\phi + z) = \dfrac{q}{m}\sin\phi = \dfrac{q}{n}\cos\phi \end{array}\right\} \quad (324)$$

and in this form it will be unnecessary to find k.*

EXAMPLE.

To find z from (321) when $m = -1{\cdot}0498332$, $n = +0{\cdot}7466898$, and $q = -0{\cdot}4316893$.

By (322) and (323).		By (324).	
$\log m = \log k \sin\phi$	$-0{\cdot}0211203$	$\log m$	$-0{\cdot}0211203$
$\log n = \log k \cos\phi$	$+9{\cdot}8731402$	$\log n$	$+9{\cdot}8731402$
$\log\tan\phi$	$-0{\cdot}1479801$	$\log\tan\phi$	$-0{\cdot}1479801$
ϕ	305° 25′ 20″	ϕ	125° 25′ 20″
$\log\sin\phi$	$-9{\cdot}9111059$	$\log\sin\phi$	$+9{\cdot}9111059$
$\log k$	$+0{\cdot}1100144$	$\log q$	$-9{\cdot}6351713$
$\log q$	$-9{\cdot}6351713$	ar co $\log m$	$-9{\cdot}9788797$
$\log\sin(\phi + z)$	$-9{\cdot}5251569$	$\log\sin(\phi + z)$	$+9{\cdot}5251569$
$\phi + z$	199° 34′ 40″ or 340° 25′ 20″	$\phi + z$	19° 34′ 40″ or 160° 25′ 20″
z	− 105° 50′ 40″ or 35° 0′ 0′	z	− 105° 50′ 40″ or 35° 0′ 0″

To avoid the negative value of z, in the first of these solutions, we may take for the first value of

$$\phi + z,\ 360° + 199°\ 34'\ 40'' = 559°\ 34'\ 40''$$

whence $z = 559°\ 34'\ 40'' - 305°\ 25'\ 20'' = 254°\ \ 9'\ 20''$. The second solution gives a like result.

If we suppose ϕ in (324) to be limited to acute values positive or negative, we take $\phi = -54°\ 34'\ 40''$, which gives $\phi + z = 199°\ 34'\ 40''$, or 340° 25′ 20″, whence the same values of z as before.

We may repeat the latter part of the work with $\cos\phi$ for verification.

---

* The solution is, by (323), impossible when $\frac{q}{k}$ is greater than unity; and by adding the squares of (322), $k^2 = m^2 + n^2$; therefore the solution is impossible when $q^2 > m^2 + n^2$.

176. *To solve the equation*

$$a \sin(\alpha + z) + b \sin(\beta + z) + c \sin(\gamma + z) + \&c. = q \tag{325}$$

Developing by (36) and putting

$$a \sin \alpha + b \sin \beta + c \sin \gamma + \&c. = m$$
$$a \cos \alpha + b \cos \beta + c \cos \gamma + \&c. = n$$

this becomes

$$m \cos z + n \sin z = q$$

which is solved in the preceding article. The same process applies if any or all of the terms contain cosines.

177. *To find k and z from the equations*

$$\left.\begin{aligned} k \sin(\alpha + z) &= m \\ k \sin(\beta + z) &= n \end{aligned}\right\} \tag{326}$$

The sum and difference of these equations are, by (105) and (106),

$$2\,k \sin[\tfrac{1}{2}(\alpha + \beta) + z] \cos \tfrac{1}{2}(\alpha - \beta) = m + n$$
$$2\,k \cos[\tfrac{1}{2}(\alpha + \beta) + z] \sin \tfrac{1}{2}(\alpha - \beta) = m - n$$

whence

$$\left.\begin{aligned} 2\,k \sin[\tfrac{1}{2}(\alpha + \beta) + z] &= \frac{m + n}{\cos \frac{1}{2}(\alpha - \beta)} \\ 2\,k \cos[\tfrac{1}{2}(\alpha + \beta) + z] &= \frac{m - n}{\sin \frac{1}{2}(\alpha - \beta)} \end{aligned}\right\} \tag{327}$$

from which $2\,k$ and $\frac{1}{2}(\alpha + \beta) + z$ are determined by Art. 174. The logs. of the second members of these equations should be computed separately, for the purpose of readily discovering the signs of the sine and cosine in the first members. The solution is determinate (according to Art. 174) when the sign of k is given.

From (327) we find, by division,

$$\tan[\tfrac{1}{2}(\alpha + \beta) + z] = \frac{m + n}{m - n} \tan \tfrac{1}{2}(\alpha - \beta) \tag{328}$$

which requires a less number of logs. than the separate computation of (327), but we are obliged to refer to (327) to determine (by an inspection of the second members) the signs of the sine and cosine.

If we assume

$$\tan \phi = \frac{n}{m}$$

we may compute (328) by the formula

$$\tan[\tfrac{1}{2}(\alpha + \beta) + z] = \tan(45° + \phi) \tan \tfrac{1}{2}(\alpha - \beta) \tag{329}$$

EXAMPLE.

In (326) given $\alpha = 200°$, $\beta = 140°$, $m = -0{\cdot}42345$ and $n = -0{\cdot}20123$, to find z and k, k being positive.

	By (327).	
	$m + n$	$-0{\cdot}62468$
	$m - n$	$-0{\cdot}22222$
	$\frac{1}{2}(\alpha + \beta)$	$170°$
	$\frac{1}{2}(\alpha - \beta)$	$30°$
	$\log(m + n)$	$-9{\cdot}7956576$
	$\log\cos\frac{1}{2}(\alpha - \beta)$	$+9{\cdot}9375306$
(a)	$\log 2k\sin[\frac{1}{2}(\alpha + \beta) + z]$	$-9{\cdot}8581270$
	$\log(m - n)$	$-9{\cdot}3467831$
	$\log\sin\frac{1}{2}(\alpha - \beta)$	$+9{\cdot}6989700$
(b)	$\log 2k\cos[\frac{1}{2}(\alpha + \beta) + z]$	$-9{\cdot}6478131$
(a) — (b)	$\log\tan[\frac{1}{2}(\alpha + \beta) + z]$	$+0{\cdot}2103140$
	$\frac{1}{2}(\alpha + \beta + z)$	$238°\,21'\,38''{\cdot}6$
	z	$68°\,21'\,38''{\cdot}6$
(c)	$\log\sin[\frac{1}{2}(\alpha + \beta) + z]$	$-9{\cdot}9301171$
(a) — (c)	$\log 2k$	$+9{\cdot}9280099$
	$2k$	$0{\cdot}8472467$
	k	$0{\cdot}4236234$

178. A more general solution of (326) is the following.* Let γ be any angle assumed at pleasure, and in (171) let

$$x = \alpha + z, \qquad y = \beta + z, \qquad z' = \gamma + z$$

(distinguishing the z of (171) by an accent); then we shall find

$$\sin(\alpha - \beta)\sin(\gamma + z) = \sin(\alpha - \gamma)\sin(\beta + z) - \sin(\beta - \gamma)\sin(\alpha + z)$$

In this let γ (whose value is arbitrary) be exchanged for $\gamma + 90°$; then

$$\sin(\alpha - \beta)\cos(\gamma + z) = -\cos(\alpha - \gamma)\sin(\beta + z) + \cos(\beta - \gamma)\sin(\alpha + z)$$

Multiplying these equations by k and substituting m and n from (326)

$$\left.\begin{aligned} k\sin(\alpha - \beta)\sin(\gamma + z) &= m\sin(\gamma - \beta) - n\sin(\gamma - \alpha) \\ k\sin(\alpha - \beta)\cos(\gamma + z) &= m\cos(\gamma - \beta) - n\cos(\gamma - \alpha) \end{aligned}\right\} \quad (330)$$

which (γ being assumed at pleasure), determine k and $\gamma + z$.

If we take $\gamma = 0$, we find

$$\tan z = \frac{-m\sin\beta + n\sin\alpha}{m\cos\beta - n\cos\alpha}$$

If $\gamma = \alpha$,

$$\left.\begin{aligned} k\sin(\alpha + z) &= m \\ k\cos(\alpha + z) &= \frac{m\cos(\alpha - \beta) - n}{\sin(\alpha - \beta)} \end{aligned}\right\} \quad (331)$$

If $\gamma = \beta$, we have a similar result.

If $\gamma = \frac{1}{2}(\alpha + \beta)$ we obtain the solution of the preceding article.

If k is required, without first finding z, we have, by adding the squares of (330)

$$k\sin(\alpha - \beta) = \sqrt{[m^2 + n^2 - 2mn\cos(\alpha - \beta)]} \quad (332)$$

---

* GAUSS. *Theoria Motus Corporum Cœlestium*, Art. 78.

179. *To find k and z from the equations*

$$\left.\begin{aligned} k\cos(\alpha+z) &= m \\ k\cos(\beta+z) &= n \end{aligned}\right\} \quad (333)$$

These are reduced to the form (326) by substituting $90°+\alpha$ and $90°+\beta$ for α and β. We find, however, by a process similar to that of Art. 177,

$$\left.\begin{aligned} 2k\sin\left[\tfrac{1}{2}(\alpha+\beta)+z\right] &= \frac{n-m}{\sin\frac{1}{2}(\alpha-\beta)} \\ 2k\cos\left[\tfrac{1}{2}(\alpha+\beta)+z\right] &= \frac{n+m}{\cos\frac{1}{2}(\alpha-\beta)} \end{aligned}\right\} \quad (334)$$

or

$$\left.\begin{aligned} \tan\phi &= \frac{m}{n} \\ \cot\left[\tfrac{1}{2}(\alpha+\beta)+z\right] &= \tan(45°+\phi)\tan\tfrac{1}{2}(\alpha-\beta) \end{aligned}\right\} \quad (335)$$

EXAMPLE. In (333) given $\alpha = 280°\ 16'$, $\beta = 200°\ 10'$, $m = -0{\cdot}62342$, and $n = 0{\cdot}69725$, find z and k, k being positive.

Ans. $z = 207°\ 5'\ 34''{\cdot}4 \quad k = 1{\cdot}0273643$

180. The more general solution of (333) may be found directly from (172), but it will be simpler to obtain it from (330) by substituting $90°+\alpha$ for α, and $90°+\beta$ for β, whence

$$\left.\begin{aligned} k\sin(\alpha-\beta)\sin(\gamma+z) &= -m\cos(\gamma-\beta)+n\cos(\gamma-\alpha) \\ k\sin(\alpha-\beta)\cos(\gamma+z) &= m\sin(\gamma-\beta)-n\sin(\gamma-\alpha) \end{aligned}\right\} \quad (336)$$

γ being arbitrary as before.

If $\gamma = 0$, we find

$$\tan z = \frac{-m\cos\beta+n\cos\alpha}{-m\sin\beta+n\sin\alpha}$$

If $\gamma = \alpha$,

$$\left.\begin{aligned} k\sin(\alpha+z) &= \frac{-m\cos(\alpha-\beta)+n}{\sin(\alpha-\beta)} \\ k\cos(\alpha+z) &= m \end{aligned}\right\} \quad (337)$$

If $\gamma = \frac{1}{2}(\alpha+\beta)$, we obtain the solution (334).

If k is required directly, the sum of the squares of (336) gives

$$k\sin(\alpha-\beta) = \sqrt{[m^2+n^2-2mn\cos(\alpha-\beta)]}$$

as in Art. 178.

181. The solutions of the preceding articles may be applied to a single equation of the form

$$n\sin(\alpha+z) = m\sin(\beta+z)$$

which is a more general form of (309). For if we assume

$$k\sin(\alpha+z) = m$$

we have

$$k\sin(\beta+z) = n$$

whence k and z are found by Arts. 177, 178.

182. In like manner, if the proposed equation is

$$n \cos(\alpha + z) = m \cos(\beta + z)$$

we assume

$$k \cos(\alpha + z) = m$$

whence

$$k \cos(\beta + z) = n$$

and k and z are found by Arts. 179, 180. As the sign of k (in this and the preceding article) may be arbitrarily assumed, there will be two solutions.

Numerical Equations of the Second and Third Degrees.

183. *To solve the equation*

$$x^2 + px + q = 0 \tag{338}$$

when q is essentially positive, and p either positive or negative.

We have from (144), exchanging x for ϕ,

$$\tan^2 \tfrac{1}{2}\phi - 2 \operatorname{cosec}\phi \tan \tfrac{1}{2}\phi + 1 = 0 \tag{339}$$

and (338) may be reduced to this form by substituting

$$x = z\sqrt{q}$$

in which we may take the radical only with the positive sign, since we may assume x and z to have the same sign. We thus reduce (338) to

$$z^2 + \frac{p}{\sqrt{q}}z + 1 = 0$$

which compared with (339) gives

$$-2 \operatorname{cosec}\phi = \frac{p}{\sqrt{q}}, \qquad z = \tan \tfrac{1}{2}\phi$$

or

$$\sin\phi = -\frac{2\sqrt{q}}{p}, \qquad x = \sqrt{q}\tan \tfrac{1}{2}\phi \tag{340}$$

which gives two values of ϕ less than 360° and consequently two values of x. If θ be the least of these two values of ϕ less than 360° ($= 2\pi$), all the values of ϕ which have the same sine are

$$\theta, \qquad \pi - \theta, \qquad 2\pi + \theta, \qquad 3\pi - \theta, \text{ \&c.}$$

and all the values of $\tan \frac{1}{2}\phi$ are

$$\tan \tfrac{1}{2}\theta, \qquad \cot \tfrac{1}{2}\theta, \qquad \tan \tfrac{1}{2}\theta, \qquad \cot \tfrac{1}{2}\theta, \text{ \&c.}$$

Hence the two roots of (338) are found by the formulæ

$$\sin\theta = -\frac{2\sqrt{q}}{p}, \qquad x_1 = \sqrt{q}\tan \tfrac{1}{2}\theta, \qquad x_2 = \sqrt{q}\cot \tfrac{1}{2}\theta \tag{341}$$

in which θ may be always taken $< 90°$ with the sign of its sine, and $\sqrt{q}$ is to be regarded as a positive quantity.

As long as $2\sqrt{q}$ is not greater than p, this solution is possible, but when $2\sqrt{q} > p$, $\sin\theta$ is not possible, and both roots are imaginary; which agrees with what is shown in algebra.

184. *To solve the equation*

$$x^2 + px - q = 0 \tag{342}$$

when $-q$ *is essentially negative,* p *being either positive or negative.*

We have, by (143)

$$\tan^2 \tfrac{1}{2}\phi + 2 \cot \phi \tan \tfrac{1}{2}\phi - 1 = 0$$

and (342) is reduced to this form by substituting

$$x = z\sqrt{q}$$

whence

$$z^2 + \frac{p}{\sqrt{q}} z - 1 = 0$$

The required solution is therefore

$$2 \cot \phi = \frac{p}{\sqrt{q}}, \qquad z = \tan \tfrac{1}{2}\phi$$

or

$$\tan \phi = \frac{2\sqrt{q}}{p}, \qquad x = \sqrt{q} \tan \tfrac{1}{2}\phi \tag{343}$$

If θ is the least value of ϕ, all the values of ϕ which have the same tangent are

$$\theta, \qquad \pi + \theta, \qquad 2\pi + \theta, \qquad 3\pi + \theta, \text{ \&c.}$$

and all the values of $\tan \frac{1}{2}\phi$ are

$$\tan \tfrac{1}{2}\theta, \qquad -\cot \tfrac{1}{2}\theta, \qquad \tan \tfrac{1}{2}\theta, \qquad -\cot \tfrac{1}{2}\theta, \text{ \&c.}$$

Therefore the two roots of (342) are found by the formulæ

$$\tan \theta = \frac{2\sqrt{q}}{p}, \qquad x_1 = \sqrt{q} \tan \tfrac{1}{2}\theta, \qquad x_2 = -\sqrt{q} \cot \tfrac{1}{2}\theta \tag{344}$$

in which, as before, the radical is to be taken as positive, and $\theta < 90°$ with the sign of its tangent.

In this case both roots are real, since $\tan \theta$ is always possible.

185. *To solve a numerical equation of the third degree.* It is shown in algebra that any equation of the third degree may be reduced to one in which the 2d term is wanting; we need consider therefore only the form

$$x^3 + ax + b = 0 \tag{345}$$

To resolve this, put

$$x = y + z$$

we find

$$(3yz + a)(y + z) + y^3 + z^3 + b = 0$$

Now x may be decomposed into two parts, y and z, in an infinite variety of ways, and we may therefore suppose that y and z are such as to satisfy the condition

$$3yz + a = 0$$

which reduces the first term of the preceding equation to 0, and gives the two conditions

$$yz = -\frac{a}{3}, \qquad y^3 + z^3 = -b$$

Put $y^3 = t_1$, $z^3 = t_2$; then we have

$$t_1 t_2 = -\frac{a^3}{27}, \qquad t_1 + t_2 = -b$$

so that, by the theory of equations, t_1 and t_2 are the two roots of an equation of the second degree in which the absolute term is $-\frac{a^3}{27}$ and the coefficient of the second term is b; that is, they are the roots of the equation

$$t^2 + bt - \frac{a^3}{27} = 0 \qquad (m)$$

If then we find the two roots t_1 and t_2 of (m) by the preceding methods we shall have

$$x = y + z = \sqrt[3]{t_1} + \sqrt[3]{t_2} \qquad (n)$$

It will be necessary to consider the sign of a in the equation (m).

1st. *When a is positive,* (m) comes under the form (342) and the solution by (344) gives

$$\tan\theta = \frac{2}{b}\sqrt{\frac{a^3}{27}}, \qquad t_1 = \sqrt{\frac{a^3}{27}}\tan\tfrac{1}{2}\theta, \qquad t_2 = -\sqrt{\frac{a^3}{27}}\cot\tfrac{1}{2}\theta$$

and by (n)

$$x = \sqrt{\frac{a}{3}}\left(\sqrt[3]{\tan\tfrac{1}{2}\theta} - \sqrt[3]{\cot\tfrac{1}{2}\theta}\right)$$

and if we assume

$$\tan\tfrac{1}{2}\phi = \sqrt[3]{\tan\tfrac{1}{2}\theta}$$

this becomes, by (142)

$$x = -2\sqrt{\frac{a}{3}}\cot\phi$$

Collecting these results we have, for the solution of (345), *when a is positive,*

$$\left.\begin{aligned} \tan\theta = \frac{2}{b}\sqrt{\frac{a^3}{27}}, \qquad \tan\tfrac{1}{2}\phi = \sqrt[3]{\tan\tfrac{1}{2}\theta} \\ x = -2\sqrt{\frac{a}{3}}\cot\phi \end{aligned}\right\} \qquad (346)$$

in which the radicals $\sqrt{\frac{a^3}{27}}$ and $\sqrt{\frac{a}{3}}$ are to be considered positive, and θ is to be taken $< 90°$ with the sign of the tangent. But two of the three values of $\sqrt[3]{\tan\frac{1}{2}\theta}$ being imaginary, the given equation has but one real root.*

* If r represent the real value of $\sqrt[3]{\tan\frac{1}{2}\theta}$, and a_1, a_2 the two imaginary roots of unity, the real value of x is

$$x_1 = \sqrt{\frac{a}{3}}\left(r - \frac{1}{r}\right)$$

and the imaginary values are

$$x_2 = \sqrt{\frac{a}{3}}\left(ra_1 - \frac{1}{ra_1}\right) \qquad x_3 = \sqrt{\frac{a}{3}}\left(ra_2 - \frac{1}{ra_2}\right)$$

or since $a_1 a_2 = 1$

$$x_2 = \sqrt{\frac{a}{3}}\left(ra_1 - \frac{a_2}{r}\right) \qquad x_3 = \sqrt{\frac{a}{3}}\left(ra_2 - \frac{a_1}{r}\right)$$

2d. *When a is negative and* $-4a^3 < 27b^2$. Equation (m) becomes

$$t^2 + bt + \left(-\frac{a^3}{27}\right) = 0$$

and is of the form (338); its roots are therefore found by (341) which gives

$$\sin\theta = -\frac{2}{b}\sqrt{-\frac{a^3}{27}} = -\sqrt{-\frac{4a^3}{27b^2}}$$

$$t_1 = \sqrt{-\frac{a^3}{27}}\tan\tfrac{1}{2}\theta \qquad t_2 = \sqrt{-\frac{a^3}{27}}\cot\tfrac{1}{2}\theta$$

and by (n)

$$x = \sqrt{-\frac{a}{3}}\left(\sqrt[3]{\tan\tfrac{1}{2}\theta} + \sqrt[3]{\cot\tfrac{1}{2}\theta}\right)$$

or if we put, as before, $\tan\frac{1}{2}\phi = \sqrt[3]{\tan\frac{1}{2}\theta}$, the solution of (345), *when a is negative*, is

$$\left.\begin{aligned} &\sin\theta = -\frac{2}{b}\sqrt{-\frac{a^3}{27}} \qquad \tan\tfrac{1}{2}\phi = \sqrt[3]{\tan\tfrac{1}{2}\theta} \\ &x = 2\sqrt{-\frac{a}{3}}\operatorname{cosec}\phi \end{aligned}\right\} \quad (347)$$

which gives one real root, (the other two being imaginary, as above), when $\sin\theta$ is possible, i. e. when $-4a^3 < 27b^2$.*

Substituting the values of a_1 and a_2

$$a_1 = \frac{-1+\sqrt{-3}}{2} \qquad a_2 = \frac{-1-\sqrt{-3}}{2}$$

and also

$$r = \tan\tfrac{1}{2}\phi \qquad \frac{1}{r} = \cot\tfrac{1}{2}\phi$$

we find

$$x_2 = \sqrt{\frac{a}{3}}\,(\cot\phi - \operatorname{cosec}\phi\sqrt{-3})$$

$$x_3 = \sqrt{\frac{a}{3}}\,(\cot\phi + \operatorname{cosec}\phi\sqrt{-3})$$

or finally, x_1 being the real root, the imaginary roots are

$$x_2 = -\frac{x_1}{2} - \frac{x_1\sqrt{3}}{2}\sec\phi\sqrt{-1}$$

$$x_3 = -\frac{x_1}{2} + \frac{x_1\sqrt{3}}{2}\sec\phi\sqrt{-1}$$

* The two imaginary roots will be found, by a process similar to that employed in the preceding note, to be

$$x_2 = -\frac{x_1}{2} - \frac{x_1\sqrt{3}}{2}\cos\phi\sqrt{-1}$$

$$x_3 = -\frac{x_1}{2} + \frac{x_1\sqrt{3}}{2}\cos\phi\sqrt{-1}$$

in which x_1 is the real root found by (347).

3d. *When a is negative and* $-4a^3 > 27b^2$. In this case $\sin\theta$, in (347), is impossible and the preceding solution fails. This is the *irreducible case* of Cardan's rule, the roots appearing under imaginary forms, although it is known that they are all three real. It is, however, readily solved trigonometrically.

In Art. 77, putting ϕ for x, we have

$$\sin^3\phi - \tfrac{3}{4}\sin\phi + \tfrac{1}{4}\sin 3\phi = 0 \qquad (m')$$

and (345) may be reduced to this form by substituting

$$x = kz$$

whence

$$z^3 + \frac{a}{k^2}z + \frac{b}{k^3} = 0 \qquad (n')$$

so that we must have

$$\frac{a}{k^2} = -\frac{3}{4} \quad \text{or} \quad k = 2\sqrt{-\frac{a}{3}}$$

in which the radical is to be taken positive, so that x and z shall have the same sign.

Comparing (m') and (n') we have also

$$\tfrac{1}{4}\sin 3\phi = \frac{b}{k^3} \quad \text{or} \quad \sin 3\phi = \frac{b}{2}\sqrt{-\frac{27}{a^3}} = \sqrt{-\frac{27b^2}{4a^3}}$$

which is a possible sine in the present case. We may therefore take

$$z = \sin\phi$$

and the solution is

$$\sin 3\phi = \frac{b}{2}\sqrt{-\frac{27}{a^3}} \qquad x = 2\sin\phi\sqrt{-\frac{a}{3}} \qquad (348)$$

which gives three real roots by the different values of 3ϕ, which have the same sine.

If θ is the least of these values, all the values of 3ϕ are expressed by

$$2n\pi + \theta \quad \text{and} \quad (2n+1)\pi - \theta$$

n being any integer or 0; and all the values of ϕ are expressed by

$$\phi = \frac{2n}{3}\pi + \tfrac{1}{3}\theta, \qquad \phi = \frac{2n+1}{3}\pi - \theta$$

Now all integers are included in the forms $3m$, $3m+1$ and $3m-1$.

If $n = 3m$, the above values of ϕ are

$$\phi = 2m\pi + \tfrac{1}{3}\theta, \qquad \phi = 2m\pi + \tfrac{1}{3}\pi - \tfrac{1}{3}\theta$$

whence

$$\sin\phi = \sin\tfrac{1}{3}\theta, \qquad \sin\phi = \sin\tfrac{1}{3}(\pi - \theta)$$

If $n = 3m + 1$, we find in the same way

$$\sin\phi = \sin\tfrac{1}{3}(\pi - \theta), \qquad \sin\phi = \tfrac{1}{3}\theta$$

the same as before.

If $n = 3m - 1$, we find both values to be

$$\sin\phi = -\sin\tfrac{1}{3}(\pi + \theta)$$

so that there are but three different values of sin ϕ. Substituting these in (348), the three roots of (345), *when a is negative and* $-4a^3 > 27b^2$, are found by

$$\left.\begin{aligned} \sin\theta &= \frac{b}{2}\sqrt{-\frac{27}{a^3}} \\ x_1 &= 2\sqrt{-\frac{a}{3}}\sin\tfrac{1}{3}\theta \\ x_2 &= 2\sqrt{-\frac{a}{3}}\sin\tfrac{1}{3}(\pi-\theta) = 2\sqrt{-\frac{a}{3}}\sin(60^\circ-\tfrac{1}{3}\theta) \\ x_3 &= -2\sqrt{-\frac{a}{3}}\sin\tfrac{1}{3}(\pi+\theta) = -2\sqrt{-\frac{a}{3}}\sin(60^\circ+\tfrac{1}{3}\theta) \end{aligned}\right\} \quad (349)$$

in which $\theta < 90^\circ$ with the sign of its sine, and the radicals are taken with the positive sign.

Examples.

1. Solve (345) when $a = -6{\cdot}101315$, $b = -5{\cdot}766578$. We find

$$\log\frac{2}{b}\sqrt{-\frac{a^3}{27}} = -0{\cdot}002651*$$

which being greater than any log. sine, we take its arithmetical complement and proceed by (349). Then

$$\log\sin\theta = -9{\cdot}9974349$$
$$\theta = -83^\circ\,46'\,44''$$

$\frac{1}{3}\theta$	$-27^\circ\,55'\,34''{\cdot}7$	$60^\circ-\frac{1}{3}\theta$	$87^\circ\,55'\,34''{\cdot}7$	$-(60^\circ+\frac{1}{3}\theta)$	$-32^\circ 4' 25''{\cdot}3$
log sin	$-9{\cdot}6705571$		$9{\cdot}9997155$		$-9{\cdot}7251024$
$\log 2\sqrt{-\frac{a}{3}}$	$0{\cdot}4551811$		$0{\cdot}4551811$		$0{\cdot}4551811$
$\log x_1$	$-0{\cdot}1257382$	$\log x_2$	$0{\cdot}4548966$	$\log x_3$	$-0{\cdot}1802835$
$x_1 =$	$-1{\cdot}335790$	$x_2 =$	$2{\cdot}850339$	$x_3 =$	$-1{\cdot}514549$

2. Solve (345) when $a = -7$, and $b = 7$.
Ans. $x_1 = 1{\cdot}356896$, $x_2 = 1{\cdot}692021$, $x_3 = -3{\cdot}048917$.

3. Solve (345) when $a = 1{\cdot}5$, and $b = -45$.
Ans. The real root $= 3{\cdot}4163975$.

It may be observed that the algebraic sum of the three roots is always zero, in consequence of the absence of the term in x^2 from the given equation. This is easily shown from (349) where there are three real roots, and from the forms in the notes p. 98, where there are imaginary roots. This principle furnishes a simple verification of the values found by (349).

* The sign — here belongs to the number of which this is the logarithm.

CHAPTER XI.

DIFFERENCES AND DIFFERENTIALS OF THE TRIGONOMETRIC FUNCTIONS.

186. In the applications of trigonometry, it is often required to compute a function of one angle from that of an angle which differs from the first by a small quantity. In such cases it is generally most convenient to compute the *difference* of the two functions, which may be applied to either to obtain the other.

187. *To find the increment of the sine or cosine of an angle, corresponding to a given increment of the angle.*

Let the angle x be increased by Δx, (this notation signifying *difference*, or *increment* of x), and let the corresponding difference or increment of the sine be expressed by $\Delta \sin x$ and of the cosine by $\Delta \cos x$; we have, by this notation,

$$\Delta \sin x = \sin(x + \Delta x) - \sin x$$
$$\Delta \cos x = \cos(x + \Delta x) - \cos x$$

and by (106) and (108)

$$\Delta \sin x = \quad 2 \cos(x + \tfrac{1}{2} \Delta x) \sin \tfrac{1}{2} \Delta x \tag{350}$$
$$\Delta \cos x = -2 \sin(x + \tfrac{1}{2} \Delta x) \sin \tfrac{1}{2} \Delta x \tag{351}$$

which are the required formulæ.

We here consider the difference always as an *increment*, i. e. an increase, and give it the positive (algebraic) sign; its essential sign may, however, be negative, and it will then be in fact a decrement. Thus, in (351) the second member will be negative so long as $x < 180°$, and therefore the increment of the cosine is negative; that is, from 0° to 180° the cosine decreases as the angle increases. In like manner $\Delta \sin x$ is negative when $x > 90°$, and $< 270°$.

188. *To find the increment of the tangent and cotangent.* We have

$$\Delta \tan x = \tan(x + \Delta x) - \tan x$$
$$\Delta \cot x = \cot(x + \Delta x) - \cot x$$

and by (116) and (119)

$$\Delta \tan x = \frac{\sin \Delta x}{\cos(x + \Delta x) \cos x} = \sec(x + \Delta x) \sec x \sin \Delta x \tag{352}$$

$$\Delta \cot x = \frac{-\sin \Delta x}{\sin(x + \Delta x) \sin x} = -\operatorname{cosec}(x + \Delta x) \operatorname{cosec} x \sin \Delta x \tag{353}$$

189. *To find the increment of the secant and cosecant.* We have

$$\Delta \sec x = \sec(x + \Delta x) - \sec x$$
$$\Delta \operatorname{cosec} x = \operatorname{cosec}(x + \Delta x) - \operatorname{cosec} x$$

or by (130) and (132)

$$\Delta \sec x = \frac{2 \sin(x + \frac{1}{2}\Delta x) \sin \frac{1}{2}\Delta x}{\cos(x + \Delta x)\cos x} \tag{354}$$

$$\Delta \operatorname{cosec} x = \frac{-2 \cos(x + \frac{1}{2}\Delta x) \sin \frac{1}{2}\Delta x}{\sin(x + \Delta x)\sin x} \tag{355}$$

190. *To find the increments of the squares of the trigonometric functions corresponding to a given increment of the angle.*

We have

$$\Delta \sin^2 x = \sin^2(x + \Delta x) - \sin^2 x$$
$$= \cos^2 x - \cos^2(x + \Delta x)$$

whence by (133)

$$\Delta \sin^2 x = -\Delta \cos^2 x = \sin(2x + \Delta x)\sin \Delta x \tag{356}$$

From (115), (116), and (119) we deduce

$$\tan^2 x - \tan^2 y = \frac{\sin(x+y)\sin(x-y)}{\cos^2 x \cos^2 y}$$

$$\cot^2 x - \cot^2 y = \frac{-\sin(x+y)\sin(x-y)}{\sin^2 x \sin^2 y}$$

whence

$$\Delta \tan^2 x = \frac{\sin(2x + \Delta x)\sin \Delta x}{\cos^2(x + \Delta x)\cos^2 x} \tag{357}$$

$$\Delta \cot^2 x = \frac{-\sin(2x + \Delta x)\sin \Delta x}{\sin^2(x + \Delta x)\sin^2 x} \tag{358}$$

From (16) we have

$$\sec^2(x + \Delta x) = \tan^2(x + \Delta x) + 1$$
$$\sec^2 x = \tan^2 x + 1$$

the difference of which gives

$$\Delta \sec^2 x = \Delta \tan^2 x \tag{359}$$

and in the same manner, from (17),

$$\Delta \operatorname{cosec}^2 x = \Delta \cot^2 x \tag{360}$$

and the values of $\Delta \tan^2 x$, $\Delta \cot^2 x$, may be substituted in (359) and (360).

191. When the increment of an angle, or arc, is *infinitely small*, it is called the *differential* of the angle, or arc; and the corresponding increments of the trigonometric functions are the differentials of these functions.

The differential of x is denoted by dx; of $\sin x$ by $d\ \sin x$, &c.

192. *To find the differentials of the trigonometric functions from the differential of the angle.*

Let the angle x and its increment Δx be expressed in the unit of Art. 11; or, which is equivalent, let x and Δx be the arcs which measure the angle and its increment in the circle whose radius $= 1$. It is evident that the less the arc, the more nearly does it coincide with its sine or tangent; therefore, when Δx is infinitely small, or becomes dx,

$$\sin dx = dx \qquad \sin \tfrac{1}{2}\, dx = \tfrac{1}{2}\, dx$$

This may be demonstrated more rigorously thus. When dx is infinitely small, we have $\cos dx = 1$, whence

$$\frac{\sin dx}{\tan dx} = \cos dx = 1$$

$$\sin dx = \tan dx$$

but the arc cannot be less than the sine, nor greater than the tangent, and therefore

$$dx = \sin dx = \tan dx$$

Again, when Δx is infinitely small, or becomes dx, we must, according to the principles of the differential calculus, reject it when connected with finite quantities by the signs $+$ or $-$; thus we must substitute x for $x + dx$, or for $x + \frac{1}{2}\, dx$.

Upon these principles we find the differentials directly from the finite differences (350), (351), (352), (353), (354) and (355) as follows:

$$d \sin x = \cos x\, dx \tag{361}$$

$$d \cos x = -\sin x\, dx \tag{362}$$

$$d \tan x = \sec^2 x\, dx = (1 + \tan^2 x)\, dx \tag{363}$$

$$d \cot x = -\operatorname{cosec}^2 x\, dx = -(1 + \cot^2 x)\, dx \tag{364}$$

$$d \sec x = \tan x \sec x\, dx \tag{365}$$

$$d \operatorname{cosec} x = - \cot x \operatorname{cosec} x\, dx \tag{366}$$

193. In the same manner the equations (356), (357), (358), (359) and (360) give

$$d \sin^2 x = -\, d \cos^2 x = \sin 2\, x\, dx \tag{367}$$

$$d \tan^2 x = \quad d \sec^2 x = \frac{\sin 2\, x}{\cos^4 x}\, dx \tag{368}$$

$$= \frac{2 \sin x}{\cos^3 x}\, dx = \frac{2 \tan x}{\cos^2 x}\, dx \tag{369}$$

$$d \cot^2 x = d \operatorname{cosec}^2 x = \frac{-\sin 2\, x}{\sin^4 x}\, dx \tag{370}$$

$$= \frac{-2 \cos x}{\sin^3 x}\, dx = \frac{-2 \cot x}{\sin^2 x}\, dx \tag{371}$$

194. Although the equations (361), (362), (363), (364), (365) and (366), are rigorously true only when dx is infinitely small, they may be used when dx is a finite difference, instead of the equations, (350), (351), (352), (353), (354) and (355), provided dx is sufficiently small to be considered equal to its sine without sensible error, and is also very small in comparison with x. This is very frequently the case in practice, and the differential equations are then preferred on account of their simplicity. It is only necessary to observe that dx must be expressed *in arc*, i. e. in terms of the unit radius; if it is given in seconds, it may be reduced to arc by Art. 9.

195. *To find the differential of an angle from the differentials of its functions.*

From (361) we have

$$dx = \frac{d \sin x}{\cos x} \tag{372}$$

but it is convenient in this case to employ the notation of inverse functions, Art. 87. Thus, if $y = \sin x$, $x = \sin^{-1} y$, and the preceding equation becomes

$$d \sin^{-1} y = \frac{dy}{\sqrt{(1 - y^2)}} \tag{373}$$

In the same manner from (362), &c., we find

$$d \cos^{-1} y = \frac{-dy}{\sqrt{(1 - y^2)}} \tag{374}$$

$$d \tan^{-1} y = \frac{dy}{1 + y^2} \tag{375}$$

$$d \cot^{-1} y = \frac{-dy}{1 + y^2} \tag{376}$$

$$d \sec^{-1} y = \frac{dy}{y\sqrt{(y^2 - 1)}} \tag{377}$$

$$d \operatorname{cosec}^{-1} y = \frac{-dy}{y\sqrt{(y^2 - 1)}} \tag{378}$$

CHAPTER XII.

DIFFERENCES AND DIFFERENTIALS OF PLANE TRIANGLES.

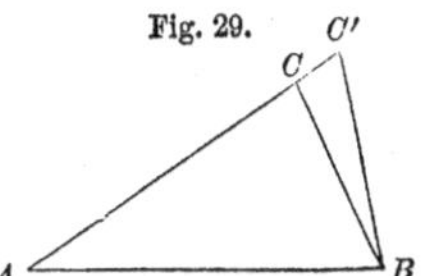

Fig. 29.

196. In trigonometrical investigations it is often necessary to determine the effect of a small change in one of the data, upon the computed parts. Thus, Fig. 29, if A, AB and AC, of the plane triangle ABC, are the data, and AC is subject to an error of CC', the required parts will be subject to errors which are respectively, the differences between ACB and $AC'B$, ABC and ABC', BC and BC'. In the same figure, the data may be supposed to be A, AB and ABC, and the angle ABC may be regarded as subject to the error CBC' which produces the corresponding errors in the remaining parts. In the same manner, the data may be A, AB, and ACB, ACB being variable; or, A, AB, and BC, BC being variable. In all these instances, A and AB are *constant*, while the remaining four parts are *variable*, and may be considered as receiving, simultaneously, certain increments which are related to each other. We propose, then, to solve the general problem:

In a plane triangle, any two parts being constant, and the rest variable, to determine the relations between the increments of the variable parts.

It is evident that the solution of this problem resolves itself into an investigation of the *differences of two triangles which have two parts in common.* We shall consider the several cases successively; distinguishing the triangle formed from the given one by the application of the increments, as the *derived triangle.*

197. Case I. *A and c constant.* The six parts of the given triangle, ABC, Fig. 29, being A, B, C, a, b, c, those of the derived triangle formed by varying all but A and c, are A, $B + \Delta B$, $C + \Delta C$, $a + \Delta a$, $b + \Delta b$, and c. In these two triangles we have

$$A + B + C = 180°$$

$$A + B + \Delta B + C + \Delta C = 180°$$

whence $$\Delta B + \Delta C = 0, \qquad \Delta B = -\Delta C \tag{379}$$

Also in the two triangles we have

$$a = c \sin A \operatorname{cosec} C \tag{m}$$

$$a + \Delta a = c \sin A \operatorname{cosec} (C + \Delta C) \tag{n}$$

the half difference of which by (355) is

$$\tfrac{1}{2} \Delta a = - \frac{c \sin A \cos (C + \frac{1}{2} \Delta C) \sin \frac{1}{2} \Delta C}{\sin C \sin (C + \Delta C)} \tag{p}$$

$$\frac{\frac{1}{2} \Delta a}{\sin \frac{1}{2} \Delta B} = - \frac{\frac{1}{2} \Delta a}{\sin \frac{1}{2} \Delta C} = \frac{a \cos (C + \frac{1}{2} \Delta C)}{\sin (C + \Delta C)} \tag{380}$$

The half sum of (m) and (n) by (131) is

$$a + \tfrac{1}{2} \Delta a = \frac{c \sin A \sin (C + \frac{1}{2} \Delta C) \cos \frac{1}{2} \Delta C}{\sin C \sin (C + \Delta C)}$$

which combined with (p) gives

$$\frac{\frac{1}{2} \Delta a}{\tan \frac{1}{2} \Delta B} = - \frac{\frac{1}{2} \Delta a}{\tan \frac{1}{2} \Delta C} = \frac{a + \frac{1}{2} \Delta a}{\tan (C + \frac{1}{2} \Delta C)} \tag{381}$$

From (262) we have

$$\tan C = \frac{c \sin A}{b - c \cos A}$$

whence

$$b - c \cos A = c \sin A \cot C$$

$$b + \Delta b - c \cos A = c \sin A \cot (C + \Delta C)$$

the difference of which by (353) is

$$\Delta b = - \frac{c \sin A \sin \Delta C}{\sin C \sin (C + \Delta C)}$$

therefore

$$\frac{\Delta b}{\sin \Delta B} = - \frac{\Delta b}{\sin \Delta C} = \frac{a}{\sin (C + \Delta C)} \tag{382}$$

This equation gives by (135)

$$-\frac{\frac{1}{2}\Delta b}{\sin\frac{1}{2}\Delta C\cos\frac{1}{2}\Delta C}=\frac{a}{\sin(C+\Delta C)}$$

and dividing (380) by this

$$\frac{\Delta a}{\Delta b}=\frac{\cos(C+\frac{1}{2}\Delta C)}{\cos\frac{1}{2}\Delta C} \tag{383}$$

It is to be observed that the increments (or half increments) of the angles must be deduced from their sines or tangents, since it is only by these functions that a small angle can be accurately determined. Moreover, a small arc being nearly equal to its sine or tangent, the equations (380), (381) and (382) express very nearly the ratios of the increments of the sides to the increments of the angles, or rather to those increments reduced to arc by Art. 9, or Art. 54.

198. CASE II. *A and a constant.* We have as in the preceding case $\Delta B=-\Delta C$; and in the two triangles

$$b\sin A=a\sin B$$

$$(b+\Delta b)\sin A=a\sin(B+\Delta B)$$

the difference and sum of which give

$$\tfrac{1}{2}\Delta b\sin A=a\cos(B+\tfrac{1}{2}\Delta B)\sin\tfrac{1}{2}\Delta B \tag{p}$$

$$(b+\tfrac{1}{2}\Delta b)\sin A=a\sin(B+\tfrac{1}{2}\Delta B)\cos\tfrac{1}{2}\Delta B$$

whence by division

$$\frac{\frac{1}{2}\Delta b}{\tan\frac{1}{2}\Delta B}=-\frac{\frac{1}{2}\Delta b}{\tan\frac{1}{2}\Delta C}=\frac{b+\frac{1}{2}\Delta b}{\tan(B+\frac{1}{2}\Delta B)} \tag{384}$$

In the same way

$$\frac{\frac{1}{2}\Delta c}{\tan\frac{1}{2}\Delta C}=-\frac{\frac{1}{2}\Delta c}{\tan\frac{1}{2}\Delta B}=\frac{c+\frac{1}{2}\Delta c}{\tan(C+\frac{1}{2}\Delta C)} \tag{385}$$

From the equations

$$c\sin A=a\sin C$$

$$(c+\Delta c)\sin A=a\sin(C+\Delta C)$$

we find $$\tfrac{1}{2}\Delta c\sin A=a\cos(C+\tfrac{1}{2}\Delta C)\sin\tfrac{1}{2}\Delta C \tag{q}$$

which combined with the equation (p) gives, since $\sin \frac{1}{2}\Delta C = -\sin\frac{1}{2}\Delta B$,

$$\frac{\Delta b}{\Delta c} = -\frac{\cos\,(B + \frac{1}{2}\,\Delta B)}{\cos\,(C + \frac{1}{2}\,\Delta C)} \tag{386}$$

From (p) we also have

$$\frac{\frac{1}{2}\,\Delta b}{\sin\frac{1}{2}\,\Delta B} = -\frac{\frac{1}{2}\,\Delta b}{\sin\frac{1}{2}\,\Delta C} = \frac{b\cos\,(B + \frac{1}{2}\,\Delta B)}{\sin B} \tag{387}$$

which, when $\Delta\,b$ is to be found from ΔB, is more convenient than (384). In the same way from (q)

$$\frac{\frac{1}{2}\,\Delta c}{\sin\frac{1}{2}\,\Delta C} = -\frac{\frac{1}{2}\,\Delta c}{\sin\frac{1}{2}\,\Delta B} = \frac{c\cos\,(C + \frac{1}{2}\,\Delta C)}{\sin C} \tag{388}$$

199. Case III. *b and c constant.* We have

$$c\sin B = b\sin C$$
$$c\sin\,(B + \Delta B) = b\sin\,(C + \Delta C)$$

the sum and difference of which give

$$c\sin\,(B + \tfrac{1}{2}\,\Delta B)\cos\tfrac{1}{2}\,\Delta B = b\sin\,(C + \tfrac{1}{2}\,\Delta C)\cos\tfrac{1}{2}\,\Delta C \qquad (p)$$
$$c\cos\,(B + \tfrac{1}{2}\,\Delta B)\sin\tfrac{1}{2}\,\Delta B = b\cos\,(C + \tfrac{1}{2}\,\Delta C)\sin\tfrac{1}{2}\,\Delta C \qquad (q)$$

the quotient of these gives

$$\frac{\tan\frac{1}{2}\,\Delta B}{\tan\frac{1}{2}\,\Delta C} = \frac{\tan\,(B + \frac{1}{2}\,\Delta B)}{\tan\,(C + \frac{1}{2}\,\Delta C)} \tag{389}$$

By (224) we have

$$a = b\cos C + c\cos B$$
$$a + \Delta a = b\cos\,(C + \Delta C) + c\cos\,(B + \Delta B)$$

the sum and difference of which give

$$a + \tfrac{1}{2}\,\Delta a = b\cos\,(C + \tfrac{1}{2}\,\Delta C)\cos\tfrac{1}{2}\,\Delta C + c\cos\,(B + \tfrac{1}{2}\,\Delta B)\cos\tfrac{1}{2}\,\Delta B$$
$$-\tfrac{1}{2}\,\Delta a = b\sin\,(C + \tfrac{1}{2}\,\Delta C)\sin\tfrac{1}{2}\,\Delta C + c\sin\,(B + \tfrac{1}{2}\,\Delta B)\sin\tfrac{1}{2}\,\Delta B$$

These expressions are reduced by (p) and (q) to

$$a + \tfrac{1}{2}\,\Delta a = c\cos\,(B + \tfrac{1}{2}\,\Delta B)\cos\tfrac{1}{2}\,\Delta B\cot\tfrac{1}{2}\,\Delta C\,(\tan\tfrac{1}{2}\Delta B + \tan\tfrac{1}{2}\,\Delta C) \qquad (r)$$
$$-\tfrac{1}{2}\,\Delta a = c\sin\,(B + \tfrac{1}{2}\,\Delta B)\cos\tfrac{1}{2}\,\Delta B\,(\tan\tfrac{1}{2}\,\Delta B + \tan\tfrac{1}{2}\,\Delta C) \qquad (s)$$

and by division

$$\frac{\frac{1}{2}\,\Delta a}{\tan\frac{1}{2}\,\Delta\,C} = -\frac{a + \frac{1}{2}\,\Delta a}{\cot\,(B + \frac{1}{2}\,\Delta B)} \tag{390}$$

In the same way we have

$$\frac{\frac{1}{2}\Delta a}{\tan\frac{1}{2}\Delta B} = -\frac{a+\frac{1}{2}\Delta a}{\cot(C+\frac{1}{2}\Delta C)} \qquad (391)$$

Since the sum of the three angles is constant,

$$\Delta A + \Delta B + \Delta C = 0$$

$$\tfrac{1}{2}(\Delta B + \Delta C) = -\tfrac{1}{2}\Delta A$$

therefore by (115)

$$\tan\tfrac{1}{2}\Delta B + \tan\tfrac{1}{2}\Delta C = \frac{\sin\frac{1}{2}(\Delta B + \Delta C)}{\cos\frac{1}{2}\Delta B\cos\frac{1}{2}\Delta C}$$

$$= -\frac{\sin\frac{1}{2}\Delta A}{\cos\frac{1}{2}\Delta B\cos\frac{1}{2}\Delta C} \qquad (t)$$

which substituted in (s) gives

$$\frac{\frac{1}{2}\Delta a}{\sin\frac{1}{2}\Delta A} = \frac{c\sin(B+\frac{1}{2}\Delta B)}{\cos\frac{1}{2}\Delta C} \qquad (392)$$

and in the same manner

$$\frac{\frac{1}{2}\Delta a}{\sin\frac{1}{2}\Delta A} = \frac{b\sin(C+\frac{1}{2}\Delta C)}{\cos\frac{1}{2}\Delta B} \qquad (393)$$

Substituting (t) in (r) we find

$$\frac{\sin\frac{1}{2}\Delta C}{\sin\frac{1}{2}\Delta A} = -\frac{c\cos(B+\frac{1}{2}\Delta B)}{a+\frac{1}{2}\Delta a} \qquad (394)$$

whence also

$$\frac{\sin\frac{1}{2}\Delta B}{\sin\frac{1}{2}\Delta A} = -\frac{b\cos(C+\frac{1}{2}\Delta C)}{a+\frac{1}{2}\Delta a} \qquad (395)$$

By differencing the equation

$$a^2 = b^2 + c^2 - 2bc\cos A$$

we find instead of (392) and (393)

$$\frac{\frac{1}{2}\Delta a}{\sin\frac{1}{2}\Delta A} = \frac{bc\sin(A+\frac{1}{2}\Delta A)}{a+\frac{1}{2}\Delta a} \qquad (396)$$

200. Case IV. *A and B constant.* We have

$$b = \frac{\sin B}{\sin A} a$$

$$b + \Delta b = \frac{\sin B}{\sin A} (a + \Delta a)$$

whence
$$\Delta b = \frac{\sin B}{\sin A} \Delta a$$

In this case the third angle is also constant and there are but three variables related by the equation

$$\frac{\Delta a}{\sin A} = \frac{\Delta b}{\sin B} = \frac{\Delta c}{\sin C} \tag{397}$$

This case is not strictly included in the general problem as stated in Art. 196, since the two triangles have not two parts in common.

201. The second members of the equations (380), (381), (382), (383), (384), (385), (386), (387), (388), (389), (390), (391), (392), (393), (394), (395), (396), involve the increments themselves, which are the quantities sought. It is therefore necessary, in many cases, to solve these equations by *successive approximations.*

For a *first approximation* we consider the increments in the second member to be $= 0$, employing B for $B + \frac{1}{2}\Delta B$, &c., and taking $\cos \frac{1}{2}\Delta B = 1$, &c. This will evidently produce but a slight error so long as the increments are small as compared with the entire parts of the triangle. We then obtain a *second approximation*, by recomputing the equation in its complete form, employing in the second members the approximate values of the increments. With these second values we may, in the same way, obtain a third approximation, &c. Theoretically, it requires an infinite number of such approximations to arrive at a perfect result; but in practice, the tenths or hundredths of seconds being the limits of accuracy, it is rare that more than a second approximation is necessary.

It is also to be observed that in computing the values of small quantities such as the increments in question, we may employ logarithms of only four or five decimal places and take the angles to the nearest minute. This is in fact one of the chief advantages of computing by differential formulæ, rather than by the direct formulæ applied to each of the two triangles successively.

EXAMPLE.

In a plane triangle whose parts are

$A = 58° \, 41' \, 48''.9 \quad B = 35° \, 11' \, 3''.4 \quad C = 86° \, 7' \, 7''.7$

$a = 6053 \quad b = 4082 \quad c = 7068$

let A and a be constant while b is *diminished* by 50·5; to find the change in the angle B.

We have in this case $\Delta b = - 50.5$; and by (387)

$$\sin \tfrac{1}{2} \Delta B = \frac{\tfrac{1}{2} \Delta b \sin B}{b \cos (B + \tfrac{1}{2} \Delta B)}$$

	1st Approx.	2d Approx.
$\frac{1}{2} \Delta b$	− 25·25	
b	4082	
B	35° 11′	35° 11′
$\frac{1}{2} \Delta B$	0	− 15′
$B + \frac{1}{2} \Delta B$	35° 11′	34° 56′
$\log \frac{1}{2} \Delta b$	− 1·4023	}
ar. co. log. b	6·3891	} − 7·5520
log sin B	9·7606	}
ar. co. l. cos $(B + \frac{1}{2} \Delta B)$	0·0876	0·0863
log. sin $\frac{1}{2} \Delta B$	− 7·6396	− 7·6383
$\frac{1}{2} \Delta B$	− 15′ 0″	− 14′ 56″·8

It is evident that changing the angle $B + \frac{1}{2} \Delta B$ by only three seconds would not affect the fourth place of its cosine; a third approximation is therefore unnecessary, and we have finally $\Delta B = - 29' \, 53''.6$. As the log. sines of small angles do not vary proportionally with the angles, it will conduce to accuracy to employ the methods explained in Art. 115.

DIFFERENTIAL VARIATIONS OF PLANE TRIANGLES.

202. The equations (380), (381), (382), (383), (384), (385), (386), (387), (388), (389), (390), (391), (392), (393), (394), (395), (396) and (397) become *differential* by making the increments infinitely small, that is, by omitting the increments when connected with finite quanti-

ties by the signs + or —, and substituting the increment itself for its sine or tangent, and unity for its cosine, (Art. 192.) The character d must also be substituted for Δ. These changes being made, we easily deduce the following differential relations.

CASE I. *A and c constant.*

$$\left.\begin{array}{c} dB = -dC \\ \dfrac{da}{dB} = -\dfrac{da}{dC} = a\cot C \\ \dfrac{db}{dB} = -\dfrac{db}{dC} = \dfrac{a}{\sin C} \\ \dfrac{da}{db} = \cos C \end{array}\right\} \quad (398)$$

CASE II. *A and a constant.*

$$\left.\begin{array}{c} dB = -dC \\ \dfrac{db}{dB} = -\dfrac{db}{dC} = b\cot B \\ \dfrac{dc}{dC} = -\dfrac{dc}{dB} = c\cot C \\ \dfrac{db}{dc} = -\dfrac{\cos B}{\cos C} \end{array}\right\} \quad (399)$$

CASE III. *b and c constant.*

$$\left.\begin{array}{c} dA + dB + dC = 0 \\ \dfrac{dB}{dC} = \dfrac{\tan B}{\tan C} \\ \dfrac{da}{dC} = -a\tan B, \qquad \dfrac{da}{dB} = -a\tan C \\ \dfrac{da}{dA} = c\sin B = b\sin C \\ \dfrac{dC}{dA} = -\dfrac{c}{a}\cos B, \qquad \dfrac{dB}{dA} = -\dfrac{b}{a}\cos C \end{array}\right\} \quad (400)$$

CASE IV. *The angles, A, B, C, constant.*

$$\frac{da}{\sin A} = \frac{db}{\sin B} = \frac{dc}{\sin C} \tag{401}$$

203. These differential relations are often employed when the increments are very small, instead of the equations of finite differences. We have already seen that the equation of differences often requires to be solved by successive approximations, the first approximation being in fact obtained by employing the corresponding differential equation. In all cases therefore where a second approximation in the use of finite differences could not alter the result of the first, it is plain that the differential equation is sufficiently accurate.

The increments of the angles must generally be expressed in arc. Thus if dB is given in seconds we must divide it by $R'' = 206264''.8$, or substitute $dB \sin 1''$ for dB.

But in such fractions as $\frac{dA}{dB}$, this substitution is evidently unnecessary provided the two increments are always expressed in the *same unit*, as minutes, seconds, &c.

EXAMPLE.

In a plane triangle whose parts are

$A = 58° \, 41' \, 48''.9$ $\quad B = 35° \, 11' \, 3''.4$ $\quad C = 86° \, 7' \, 7''.7$

$a = 6053$ $\quad b = 4082$ $\quad c = 7068$

suppose b and c to be constant and the angle A to receive the increment $dA = 20''.6$; find da and dC.

From (400) we have

$$da = dA \sin 1'' \, c \sin B$$

$$dC = \frac{-\,dA \; c \cos B}{a}$$

$\log dA$	1·3139	$\log(-dA)$	$-$ 1·3139
$\log \sin 1''$	4·6856	$\log c$	3·8493
$\log c$	3·8493	$\log \cos B$	9·9124
$\log \sin B$	9·7606	ar. co. $\log a$	6·2180
$\log da$	9·6094	$\log dC$	$-$ 1·2936
da	0·407	dC	$-$ 19″·7

204. The error of employing the differentials in any case may be determined approximately by developing the equation of finite differences and comparing it with the corresponding differential equation. We shall select a simple example.

We have from (387) and its corresponding differential equation in (399)

$$\tfrac{1}{2}\,\Delta b = \frac{b \cos\left(B + \frac{1}{2}\,\Delta B\right)}{\sin B} \sin \tfrac{1}{2}\,\Delta B$$

$$\Delta b = b \cot B\,\Delta B \sin 1''$$

the first of which when developed gives

$$\tfrac{1}{2}\,\Delta b = b \cot B \sin \tfrac{1}{2}\,\Delta B - \frac{2\,b \sin\left(B + \frac{1}{4}\,\Delta B\right)}{\sin B} \sin \tfrac{1}{2}\,\Delta B \sin \tfrac{1}{4}\,\Delta B$$

or substituting $\sin \frac{1}{2}\,\Delta B = \frac{1}{2}\,\Delta B \sin 1''$, $\sin \frac{1}{4}\,\Delta B = \frac{1}{4}\,\Delta B \sin 1''$, and also B for $B + \frac{1}{4}\,\Delta B$ in the second term, which will affect so small a term but slightly,

$$\Delta b = b \cot B\,\Delta B \sin 1'' - \frac{b}{2}\,(\Delta B \sin 1'')^2$$

Comparing this with the differential equation above, the error of employing the latter is approximately

$$-\frac{b}{2}\,(\Delta B \sin 1'')^2$$

which for $\Delta B = 1°$ is $-\,{\cdot}000015\,b$.

It appears from this example that the error is expressed by a term involving the *square* of the increment; and if we develop all the equations of finite differences we shall find that they differ from the corresponding differential equations by terms involving the squares and higher powers of the increment. Hence, *employing the differentials instead of the finite differences amounts to neglecting the terms involving the squares and higher powers of the increments.*

205. The differential relations above obtained could have been deduced more directly from the formulæ of plane triangles by differentiation, employing the values of the differentials given in Art. 192. Thus in Case I, A and c being constant, if we differentiate the equation

$$a = c \sin A \operatorname{cosec} C$$

we have

$$\begin{aligned} d\,a &= c \sin A \; d \operatorname{cosec} C \\ &= -\,c \sin A \cot C \operatorname{cosec} C\,d\,C \\ &= -\,a \cot C\,d\,C \end{aligned}$$

as in (398).

The student may exercise himself by deducing the other relations of (398), (399), and (400) in a similar manner.

CHAPTER XIII.

TRIGONOMETRIC SERIES. DEVELOPMENTS OF THE FUNCTIONS OF AN ARC IN TERMS OF THE ARC, AND RECIPROCALLY.*

206. THE investigation of trigonometric series is most readily carried on with the aid of a few elementary principles of the Differential Calculus. All that will be required here will be no more than is generally given in the first chapter of a treatise on that subject, namely, the differentiation of simple algebraic functions, and Taylor's Theorem. We shall employ the following expression of this theorem:

$$f(y+h) = fy + \frac{d.fy}{dy}\cdot\frac{h}{1} + \frac{d^2.fy}{dy^2}\cdot\frac{h^2}{1\cdot 2} + \frac{d^3.fy}{dy^3}\cdot\frac{h^3}{1\cdot 2\cdot 3} + \&c. \quad (402)$$

in which fy denotes what $f(y+h)$ becomes when $h = 0$ and $\frac{d.fy}{dy}, \frac{d^2.fy}{dy^2}$, &c., are the successive differential coefficients, or derivatives of fy.

207. *To develop* $\sin x$ *and* $\cos x$ *in terms of* x.

We shall first develop $\sin(y+x)$ and $\cos(y+x)$ by (402). By (361) and (362), if

$$fy = \sin y$$

we have

$$\frac{d.fy}{dy} = \frac{d\sin y}{dy} = \cos y$$

$$\frac{d^2.fy}{dy^2} = \frac{d\cos y}{dy} = -\sin y$$

$$\frac{d^3.fy}{dy^3} = -\frac{d\sin y}{dy} = -\cos y$$

$$\frac{d^4.fy}{dy^4} = -\frac{d\cos y}{dy} = \sin y$$

---

* The leading results of this Chapter being of very general utility and constant application are printed in the larger type, but as they are not referred to in the subsequent large print of this work, and moreover require a limited acquaintance with the Differential Calculus, the student can omit them at the first perusal, and pass directly to Part II.

so that the values of the coefficients of the series (402) recur in the order $+\sin y$, $+\cos y$, $-\sin y$, $-\cos y$, and therefore $f(y+x)=$

$$\sin(y+x)=\sin y+\cos y\frac{x}{1}-\sin y\frac{x^2}{1\cdot 2}-\cos y\frac{x^3}{1\cdot 2\cdot 3}+\&c. \quad (403)$$

If we commence with

$$fy=\cos y$$

the coefficients will recur in the order $+\cos y$, $-\sin y$, $-\cos y$, $+\sin y$, and (402) will give

$$\cos(y+x)=\cos y-\sin y\frac{x}{1}-\cos y\frac{x^2}{1\cdot 2}+\sin y\frac{x^3}{1\cdot 2\cdot 3}-\&c. \quad (404)$$

If now we put $y=0$ in (403) and (404), $\sin y=0$, $\cos y=1$, the alternate terms of the series vanish, and we have

$$\sin x=\frac{x}{1}-\frac{x^3}{1\cdot 2\cdot 3}+\frac{x^5}{1\cdot 2\cdot 3\cdot 4\cdot 5}-\frac{x^7}{1\cdot 2\cdot 3\cdot 4\cdot 5\cdot 6\cdot 7}+\&c. \quad (405)$$

$$\cos x=1-\frac{x^2}{1\cdot 2}+\frac{x^4}{1\cdot 2\cdot 3\cdot 4}-\frac{x^6}{1\cdot 2\cdot 3\cdot 4\cdot 5\cdot 6}+\&c. \quad (406)$$

It may be observed that (406) can be deduced from (405) by differentiation.

208. The series (405) and (406) are directly available for the construction of the trigonometric table. For this purpose x in the series must be expressed in arc, since (361) and (362), upon which the preceding demonstration rests, require x to be in arc, Art. 9.

Example.

Find cos. 10°. Reducing 10° to arc, by Art. 9, we have

$$x=10\times\cdot 01745329=\cdot 1745329$$

and computing separately the positive and negative terms of (406),

$$1=1\cdot \qquad\qquad -\frac{x^2}{1\cdot 2}=-\cdot 01523086$$

$$\frac{x^4}{1\cdot 2\cdot 3\cdot 4}=\cdot 00003866 \qquad\qquad -\frac{x^6}{1\ldots 6}=-\cdot 00000004$$

$$1\cdot 00003866 \qquad\qquad -\cdot 01523090$$

$$-\ \cdot 01523090$$

$$\cos 10^\circ=\cdot 98480776$$

agreeing with the tables, which give ·9848078. The student may, for practice, verify any other sine or cosine of his table.

209. *To develop* $\tan x$ *in terms of* x.

Representing the coefficients in the series (405) and (406) by letters, we have

$$\tan x = \frac{x - a_3 x^3 + a_5 x^5 - a_7 x^7 + \&c.}{1 - a_2 x^2 + a_4 x^4 - a_6 x^6 + \&c.} \tag{407}$$

in which $\qquad a_2 = \dfrac{1}{1\cdot 2} \qquad a_3 = \dfrac{1}{1\cdot 2\cdot 3}$ &c.

If we perform the division of the numerator by the denominator, we perceive that the result will be a series containing only *odd* powers of x, and commencing with the term x. But as the law for the successive formation of the coefficients is not easily shown in this way, we shall resort to the following process. Assume the series to be

$$\tan x = c_1 x + c_3 x^3 + c_5 x^5 + \&c. \tag{m}$$

and differentiate it; we find, by (363), after dividing by dx,

$$1 + \tan^2 x = c_1 + 3 c_3 x^2 + 5 c_5 x^4 + \&c.$$

or, since from the division of (407) we know that $c_1 = 1$,

$$\tan^2 x = 3 c_3 x^2 + 5 c_5 x^4 + 7 c_7 x^6 + 9 c_9 x^8 + \&c. \tag{n}$$

The square of (m) is

$$\tan^2 x = c_1 c_1 x^2 + \begin{array}{l|l|l|l} c_1 c_3 & x^4 + c_1 c_5 & x^6 + c_1 c_7 & x^8 + \&c. \\ + c_3 c_1 & \quad + c_3 c_3 & \quad + c_3 c_5 & \quad + \&c. \\ & \quad + c_5 c_1 & \quad + c_5 c_3 & \quad + \&c. \\ & & \quad + c_7 c_1 & \quad + \&c. \end{array}$$

which compared with (n) gives

$$c_3 = \frac{1}{3} c_1 c_1$$

$$c_5 = \frac{1}{5} (c_1 c_3 + c_3 c_1)$$

$$c_7 = \frac{1}{7} (c_1 c_5 + c_3 c_3 + c_5 c_1)$$

$$c_9 = \frac{1}{9} (c_1 c_7 + c_3 c_5 + c_5 c_3 + c_7 c_1)$$

$$\&c. \qquad \&c.$$

where the law of derivation is obvious. We have preserved the factor c_1, although it is equal to unity, in order to render this law more apparent.

Since the first and last terms of these expressions are equal, as also the terms equally distant from them, we may write them as follows:

$$c_1 = 1$$

$$c_3 = \frac{1}{3} (c_1 c_1)$$

$$c_5 = \frac{1}{5} (2 c_3 c_1)$$

$$c_7 = \frac{1}{7} (2 c_5 c_1 + c_3 c_3)$$

$$c_9 = \frac{1}{9} (2 c_7 c_1 + 2 c_5 c_3)$$

$$c_{11} = \frac{1}{11} (2 c_9 c_1 + 2 c_7 c_3 + c_5 c_5)$$

$$\&c. \qquad \&c.$$

in which form any coefficient c_{2n+1} *when n is even*, is expressed by $\frac{n}{2}$ terms all of whose coefficients are $= 2$; and *when n is odd*, by $\frac{n+1}{2}$ terms all of whose coefficients are 2 except the last, which is 1.*

If we now substitute the value of $c_1 = 1$, and deduce the numerical values of the coefficients successively, we shall find

$$\tan x = x + \frac{x^3}{3} + \frac{2\,x^5}{3 \cdot 5} + \frac{17\,x^7}{3^2 \cdot 5 \cdot 7} + \frac{62\,x^9}{3^2 \cdot 5 \cdot 7 \cdot 9} + \frac{1382\,x^{11}}{3^2 \cdot 5^2 \cdot 7 \cdot 9 \cdot 11} + \text{\&c.} \qquad (408)$$

210. *To develop* $\cot x$ *in terms of* x.

If we invert (407) we have

$$\cot x = \frac{1 - a_2 x^2 + a_4 x^4 - \text{\&c.}}{x - a_3 x^3 + a_5 x^5 - \text{\&c.}} \qquad (409)$$

and the first term of the actual division is $\frac{1}{x}$, the second term $-(a_2 - a_3)\,x$, and the succeeding terms evidently involve only the odd powers of x. Therefore let

$$\cot x = \frac{1}{x} - d_1 x - d_3 x^3 - d_5 x^5 - \text{\&c.} \qquad (o)$$

The coefficients cannot be determined by the method of the preceding article in consequence of the negative exponent in the first term; but they are directly deducible from those of the series for $\tan x$. We have by (142)

$$\tan x = \cot x - 2 \cot 2\,x \qquad (p)$$

Now the series (o) being true for any value of x will give $\cot 2\,x$ by substituting $2\,x$ for x, whence

$$2 \cot 2\,x = \frac{1}{x} - 2^2 d_1 x - 2^4 d_3 x^3 - 2^6 d_5 x^5 - \text{\&c.}$$

Subtracting this from (o) we have by (p)

$$\tan x = (2^2 - 1)\, d_1 x + (2^4 - 1)\, d_3 x^3 + (2^6 - 1)\, d_5 x^5 + \text{\&c.}$$

Designating the coefficients of (408) by c_1, c_3, c_5, &c. we have also

$$\tan x = c_1 x + c_3 x^3 + c_5 x^5 + \text{\&c.}$$

and the comparison of these two values of $\tan x$ gives

$$d_1 = \frac{c_1}{2^2 - 1} = \frac{c_1}{(2-1)(2+1)} = \frac{c_1}{1 \cdot 3}$$

$$d_3 = \frac{c_3}{2^4 - 1} = \frac{c_3}{(2^2-1)(2^2+1)} = \frac{c_3}{3 \cdot 5}$$

$$d_5 = \frac{c_5}{2^6 - 1} = \frac{c_5}{(2^3-1)(2^3+1)} = \frac{c_5}{7 \cdot 9}$$

&c. &c.

$$d_n = \frac{c_n}{2^{n+1} - 1}$$

---

* Euler, and after him Cagnoli and others, make these coefficients depend upon those of the series $\sin x$ and $\cos x$, but the number of given quantities by which each coefficient is expressed is double the number required in the method of the text.

Substituting the values from (408)

$$c_1 = 1 \qquad c_3 = \frac{1}{3} \qquad c_5 = \frac{2}{3\cdot 5} \text{ \&c.}$$

and reducing the coefficients to their simplest forms, we find the series (o) to be

$$\cot x = \frac{1}{x} - \frac{x}{3} - \frac{x^3}{3^2\cdot 5} - \frac{2x^5}{3^3\cdot 5\cdot 7} - \frac{x^7}{3^3\cdot 5^2\cdot 7} - \frac{2x^9}{3^3\cdot 5\cdot 7\cdot 9\cdot 11} - \text{\&c.} \qquad (410)$$

211. By a process similar to that of Art. 209, but which we leave to the student, we find

$$\sec x = 1 + \frac{x^2}{2} + \frac{5x^4}{2^3\cdot 3} + \frac{61x^6}{2^4\cdot 3^2\cdot 5} + \frac{277x^8}{2^7\cdot 3^2\cdot 7} + \text{\&c.} \qquad (411)$$

And from (408) and (410) by means of the formula

$$\operatorname{cosec} x = \tfrac{1}{2}(\cot \tfrac{1}{2}x + \tan \tfrac{1}{2}x)$$

we find

$$\operatorname{cosec} x = \frac{1}{x} + \frac{x}{2\cdot 3} + \frac{7x^3}{2^3\cdot 3^2\cdot 5} + \frac{31x^5}{2^4\cdot 3^3\cdot 5.7} + \frac{127x^7}{2^7\cdot 3^3\cdot 5^2\cdot 7} + \text{\&c.} \qquad (412)$$

212. *To develop* $\sin^{-1} y$ *in terms of* y. (See Art. 87).

Let $x = \sin^{-1} y$ (or $\sin x = y$); then by (373)

$$\frac{dx}{dy} = \frac{1}{\sqrt{(1-y^2)}} = (1-y^2)^{-\frac{1}{2}}$$

Developing the second member by the Binomial Theorem,

$$\frac{dx}{dy} = 1 + \tfrac{1}{2}y^2 + \frac{1\cdot 3}{2\cdot 4}y^4 + \frac{1\cdot 3\cdot 5}{2\cdot 4\cdot 6}y^6 + \text{\&c.} \qquad (m)$$

As this contains only even powers of y, the series from which it would be obtained by differentiation must contain only odd powers of y; therefore, let

$$x = a_1 y + a_3 y^3 + a_5 y^5 + a_7 y^7 + \text{\&c.} \qquad (n)$$

There will be no term independent of y if we limit x to values between 0 and $\pm 90°$, for then when $y = 0$ we must also have $x = 0$.*
Differentiating, we have

$$\frac{dx}{dy} = a_1 + 3a_3y^2 + 5a_5y^4 + 7a_7y^6 + \text{\&c.}$$

which compared with (m) gives

$$a_1 = 1 \qquad 3a_3 = \frac{1}{2} \qquad 5a_5 = \frac{1\cdot 3}{2\cdot 4} \qquad 7a_7 = \frac{1\cdot 3\cdot 5}{2\cdot 4\cdot 6} \text{ \&c.}$$

---

* The series (413) obtained under this limitation expresses but one of the values of $\sin^{-1} y$, but if we denote the series by s, we shall have by (95) the following expression, including all the values,

$$\sin^{-1} y = n\pi + (-1)^n s$$

n being an integer, positive or negative, or zero.

therefore (n) becomes

$$x = \sin^{-1} y = y + \frac{1}{2} \cdot \frac{y^3}{3} + \frac{1 \cdot 3}{2 \cdot 4} \cdot \frac{y^5}{5} + \frac{1 \cdot 3 \cdot 5}{2 \cdot 4 \cdot 6} \cdot \frac{y^7}{7} + \&c. \quad (413)$$

It is unnecessary to develop $\cos^{-1}y$ since we have

$$\cos^{-1} y = \frac{\pi}{2} - \sin^{-1} y$$

213. *To develop* $\tan^{-1} y$. Let $x = \tan^{-1} y$, then by (375)

$$\frac{dx}{dy} = (1 + y^2)^{-1} = 1 - y^2 + y^4 - y^6 + \&c. \quad (m)$$

from which we infer, as in the preceding problem, that the required series contains only odd powers of y; therefore let

$$x = a_1 y + a_3 y^3 + a_5 y^5 + a_7 y^7 + \&c. \quad (n)$$

then
$$\frac{dx}{dy} = a_1 + 3\, a_3 y^2 + 5\, a_5 y^4 + 7\, a_7 y^6 + \&c.$$

which, compared with (m), gives

$$a_1 = 1 \qquad 3\, a_3 = -1 \qquad 5\, a_5 = 1 \qquad 7\, a_7 = -1 \ \&c.$$

so that the series is

$$x = \tan^{-1} y = y - \tfrac{1}{3} y^3 + \tfrac{1}{5} y^5 - \tfrac{1}{7} y^7 + \&c. \quad (414)$$

214. *To compute the ratio* ($= \pi$) *of the circumference of a circle to its diameter.*

We have heretofore assumed this ratio to be known from geometry, where it is found by means of circumscribed and inscribed polygons which are made to differ from the circle by as small a quantity as we please; but (414) enables us to express its value in a series. We have $\tan \frac{\pi}{4} = 1$, therefore if we make $y = 1$ in (414) we have

$$\frac{\pi}{4} = 1 - \frac{1}{3} + \frac{1}{5} - \frac{1}{7} + \frac{1}{9} - \&c. \quad (415)$$

But this series converges too slowly to be of any use. To obtain a rapidly converging series y must be a small fraction. We might employ $\tan \frac{\pi}{6} = \frac{1}{\sqrt{3}}$ (Art. 29), but in consequence of the radical, it is simpler to resolve $\frac{\pi}{4}$ into two or more arcs whose tangents are known, and to compute the value of each of these arcs by the series. To effect this let

$$\frac{\pi}{4} = \tan^{-1} t + \tan^{-1} t' \qquad (416)$$

then by (123)

$$\tan\frac{\pi}{4} = 1 = \frac{t + t'}{1 - t\,t'}$$

whence $$t' = \frac{1 - t}{1 + t} \qquad (417)$$

from which, assuming any value of t at pleasure, the corresponding value of t' is found.

If we take $t = \frac{1}{2}$, we find $t' = \frac{1}{3}$; therefore by (416) and (414)

$$\frac{\pi}{4} = \tan^{-1}\tfrac{1}{2} + \tan^{-1}\tfrac{1}{3}$$

$$= \left\{\frac{1}{2} - \frac{1}{3}\left(\frac{1}{2}\right)^3 + \frac{1}{5}\left(\frac{1}{2}\right)^5 - \&c.\right\}$$

$$+ \left\{\frac{1}{3} - \frac{1}{3}\left(\frac{1}{3}\right)^3 + \frac{1}{5}\left(\frac{1}{3}\right)^5 - \&c.\right\} \qquad (418)$$

A few terms of these series give

$$\frac{\pi}{4} = \cdot4636476 + \cdot3217506 = \cdot7853982$$

$$\pi = 3\cdot14159$$

more accurately $$\pi = 3\cdot14159\ 26535\ 89793$$

If we take $t = \frac{1}{4}$, we find $t' = \frac{3}{5}$, but the above supposition is evidently the best adapted for rendering both series sufficiently convergent.*

215. *To resolve* sin x *and* cos x *into factors.*

The series (405) shows that x is a factor of $\sin x$, and gives

$$\sin x = x\,(1 - \frac{x^2}{1\cdot2\cdot3} + \frac{x^4}{1\cdot2\cdot3\cdot4\cdot5} - \&c.) \qquad (p)$$

---

* See Note at the end of this chapter, p. 124.

and the factors of the series within the parenthesis must evidently be of the form

$$1 - \frac{x^2}{A} \qquad (q)$$

A being a constant, but having a different value in each factor. The required factors must be such as to reduce the second member of (p) to zero whenever the first member is zero. Now $\sin x$ is zero for the value $x = 0$, whence x is a factor as already seen, and also for $x = \pm n\pi$, n being any integer; therefore the general value of (q) is

$$1 - \frac{n^2 \pi^2}{A} = 0$$

whence $$A = n^2 \pi^2$$

which, substituted in (q), gives as the general factor

$$1 - \frac{x^2}{n^2 \pi^2}$$

Making n successively $= 1, 2, 3$, &c., the equation (p) becomes therefore

$$\sin x = x \left(1 - \frac{x^2}{1^2 \pi^2}\right)\left(1 - \frac{x^2}{2^2 \pi^2}\right)\left(1 - \frac{x^2}{3^2 \pi^2}\right)\cdots \qquad (419)$$

The factors of $\cos x$ in (406) must also be of the form (q); but $\cos x$ is zero for $x = \pm (2n + 1)\frac{\pi}{2}$, n being any integer or zero, and the general value of (q) is

$$1 - \frac{(2n+1)^2 \pi^2}{A \,.\, 2^2} = 0$$

whence $$A = \frac{(2n+1)^2 \pi^2}{2^2}$$

which, substituted in (q), gives the general factor

$$1 - \frac{2^2 x^2}{(2n+1)^2 \pi^2}$$

Making n successively $= 0, 1, 2, 3$, &c., we have

$$\cos x = \left(1 - \frac{2^2 x^2}{1^2 \pi^2}\right)\left(1 - \frac{2^2 x^2}{3^2 \pi^2}\right)\left(1 - \frac{2^2 x^2}{5^2 \pi^2}\right)\cdots \qquad (420)$$

216. *Logarithmic sines and cosines.* By means of (419) and (420) the logarithmic sines and cosines of the tables are readily computed.

Put $x = m\frac{\pi}{2}$, then

$$\sin m\frac{\pi}{2} = \frac{m\pi}{2}\left(1 - \frac{m^2}{2^2}\right)\left(1 - \frac{m^2}{4^2}\right)\left(1 - \frac{m^2}{6^2}\right)\cdots$$

$$\cos m\frac{\pi}{2} = \left(1 - \frac{m^2}{1^2}\right)\left(1 - \frac{m^2}{3^2}\right)\left(1 - \frac{m^2}{5^2}\right)\cdots$$

and taking the logarithms

$$\log \sin \frac{m\pi}{2} = \log \frac{\pi}{2} + \log m + \log\left(1 - \frac{m^2}{2^2}\right) + \log\left(1 - \frac{m^2}{4^2}\right) + \cdots$$

$$\log \cos \frac{m\pi}{2} = \log\left(1 - \frac{m^2}{1^2}\right) + \log\left(1 - \frac{m^2}{3^2}\right) + \log\left(1 - \frac{m^2}{5^2}\right) + \cdots$$

Developing these logs. by the known formula

$$\log(1-n) = -M\left(n + \tfrac{1}{2}n^2 + \tfrac{1}{3}n^3 + \&c.\right)$$

(in which M = modulus of common logs.) and arranging according to the powers of m, we have

$$\log \sin \frac{m\pi}{2} = \log \frac{\pi}{2} + \log m$$

$$-m^2 \cdot \frac{M}{1}\left(\frac{1}{2^2} + \frac{1}{4^2} + \frac{1}{6^2} + \&c.\right)$$

$$-m^4 \cdot \frac{M}{2}\left(\frac{1}{2^4} + \frac{1}{4^4} + \frac{1}{6^4} + \&c.\right)$$

$$-m^6 \cdot \frac{M}{3}\left(\frac{1}{2^6} + \frac{1}{4^6} + \frac{1}{6^6} + \&c.\right)$$

$$- \&c.$$

$$\log \cos \frac{m\pi}{2} = -m^2 \cdot \frac{M}{1}\left(\frac{1}{1^2} + \frac{1}{3^2} + \frac{1}{5^2} + \&c.\right)$$

$$-m^4 \cdot \frac{M}{2}\left(\frac{1}{1^4} + \frac{1}{3^4} + \frac{1}{5^4} + \&c.\right)$$

$$-m^6 \cdot \frac{M}{3}\left(\frac{1}{1^6} + \frac{1}{3^6} + \frac{1}{5^6} + \&c.\right)$$

$$- \&c.$$

ıming the constant numerical series, and substituting the value of the mo-
= ·43429 44819 and also of $\frac{\pi}{2}$, these formulæ become

$$\log \sin \frac{m\pi}{2} = 10{\cdot}19611\ 98770 + \log m$$

$-m^2 \times 0{\cdot}17859\ 64471$	$-m^{12} \times 0{\cdot}00001\ 76758$
$-m^4 \times 0{\cdot}01468\ 89690$	$-m^{14} \times 0{\cdot}00000\ 37870$
$-m^6 \times 0{\cdot}00230\ 11796$	$-m^{16} \times 0{\cdot}00000\ 08284$
$-m^8 \times 0{\cdot}00042\ 58450$	$-m^{18} \times 0{\cdot}00000\ 01841$
$-m^{10} \times 0{\cdot}00008\ 49075$	$- \&c.$

(421)

$$\log \cos \frac{m\pi}{2} = 10$$

$-m^2 \times 0{\cdot}53578\ 93412$	$-m^{12} \times 0{\cdot}07238\ 25502$
$-m^4 \times 0{\cdot}22033\ 45350$	$-m^{14} \times 0{\cdot}06204\ 20818$
$-m^6 \times 0{\cdot}14497\ 43131$	$-m^{16} \times 0{\cdot}05428\ 68115$
$-m^8 \times 0{\cdot}10859\ 04688$	$-m^{18} \times 0{\cdot}04825\ 49426$
$-m^{10} \times 0{\cdot}08686\ 03766$	$- \&c.$

(422)

In these expressions 10 is added to render the logarithms positive, as is usual in the tables.*

* See the preface to Callet's Tables, for the coefficients of these series carried to 20 decimal places, and for other forms given them by which they are rendered still more convenient.

EXAMPLE.

Compute log sin 9°. We have

$$m \times 90° = 9° \qquad m = \frac{1}{10} \qquad \log m = -1$$

and therefore by (421)

$$\begin{aligned}\log \sin 9° &= 10{\cdot}19611\,98770 - 1{\cdot} \\ &\qquad - 0{\cdot}00178\,59645 \\ &\qquad - 0{\cdot}00000\,14689 \\ &\qquad - 0{\cdot}00000\,00023 \\ &= 10{\cdot}19611\,98770 - 1{\cdot}00178\,74357 \\ \log \sin 9° &= 9{\cdot}19433\,24413\end{aligned}$$

217. If in (419) we put $x = \frac{\pi}{2}$, we have

$$\sin\frac{\pi}{2} = 1 = \frac{\pi}{2}\left(1 - \frac{1}{2^2}\right)\left(1 - \frac{1}{4^2}\right)\left(1 - \frac{1}{6^2}\right)\dots$$

$$= \frac{\pi}{2}\left(\frac{2^2-1}{2^2}\right)\left(\frac{4^2-1}{4^2}\right)\left(\frac{6^2-1}{6^2}\right)\dots$$

$$= \frac{\pi}{2}\,\frac{(2-1)(2+1)(4-1)(4+1)(6-1)(6+1)\dots}{2\;.\;2\;.\;4\;.\;4\;.\;6\;.\;6\;.\;\dots}$$

whence

$$\frac{\pi}{2} = \frac{2}{1}\cdot\frac{2}{3}\cdot\frac{4}{3}\cdot\frac{4}{5}\cdot\frac{6}{5}\cdot\frac{6}{7}\dots \tag{423}$$

which is *Wallis's* expression of π.

NOTE to page 121. *Computation of π.* Many other series besides those of Art. 214, may be given for computing π. One method of obtaining them is to resolve $\tan^{-1} t$ and $\tan^{-1} t'$ into two others, and thus make $\frac{1}{4}\pi$ to depend upon three or more arcs. From (194) we easily deduce

$$\tan^{-1}\frac{1}{m} = \tan^{-1}\frac{1}{m+n} + \tan^{-1}\frac{n}{m^2+mn+1} \tag{a}$$

$$\tan^{-1}\frac{1}{m} = \tan^{-1}\frac{1}{m-n} - \tan^{-1}\frac{n}{m^2-mn+1} \tag{b}$$

in which m being given, n may be assumed at pleasure. The numerators of the fractions in the last terms will reduce to unity when $m^2 + 1$ is divisible by n; if therefore we assume n and p so as to satisfy the condition

$$np = m^2 + 1 \tag{c}$$

we shall have

$$\tan^{-1}\frac{1}{m} = \tan^{-1}\frac{1}{m+n} + \tan^{-1}\frac{1}{m+p} \tag{d}$$

$$\tan^{-1}\frac{1}{m} = \tan^{-1}\frac{1}{m-n} - \tan^{-1}\frac{1}{p-m} \tag{e}$$

For example, let $m = 3$; then $m^2 + 1 = 10 = 1 \times 10 = 2 \times 5$, so that we may take $n = 1, p = 10$; or $n = 2, p = 5$, whence by (d) and (e)

$$\tan^{-1}\frac{1}{3} = \tan^{-1}\frac{1}{4} + \tan^{-1}\frac{1}{13}$$

$$= \tan^{-1}\frac{1}{2} - \tan^{-1}\frac{1}{7}$$

$$= \tan^{-1}\frac{1}{5} + \tan^{-1}\frac{1}{8}$$

Substituting in (418)

$$\frac{\pi}{4} = \tan^{-1}\frac{1}{2} + \tan^{-1}\frac{1}{4} + \tan^{-1}\frac{1}{13}$$

$$= 2\tan^{-1}\frac{1}{2} - \tan^{-1}\frac{1}{7}$$

$$= 2\tan^{-1}\frac{1}{3} + \tan^{-1}\frac{1}{7} \qquad (f)$$

$$= \tan^{-1}\frac{1}{2} + \tan^{-1}\frac{1}{5} + \tan^{-1}\frac{1}{8} \qquad (g)$$

The equation (f) was employed by Clausen of Germany, in computing π to 200 decimal places, and (g) was employed by Dase, also of Germany, in computing π to the same number of figures. These computations were carried on independently of each other, and the results when communicated to Schumacher, (who gives them in the Astronomische Nachrichten, No. 589), were found to agree to the last figure. They prove the value previously found by Mr. Rutherford to be erroneous beyond the 150th figure.

By means of the formulæ (a), (b), (c), (d) and (e) we may again subdivide the arcs as often as we please. Thus, it is easy to deduce

$$\frac{\pi}{4} = 2\tan^{-1}\frac{1}{5} + \tan^{-1}\frac{1}{7} + 2\tan^{-1}\frac{1}{8}$$

$$= 2\tan^{-1}\frac{1}{5} + \tan^{-1}\frac{1}{4} + \tan^{-1}\frac{1}{7} + \tan^{-1}\frac{1}{268}$$

$$= 4\tan^{-1}\frac{1}{5} - \tan^{-1}\frac{1}{18} + \tan^{-1}\frac{1}{21} + \tan^{-1}\frac{1}{268}$$

$$= 4\tan^{-1}\frac{1}{5} - \tan^{-1}\frac{3}{379} + \tan^{-1}\frac{1}{268}$$

$$= 4\tan^{-1}\frac{1}{5} - \tan^{-1}\frac{1}{239}$$

which last is known as *Machin's* formula. In deducing it we have reduced the *difference* of two arcs to a single arc by means of formula (a).

Another method is, to find by trial, or otherwise, an arc a multiple of which is nearly equal to $\frac{\pi}{4}$, and whose cotangent is a whole number; and then deduce the difference between this multiple and $\frac{\pi}{4}$. Thus it is known (from the trigonometric tables) that cot 11° 15′ = 5 nearly; therefore by the last formula of Art. 79, putting $\tan x = \frac{1}{5}$,

$$4 \tan^{-1} \frac{1}{5} = \tan^{-1} \frac{120}{119}$$

and by (194)

$$\tan^{-1} \frac{120}{119} - \frac{\pi}{4} = \tan^{-1} \frac{120}{119} - \tan^{-1} 1 = \tan^{-1} \frac{1}{239}$$

therefore

$$\frac{\pi}{4} = 4 \tan^{-1} \frac{1}{5} - \tan^{-1} \frac{1}{239}$$

as was found above.

If we resolve $\tan^{-1} \frac{1}{239}$ by means of (*c*), (*d*) and (*e*), we have $m = 239$, $m^2 + 1 = 57122 = 2 \cdot 13^4 = np$, which offers several suppositions for n and p; if we take $n = 13^2 = 169$ and $p = 2 \cdot 13^2 = 338$, we find by (*e*)

$$\frac{\pi}{4} = 4 \tan^{-1} \frac{1}{5} - \tan^{-1} \frac{1}{70} + \tan^{-1} \frac{1}{99}$$

which was employed by Rutherford.

If we take $n = 1$, $p = 57122$, we find by (*d*)

$$\frac{\pi}{4} = 4 \tan^{-1} \frac{1}{5} - \tan^{-1} \frac{1}{240} - \tan^{-1} \frac{1}{57361}$$

CHAPTER XIV.

EXPONENTIAL FORMULÆ. TRINOMIAL OR QUADRATIC FACTORS.

218. *To demonstrate Euler's formulæ*

$$\cos x = \tfrac{1}{2}\left(e^{x\sqrt{-1}} + e^{-x\sqrt{-1}}\right) \qquad (424)$$

$$\sin x = \frac{1}{2\sqrt{-1}}\left(e^{x\sqrt{-1}} - e^{-x\sqrt{-1}}\right) \qquad (425)$$

in which e *is the Naperian base of logarithms, or,*

$$e = 1 + \frac{1}{1} + \frac{1}{1 \cdot 2} + \frac{1}{1 \cdot 2 \cdot 3} + \text{\&c.}$$

It is shown in the theory of logarithms that

$$e^x = 1 + \frac{x}{(1)} + \frac{x^2}{(2)} + \frac{x^3}{(3)} + \frac{x^4}{(4)} + \text{\&c.} \qquad (426)$$

where for brevity we write

$$(1) = 1 \qquad (2) = 1 \cdot 2 \qquad (3) = 1 \cdot 2 \cdot 3, \text{ \&c.}$$

We have by (405) and (406), employing the above notation,

$$\cos x = 1 - \frac{x^2}{(2)} + \frac{x^4}{(4)} - \frac{x^6}{(6)} + \text{\&c.}$$

$$\sin x = \frac{x}{(1)} - \frac{x^3}{(3)} + \frac{x^5}{(5)} - \frac{x^7}{(7)} + \text{\&c.}$$

the terms of which are the same as those of (426), but with alternate signs. If the signs in these two series were all positive, the sum of the two would be equal to (426); and it is evident that we shall make them positive by substituting

$$x^2 = -z^2 \quad \text{or} \quad x = z\sqrt{-1}$$

which gives

$$\cos x = 1 + \frac{z^2}{(2)} + \frac{z^4}{(4)} + \frac{z^6}{(6)} + \text{\&c.}$$

$$\sin x = z\sqrt{-1}\left(1 + \frac{z^2}{(3)} + \frac{z^4}{(5)} + \frac{z^6}{(7)} + \text{\&c.}\right)$$

or $$\frac{1}{\sqrt{-1}}\sin x = \frac{z}{(1)} + \frac{z^3}{(3)} + \frac{z^5}{(5)} + \frac{z^7}{(7)} + \&c.$$

whence

$$\cos x + \frac{1}{\sqrt{-1}}\sin x = 1 + \frac{z}{(1)} + \frac{z^2}{(2)} + \frac{z^3}{(3)} + \&c. = e^z$$

But

$$\frac{1}{\sqrt{-1}} = -\frac{-1}{\sqrt{-1}} = -\sqrt{-1}, \qquad z = \frac{x}{\sqrt{-1}} = -x\sqrt{-1}$$

therefore

$$\cos x - \sqrt{-1}\sin x = e^{-x\sqrt{-1}} \tag{427}$$

If in this equation we substitute $-x$ for x, we have, by (56),

$$\cos x + \sqrt{-1}\sin x = e^{x\sqrt{-1}} \tag{428}$$

The sum and difference of these equations are

$$2\cos x = e^{x\sqrt{-1}} + e^{-x\sqrt{-1}} \tag{429}$$

$$2\sqrt{-1}\sin x = e^{x\sqrt{-1}} - e^{-x\sqrt{-1}} \tag{430}$$

whence (424) and (425).

219. The quotient of (430) divided by (429) is

$$\sqrt{-1}\tan x = \frac{e^{x\sqrt{-1}} - e^{-x\sqrt{-1}}}{e^{x\sqrt{-1}} + e^{-x\sqrt{-1}}} = \frac{e^{2x\sqrt{-1}} - 1}{e^{2x\sqrt{-1}} + 1} \tag{431}$$

220. If we put

$$y = e^{x\sqrt{-1}} = \cos x + \sqrt{-1}\sin x \tag{432}$$

we have $$y^{-1} = e^{-x\sqrt{-1}} = \cos x - \sqrt{-1}\sin x \tag{433}$$

and (429) and (430) become

$$2\cos x = y + y^{-1} \tag{434}$$

$$2\sqrt{-1}\sin x = y - y^{-1} \tag{435}$$

If mx be substituted for x in these formulæ, we have

$$y^m = e^{mx\sqrt{-1}} = \cos mx + \sqrt{-1}\sin mx \tag{436}$$

$$y^{-m} = e^{-mx\sqrt{-1}} = \cos mx - \sqrt{-1}\sin mx \tag{437}$$

$$2\cos mx = y^m + y^{-m} \tag{438}$$

$$2\sqrt{-1}\sin mx = y^m - y^{-m} \tag{439}$$

221. *Moivre's Formula.* The value of y^m from (432), compared with (436) gives

$$(\cos x + \sqrt{-1}\sin x)^m = \cos mx + \sqrt{-1}\sin mx \qquad (440)$$

which is Moivre's Formula. It shows that the involution of the expression $\cos x + \sqrt{-1}\sin x$ is effected by the multiplication of the angle.

Again, if we multiply (432) by

$$\cos x' + \sqrt{-1}\sin x' = e^{x'\sqrt{-1}}$$

we have

$$(\cos x + \sqrt{-1}\sin x)(\cos x' + \sqrt{-1}\sin x') = e^{(x+x')\sqrt{-1}}$$
$$= \cos(x+x') + \sqrt{-1}\sin(x+x')$$

which shows that factors of this form are multiplied by the addition of the angles.

We have also

$$(\cos x + \sqrt{-1}\sin x)(\cos x - \sqrt{-1}\sin x) = \cos^2 x + \sin^2 x = e^0 = 1 \quad (441)$$

222. *General form of Moivre's Formula.* As long as m is an integer, both members of (440) can have but one value; but if $m = \frac{p}{q}$ the first member becomes

$$(\cos x + \sqrt{-1}\sin x)^{\frac{p}{q}} = \sqrt[q]{(\cos x + \sqrt{-1}\sin x)^p}$$

which has q different values* in consequence of the radical of the degree q, while the second member

$$\cos\frac{p}{q}x + \sqrt{-1}\sin\frac{p}{q}x \qquad (a)$$

has but one value.

In order that both members may have the same generality, as should be the case with every analytical expression, it is necessary to suppose that we take for the arc x not merely the arc less than the circumference which has the given sine and cosine, but also all the arcs which have the same sine and cosine; that is, c denoting the circumference, all the arcs

$$x, \quad x + c, \quad x + 2c, \quad x + 3c, \text{ \&c.}$$

Now there is an infinite number of these arcs, but only q of them can give different values to (a); for all the values of the arc in (a) will be

$$\frac{p}{q}x, \quad \frac{p}{q}x + \frac{pc}{q}, \quad \frac{p}{q}x + \frac{2pc}{q}, \ldots\ldots \frac{p}{q}x + \frac{(q-1)pc}{q},$$
$$\frac{p}{q}x + \frac{qpc}{q}, \quad \frac{p}{q}x + \frac{(q+1)pc}{q}, \quad \text{\&c.} \ldots\ldots \text{\&c.}$$

---

* That is q values *real* and *imaginary;* thus it is shown in algebra that $\sqrt{a^2} = +a$ and $-a$; $\sqrt[3]{a^3} = +a$, $a\left(\frac{-1+\sqrt{-3}}{2}\right)$ and $a\left(\frac{-1-\sqrt{-3}}{2}\right)$; $\sqrt[4]{a^4} = +a, -a, +a\sqrt{-1}, -a\sqrt{-1}$; &c.

But $\frac{p}{q}x + \frac{qpc}{q} = \frac{p}{q}x + pc$ has the same sign and cosine as $\frac{p}{q}x$;

$\frac{p}{q}x + \frac{(q+1)pc}{q} = \left(\frac{p}{q}x + \frac{pc}{q}\right) + pc$ the same sine and cosine as $\frac{p}{q}x + \frac{pc}{q}$, &c.; so that after the first q terms of the above series, the same values of the sine and cosine return ad infinitum. Representing, therefore, the circumference by 2π, the equation is entirely general under the form

$$(\cos x + \sqrt{-1}\sin x)^{\frac{p}{q}} = \cos\frac{p}{q}(2n\pi + x) + \sqrt{-1}\sin\frac{p}{q}(2n\pi + x) \qquad (442)$$

in which n is any number of the series 0, 1, 2, 3 . . . . . $q-1$.

223. *Trigonometric expressions of the real and imaginary roots of unity.*

If $x = 0$ in (442) it gives

$$(1)^{\frac{p}{q}} = \cos\frac{p}{q}2n\pi + \sqrt{-1}\sin\frac{p}{q}2n\pi \qquad (443)$$

or

$$(1)^m = \cos 2mn\pi + \sqrt{-1}\sin 2mn\pi \qquad (444)$$

m being fractional or integral. If $p = 1$, (443) gives

$$\sqrt[q]{1} = \cos\frac{2n\pi}{q} + \sqrt{-1}\sin\frac{2n\pi}{q} \qquad (445)$$

which expresses the q roots of unity by making n successively 0, 1, 2, 3 . . . $q-1$.

For example, let $q = 4$, (445) gives for

$$n = 0, \quad \sqrt[4]{1} = \cos 0 + \sqrt{-1}\sin 0 = 1$$

$$n = 1, \quad \sqrt[4]{1} = \cos\frac{\pi}{2} + \sqrt{-1}\sin\frac{\pi}{2} = \sqrt{-1}$$

$$n = 2, \quad \sqrt[4]{1} = \cos\pi + \sqrt{-1}\sin\pi = -1$$

$$n = 3, \quad \sqrt[4]{1} = \cos\frac{3\pi}{2} + \sqrt{-1}\sin\frac{3\pi}{2} = -\sqrt{-1}$$

as found in algebra.

If $x = \frac{\pi}{2}$ in (442), it gives

$$(\sqrt{-1})^m = \cos\frac{m(4n+1)\pi}{2} + \sqrt{-1}\sin\frac{m(4n+1)\pi}{2} \qquad (446)$$

which shows that an imaginary term of any degree can be reduced to a binomial of the form $A + B\sqrt{-1}$.

If $x = \pi$ in (442) we find

$$(-1)^m = \cos m(2n+1)\pi + \sqrt{-1}\sin m(2n+1)\pi \qquad (447)$$

224. *To reduce an imaginary quantity of the form* $(a + b\sqrt{-1})^m$ *to the form* $A + B\sqrt{-1}$.

Let k and x be determined from the equations

$$k\cos x = a, \qquad k\sin x = b$$

by Art. 174; then, by Moivre's Formula,

$$(a + b\sqrt{-1})^m = k^m(\cos x + \sqrt{-1}\sin x)^m$$
$$= k^m(\cos mx + \sqrt{-1}\sin mx)$$

and putting $A = k^m\cos mx$, $B = k^m\sin mx$,

$$(a + b\sqrt{-1})^m = A + B\sqrt{-1}$$

Trinomial or Quadratic Factors.

225. *To find the quadratic (trinomial) factors of the expression* $z^{2m} - 2z^m \cos\phi + 1$; *m being integral.*

By (438) and (434) we have

$$y^{2m} - 2y^m \cos mx + 1 = 0$$

$$y^2 - 2y \cos x + 1 = 0$$

Therefore if we put $y = z$, $mx = 2n\pi + \phi$, or $x = \frac{2n\pi + \phi}{m}$, we have

$$z^{2m} - 2z^m \cos\phi + 1 = 0 \tag{448}$$

$$z^2 - 2z \cos\frac{2n\pi + \phi}{m} + 1 = 0 \tag{449}$$

As these two equations exist at the same time, they have common roots, and the second is therefore a divisor or factor of the first; but this factor has m values in consequence of the m values of $\cos\frac{2n\pi + \phi}{m}$ (Art. 222), found by making $n = 0, 1, 2, 3 \ldots m - 1$. Therefore the m quadratic factors of (448) are all expressed by (449), and we have

$$\begin{aligned} z^{2m} - 2z^m \cos\phi + 1 = &\left(z^2 - 2z\cos\frac{\phi}{m} + 1\right) \\ &\times\left(z^2 - 2z\cos\frac{2\pi + \phi}{m} + 1\right) \\ &\times\left(z^2 - 2z\cos\frac{4\pi + \phi}{m} + 1\right) \\ &\times \cdots \\ &\times\left(z^2 - 2z\cos\frac{2(m-1)\pi + \phi}{m} + 1\right) \end{aligned} \tag{450}$$

226. To obtain the simple factors of (448), we have only to find the two simple factors of each of the quadratic factors in (450), or to find the two factors of the general quadratic (449). Now, by the theory of equations, if z_1 and z_2 are the two roots of (449), the first member is equal to

$$(z - z_1)(z - z_2)$$

but we have by (432)

$$z = y = \cos x + \sqrt{-1}\sin x = \cos\frac{2n\pi + \phi}{m} + \sqrt{-1}\sin\frac{2n\pi + \phi}{m}$$

which gives the two values of z by the double sign belonging to $\sqrt{-1}$. Therefore the simple factors of (448) are all included in the form

$$z - \left(\cos\frac{2n\pi + \phi}{m} \pm \sqrt{-1}\sin\frac{2n\pi + \phi}{m}\right) \tag{451}$$

EXAMPLES.

1. Find the quadratic and simple factors of

$$z^4 - 2z^2 + 1$$

Here $m = 2$, $2 \cos \phi = 2$, $\cos \phi = 1$, $\phi = 0$; and by (450),

$$z^4 - 2z^2 + 1 = (z^2 - 2z \cos 0 + 1)(z^2 - 2z \cos \pi + 1)$$
$$= (z^2 - 2z + 1)(z^2 + 2z + 1)$$

by (451)

$$= [z - (\cos 0 + \sqrt{-1} \sin 0)]$$
$$\times [z - (\cos 0 - \sqrt{-1} \sin 0)]$$
$$\times [z - (\cos \pi + \sqrt{-1} \sin \pi)]$$
$$\times [z - (\cos \pi - \sqrt{-1} \sin \pi)]$$
$$= (z-1)(z-1)(z+1)(z+1)$$

2. Find the factors of $z^4 + 2z^2 + 1$. Here $m = 2$, $2 \cos \phi = -2$, $\phi = \pi$, and

$$z^4 + 2z^2 + 1 = (z^2 + 1)(z^2 + 1)$$
$$= (z - \sqrt{-1})(z + \sqrt{-1})(z - \sqrt{-1})(z + \sqrt{-1})$$

3. Find the factors of $z^4 - z^2 + 1$.

$$z^4 - z^2 + 1 = (z^2 - 2z \cos 30^\circ + 1)(z^2 + 2z \cos 30^\circ + 1)$$
$$= (z^2 - z\sqrt{3} + 1)(z^2 + z\sqrt{3} + 1)$$
$$= (z - \tfrac{1}{2}\sqrt{3} - \tfrac{1}{2}\sqrt{-1})(z - \tfrac{1}{2}\sqrt{3} + \tfrac{1}{2}\sqrt{-1})$$
$$\times (z + \tfrac{1}{2}\sqrt{3} + \tfrac{1}{2}\sqrt{-1})(z + \tfrac{1}{2}\sqrt{3} - \tfrac{1}{2}\sqrt{-1})$$

4. Find the factors of $z^6 - 2z^3 + 1$.

$$z^6 - 2z^3 + 1 = (z^2 - 2z + 1)(z^2 + z + 1)(z^2 + z + 1)$$
$$= (z - 1)^2 (z + \tfrac{1}{2} + \tfrac{1}{2}\sqrt{-3})^2 (z + \tfrac{1}{2} - \tfrac{1}{2}\sqrt{-3})^2$$

227. *To find the quadratic factors of $z^m - 1$ when m is odd.*

In (450) let $\phi = 0$, it becomes

$$(z^m - 1)^2 = (z - 1)^2 \times \left(z^2 - 2z \cos \frac{2\pi}{m} + 1 \right)$$
$$\times \left(z^2 - 2z \cos \frac{4\pi}{m} + 1 \right)$$
$$\times \cdot \cdot \cdot$$
$$\times \left(z^2 - 2z \cos \frac{2(m-1)\pi}{m} + 1 \right) \qquad (452)$$

Now m being odd, $m - 1$ is even, and the number of trinomial factors in (452), exclusive of $(z - 1)^2$, is even; but

$$\cos \frac{2(m-1)\pi}{m} = \cos \left(2\pi - \frac{2\pi}{m}\right) = \cos \frac{2\pi}{m}$$

so that the first and last of these factors are equal. In the same manner it is shown that any two of these factors equally distant from the first and last are equal;

therefore, uniting these equal factors and extracting the square root of both members, we have, *when m is odd,*

$$z^m - 1 = (z-1) \times \left(z^2 - 2z\cos\frac{2\pi}{m} + 1\right)$$
$$\times \left(z^2 - 2z\cos\frac{4\pi}{m} + 1\right)$$
$$\times \ldots .$$
$$\times \left(z^2 - 2z\cos\frac{(m-1)\pi}{m} + 1\right) \qquad (453)$$

228. *To find the quadratic factors of $z^m - 1$, when m is even.*

When m is even, $m-1$ is odd, the number of factors in (452), exclusive of $(z-1)^2$, is odd, and the middle factor will not combine with any other. This factor is the $\left(\frac{m}{2}\right)$ and contains

$$\cos\frac{2\left(\frac{m}{2}\pi\right)}{m} = \cos\pi = -1$$

and is therefore equal to

$$z^2 + 2z + 1 = (z+1)^2$$

so that uniting the remaining factors, and extracting the square root, we have, *when m is even,*

$$z^m - 1 = (z-1)(z+1) \times \left(z^2 - 2z\cos\frac{2\pi}{m} + 1\right)$$
$$\times \left(z^2 - 2z\cos\frac{4\pi}{m} + 1\right)$$
$$\times \ldots .$$
$$\times \left(z^2 - 2z\cos\frac{(m-2)\pi}{m} + 1\right) \qquad (454)$$

229. *To find the factors of $z^m + 1$, when m is odd.*

In (450) let $\phi = \pi$, it gives

$$(z^m + 1)^2 = \left(z^2 - 2z\cos\frac{\pi}{m} + 1\right)$$
$$\times \left(z^2 - 2z\cos\frac{3\pi}{m} + 1\right)$$
$$\times \ldots$$
$$\times \left(z^2 - 2z\cos\frac{(2m-1)\pi}{m} + 1\right)$$

and it is easily shown, as in the preceding articles, that the factors equally distant from the first and last are equal, and that the middle term is $z^2 + 2z + 1 = (z+1)^2$.

Hence we find, *when m is odd,*

$$z^m + 1 = (z+1) \times \left(z^2 - 2z\cos\frac{\pi}{m} + 1\right)$$
$$\times \left(z^2 - 2z\cos\frac{3\pi}{m} + 1\right)$$
$$\times \ . \ . \ .$$
$$\times \left(z^2 - 2z\cos\frac{(m-2)\pi}{m} + 1\right) \qquad (455)$$

230. *To find the factors of $z^m + 1$, when m is even.*

The same process gives

$$z^m + 1 = \left(z^2 - 2z\cos\frac{\pi}{m} + 1\right)$$
$$\times \left(z^2 - 2z\cos\frac{3\pi}{m} + 1\right)$$
$$\times \ . \ . \ . \ .$$
$$\times \left(z^2 - 2z\cos\frac{(m-1)\pi}{m} + 1\right) \qquad (456)$$

231. The simple factors of (453) and (454) are obtained from (451) by putting $\phi = 0$, and those of (455) and (456) by putting $\phi = \pi$. There will be found pairs of equal factors as in the preceding articles, but all the *different* simple factors will be found by taking only the positive sign of the radical $\sqrt{-1}$.

232. Any function of the form $z^{2m} - 2p\,z^m + q$ may also be resolved into quadratic factors. It is only necessary to reduce it to one of the preceding forms. By resolving the equation

$$z^{2m} - 2p\,z^m + q = 0 \qquad (457)$$

we shall find from its two values of z^m

$$z^{2m} - 2p\,z^m + q = \left(z^m - (p + \sqrt{p^2 - q})\right) \times \left(z^m - (p - \sqrt{p^2 - q})\right)$$

and if we put the absolute term in one of these factors $= \pm\, a^m$ (according to its sign) it becomes

$$z^m \pm a^m = a^m\left(\frac{z^m}{a^m} \pm 1\right) = a^m\,(z'^m \pm 1)$$

in which $z = az'$, and the factors of this last expression may be found by one of the preceding articles.

If, however, the values of z^m in (457) are imaginary, i. e. if $p^2 < q$, this method fails to discover the real quadratic factors, and we must proceed as follows. Put $q = a^{2m}$, then the proposed function becomes

$$a^{2m}\left(\frac{z^{2m}}{a^{2m}} - \frac{2p}{a^m}\cdot\frac{z^m}{a^m} + 1\right) = a^{2m}\left(z'^{2m} - \frac{2p}{a^m}\,.\,z'^m + 1\right)$$

in which $z = az'$; and since in the present case $p < a^m$, $\frac{p}{a^m}$ is a proper fraction, and we may put $\frac{p}{a^m} = \cos\phi$, which reduces the given function to the form (450).

CHAPTER XV.

TRIGONOMETRIC SERIES CONTINUED. MULTIPLE ANGLES.

233. THE true developments of sin mx and cos mx in series, when m is not restricted to integral values, were first obtained by *Poinsot*, and form the subject of a memoir read by him before the French Academy of Sciences, in 1823.* The following problem is the basis of these investigations.

234. *To develop* $(k + \sqrt{k^2 - 1})^m$, *in a series of ascending powers of* k. Let

$$z = (k + \sqrt{k^2 - 1})^m \qquad (a)$$

and assume

$$z = A_0 + A_1 k + A_2 k^2 + A_3 k^3 \dots + A_n k^n \dots \qquad (b)$$

Differentiating (a) and putting

$$z' = \frac{dz}{dk}$$

we find

$$z' = m\left(k + \sqrt{k^2 - 1}\right)^{m-1} \times \left(1 + \frac{k}{\sqrt{(k^2 - 1)}}\right) = \frac{mz}{\sqrt{(k^2 - 1)}} \qquad (c)$$

the square of which gives

$$m^2 z^2 - (k^2 - 1) z'^2 = 0$$

Differentiating this and putting

$$z'' = \frac{dz'}{dk}$$

we find, after dividing by z',

$$m^2 z - kz' - (k^2 - 1) z'' = 0 \qquad (d)$$

Again, differentiating (b) twice, we find,

$$\left.\begin{aligned} z' &= A_1 + 2A_2 k + 3A_3 k^2 \dots + nA_n k^{n-1} \dots \\ z'' &= 1.2A_2 + 2.3A_3 k + 3.4A_4 k^2 \dots + (n-1)nA_n k^{n-2} \dots \end{aligned}\right\} \qquad (e)$$

Substituting in (d) the values of z, z', z'', given by (b) and (e), we have

$$\begin{array}{r|r|r|rr|l} 0 = m^2 A_0 & + m^2 A_1 & k + m^2 A_2 & k^2 \dots & + m^2 A_n & k^n \dots \\ & - \quad A_1 & - \; 2A_2 & \dots & - n A_n & \dots \\ & & -1.2 A_2 & \dots & -(n-1) n A_n & \dots \\ +1.2 A_2 & +2.3 A_3 & +3.4 A_4 & \dots & +(n+1)(n+2) A_{n+2} & \dots \end{array}$$

in which each of the coefficients of the powers of k must be zero. To discover the law which governs these coefficients, it will suffice to examine that of the general term, or the coefficient of k^n, which is

$$(m^2 - n^2) A_n + (n+1)(n+2) A_{n+2} = 0$$

whence

$$A_{n+2} = -\frac{m^2 - n^2}{(n+1)(n+2)} A_n$$

* See the published memoir, "*Recherches sur l'Analyse des Sections Angulaires*," Paris, 1825.

so that from the first coefficient, A_0, we find by making $n = 0, 2, 4, 6$, &c.,

$$A_2 = -\frac{m^2}{1\cdot 2} A_0$$

$$A_4 = -\frac{m^2 - 2^2}{3\cdot 4} A_2 = \frac{m^2(m^2 - 2^2)}{1\cdot 2\cdot 3\cdot 4} A_0$$

$$A_6 = -\frac{m^2 - 4^2}{5\cdot 6} A_4 = -\frac{m^2(m^2 - 2^2)(m^2 - 4^2)}{1\cdot 2\cdot 3\cdot 4\cdot 5\cdot 6} A_0$$

&c.

and from the second coefficient, A_1, we find by making $n = 1, 3, 5$, &c.

$$A_3 = -\frac{m^2 - 1^2}{2\cdot 3} A_1$$

$$A_5 = -\frac{m^2 - 3^2}{4\cdot 5} A_3 = \frac{(m^2 - 1^2)(m^2 - 3^2)}{2\cdot 3\cdot 4\cdot 5} A_1$$

$$A_7 = -\frac{m^2 - 5^2}{6\cdot 7} A_5 = -\frac{(m^2 - 1^2)(m^2 - 3^2)(m^2 - 5^2)}{2\cdot 3\cdot 4\cdot 5\cdot 6\cdot 7} A_1$$

&c.

Therefore, if we put

$$K = 1 - \frac{m^2}{1\cdot 2} k^2 + \frac{m^2(m^2 - 2^2)}{1\cdot 2\cdot 3\cdot 4} k^4 - \frac{m^2(m^2 - 2^2)(m^2 - 4^2)}{1\cdot 2\cdot 3\cdot 4\cdot 5\cdot 6} k^6 + \&c.$$

$$K' = k - \frac{m^2 - 1^2}{2\cdot 3} k^3 + \frac{(m^2 - 1^2)(m^2 - 3^2)}{2\cdot 3\cdot 4\cdot 5} k^5 - \&c.$$

the equation (b) becomes

$$z = A_0 K + A_1 K'$$

and it only remains to find A_0 and A_1. In (a), (b), (c) and (e), put $k = 0$; we find

$$z = (\sqrt{-1})^m = A_0 \qquad z' = m(\sqrt{-1})^{m-1} = A_1$$

Therefore we have, finally,

$$z = (k + \sqrt{k^2 - 1})^m = (\sqrt{-1})^m K + (\sqrt{-1})^{m-1} m K' \tag{458}$$

235. *To develop* $(\sqrt{1 - h^2} + h\sqrt{-1})^m$ *in a series of ascending powers of* h. We have

$$(\sqrt{1 - h^2} + h\sqrt{-1})^m = (\sqrt{-1})^m (h + \sqrt{h^2 - 1})^m$$

therefore by (458), exchanging k for h,

$$(\sqrt{1 - h^2} + h\sqrt{-1})^m = (\sqrt{-1})^m [(\sqrt{-1})^m H + (\sqrt{-1})^{m-1} mH']$$

in which H and H' are what K and K' become when h is put for k. Combining the imaginary factors in the second member, observing that

$$(\sqrt{-1})^m \times (\sqrt{-1})^m = (\sqrt{1})^m = (1)^{\frac{m}{2}}$$

(which must not be put equal to unity, since m may be a fraction, and unity has imaginary roots,) and also that

$$(\sqrt{-1})^m \times (\sqrt{-1})^{m-1} = \sqrt{-1}\,(\sqrt{-1})^{m-1} \times (\sqrt{-1})^{m-1} = \sqrt{-1}\,(1)^{\frac{m-1}{2}}$$

we have

$$(\sqrt{1-h^2} + h\sqrt{-1})^m = (1)^{\frac{m}{2}}\,H + \sqrt{-1}\,(1)^{\frac{m-1}{2}}\,m\,H' \qquad (459)$$

in which

$$H = 1 - \frac{m^2}{1{\cdot}2}h^2 + \frac{m^2\,(m^2 - 2^2)}{1{\cdot}2{\cdot}3{\cdot}4}h^4 - \&c.$$

$$H' = h - \frac{m^2 - 1^2}{2{\cdot}3}h^3 + \frac{(m^2 - 1^2)\,(m^2 - 3^2)}{2{\cdot}3{\cdot}4{\cdot}5}h^5 - \&c.$$

236. *To develop the sine and cosine of the multiple angle in a series of ascending powers of the cosine of the simple angle.*

When m is an integer, this problem requires us simply to develop sin mx and cos mx in a series of powers of cos x; but when m is a fraction $= \frac{p}{q}$, the angle mx has q values which have the same sine and cosine, (Art. 222), if we consider x to represent all the angles which have the same sine and cosine as the simple angle. We shall therefore employ Moivre's Formula in its general form (442), or

$$(\cos x + \sqrt{-1}\,\sin x)^m = \cos m\,(2\,n\,\pi + x) + \sqrt{-1}\,\sin m\,(2\,n\,\pi + x)$$

Putting $k = \cos x$ we have by (458) and (446),

$$\begin{aligned}(\cos x + \sqrt{-1}\,\sin x)^m &= (k + \sqrt{k^2 - 1})^m \\ &= (\sqrt{-1})^m K + (\sqrt{-1})^{m-1}\,m\,K' \\ &= \cos\left(\frac{m\,(4\,n' + 1)\,\pi}{2}\right).\,K + \sqrt{-1}\,\sin\left(\frac{m\,(4\,n' + 1)\,\pi}{2}\right).\,K \\ &\quad + \cos\left(\frac{(m-1)\,(4\,n' + 1)\,\pi}{2}\right).\,m\,K' + \sqrt{-1}\,\sin\left(\frac{(m-1)\,(4\,n' + 1)\,\pi}{2}\right).\,mK'\end{aligned}$$

Comparing the real and imaginary terms of these two values of $(\cos x + \sqrt{-1}\,\sin x)^m$, we have

$$\cos m\,(2\,n\,\pi + x) = \cos\left(\frac{m\,(4\,n' + 1)\,\pi}{2}\right).\,K + \cos\left(\frac{(m-1)\,(4\,n' + 1)\,\pi}{2}\right).\,m\,K'$$

$$\sin m\,(2\,n\,\pi + x) = \sin\left(\frac{m\,(4\,n' + 1)\,\pi}{2}\right).\,K + \sin\left(\frac{(m-1)\,(4\,n' + 1)\,\pi}{2}\right).\,m\,K'$$

If m is a fraction $= \frac{p}{q}$, each member of these equations receives q values by taking successively for n, or n', the numbers of the series 0, 1, 2, 3, . . . $q - 1$; but we are now to show what values of n and n' correspond to each other in the two members. Let $x = \frac{\pi}{2}$, then $k = 0$, $K = 1$, $K' = 0$, and we have

$$\cos\frac{m\,(4\,n + 1)\,\pi}{2} = \cos\frac{m\,(4\,n' + 1)\,\pi}{2}$$

$$\sin\frac{m\,(4\,n + 1)\,\pi}{2} = \sin\frac{m\,(4\,n' + 1)\,\pi}{2}$$

therefore these two angles can only differ by some multiple of 2π, or we must have

$$\frac{m(4n+1)\pi}{2} = \frac{m(4n'+1)\pi}{2} + 2n''\pi$$

whence
$$m(n-n') = n''$$

but m being a fraction $\frac{p}{q}$, and n, n' numbers of the series $0, 1, 2, \ldots q-1$, we cannot have $m(n-n')$ equal to an integer n'', unless it is zero;* therefore

$$n - n' = 0, \qquad n = n'$$

and the above developments are

$$\cos m(2n\pi + x) = \cos\left(\frac{m(4n+1)\pi}{2}\right).K + \cos\left(\frac{(m-1)(4n+1)\pi}{2}\right).mK' \quad (460)$$

$$\sin m(2n\pi + x) = \sin\left(\frac{m(4n+1)\pi}{2}\right).K + \sin\left(\frac{(m-1)(4n+1)\pi}{2}\right).mK' \quad (461)$$

in which

$$K = 1 - \frac{m^2}{1 \cdot 2}\cos^2 x + \frac{m^2(m^2-2^2)}{1 \cdot 2 \cdot 3 \cdot 4}\cos^4 x - \&c.$$

$$K' = \cos x - \frac{m^2-1^2}{2 \cdot 3}\cos^3 x + \frac{(m^2-1^2)(m^2-3^2)}{2 \cdot 3 \cdot 4 \cdot 5}\cos^5 x - \&c.$$

It hence appears that, in general, it requires the combination of two series to express the cosine and sine of a multiple angle in powers of the cosine of the simple angle, when m is fractional.

237. *When m is an integer*, one of the terms of (460) and (461) will always become zero, and we shall have but a single series to express the function of the multiple angle. The first members become in all cases

$$\cos(2mn\pi + mx) = \cos mx$$
$$\sin(2mn\pi + mx) = \sin mx$$

and the second members vary according to the form of m. In (460), if

$$\left.\begin{array}{ll} m = 4m', & \cos mx = K \\ m = 4m' + 1, & \cos mx = mK' \\ m = 4m' + 2, & \cos mx = -K \\ m = 4m' + 3, & \cos mx = -mK' \end{array}\right\} \quad (462)$$

and since when m is even, the series K terminates, and when m is odd, the series K' terminates, these four equations are all finite expressions, and will give the equations of Art. 76, by making $m = 1, 2, 3$, &c.

In (461), if

$$\begin{array}{ll} m = 4m', & \sin mx = -mK' \\ m = 4m' + 1, & \sin mx = K \\ m = 4m' + 2, & \sin mx = mK' \\ m = 4m' + 3, & \sin mx = -K \end{array}$$

---

* Since $\frac{p}{q}$ is supposed to be reduced to its lowest terms, p and q are prime to each other, therefore, if $\frac{p(n-n')}{q}$ is not zero, q must divide $n - n'$; which is impossible, since the greatest value of either n or n' is $q - 1$.

In these formulæ, however, the series do not terminate, but by differentiating (462) we find for

$$\left.\begin{aligned}
m &= 4m', & \sin mx &= -m\sin x\left(\cos x - \frac{m^2-2^2}{2\cdot 3}\cos^3 x + \&c.\right)\\
m &= 4m'+1, & \sin mx &= \sin x\left(1 - \frac{m^2-1^2}{1\cdot 2}\cos^2 x + \&c.\right)\\
m &= 4m'+2, & \sin mx &= m\sin x\left(\cos x - \frac{m^2-2^2}{2\cdot 3}\cos^3 x + \&c.\right)\\
m &= 4m'+3, & \sin mx &= -\sin x\left(1 - \frac{m^2-1^2}{1\cdot 2}\cos^2 x + \&c.\right)
\end{aligned}\right\} \quad (463)$$

all of which terminate and give the equations of Art. 75.

238. *To develop the sine and cosine of the multiple angle in a series of ascending powers of the sine of the simple angle.*

We take as before

$$(\cos x + \sqrt{-1}\sin x)^m = \cos m(2n\pi + x) + \sqrt{-1}\sin m(2n\pi + x)$$

Putting $h = \sin x$, we have, by (459) and (444),

$$\begin{aligned}
(\cos x + \sqrt{-1}\sin x)^m &= (\sqrt{1-h^2} + h\sqrt{-1})^m\\
&= (1)^{\frac{m}{2}} H + \sqrt{-1}\,(1)^{\frac{m-1}{2}} mH'\\
&= \cos mn'\pi \,.\, H + \sqrt{-1}\sin mn'\pi \,.\, H\\
&\quad + \sqrt{-1}\cos (m-1)n'\pi \,.\, mH' - \sin (m-1)n'\pi \,.\, mH'
\end{aligned}$$

Comparing the real and imaginary terms of these equations,

$$\cos m(2n\pi + x) = \cos mn'\pi \,.\, H - \sin(m-1)n'\pi \,.\, mH'$$
$$\sin m(2n\pi + x) = \sin mn'\pi \,.\, H + \cos(m-1)n'\pi \,.\, mH'$$

and to find what values of n and n' correspond, let $x = 0$, then $h = \sin x = 0$, $H = 1$, $H' = 0$, and we have

$$\cos 2mn\pi = \cos mn'\pi$$
$$\sin 2mn\pi = \sin mn'\pi$$

from which we infer that $2mn\pi = mn'\pi$, or $2n = n'$, and hence

$$\cos m(2n\pi + x) = \cos 2mn\pi \,.\, H - \sin 2(m-1)n\pi \,.\, mH' \quad (464)$$
$$\sin m(2n\pi + x) = \sin 2mn\pi \,.\, H + \cos 2(m-1)n\pi \,.\, mH' \quad (465)$$

in which m being a fraction $= \frac{p}{q}$, n is any number of the series $0, 1, 2, 3, \ldots q-1$; and

$$H = 1 - \frac{m^2}{1\cdot 2}\sin^2 x + \frac{m^2(m^2-2^2)}{1\cdot 2\cdot 3\cdot 4}\sin^4 x - \&c.$$

$$H' = \sin x - \frac{m^2-1^2}{2\cdot 3}\sin^3 x + \frac{(m^2-1^2)(m^2-3^2)}{2\cdot 3\cdot 4\cdot 5}\sin^5 x - \&c.$$

239. *When m is an integer*, the first members of (464) and (465) become $\cos mx$ and $\sin mx$; and the coefficients of the second members contain only multiples of 2π; therefore we have

$$\cos mx = H \qquad\qquad \sin mx = mH'$$

But the series H terminates only when m is even, and the series H' only when m is odd, and we must also employ the derivatives of these equations to obtain finite expressions in all cases; thus we have also

$$\sin mx = -\frac{dH}{mdx} \qquad \cos mx = \frac{dH'}{dx}$$

Therefore differentiating the series H and H', we shall have, when

$$\left.\begin{array}{ll} m = 2m', & \cos mx = 1 - \frac{m^2}{1\cdot 2}\sin^2 x + \&c. \\ m = 2m' + 1, & \cos mx = \cos x\left(1 - \frac{m^2 - 1^2}{1\cdot 2}\sin^2 x + \&c.\right) \end{array}\right\} \quad (466)$$

$$\left.\begin{array}{ll} m = 2m', & \sin mx = m\cos x\left(\sin x - \frac{m^2 - 2^2}{2\cdot 3}\sin^3 x + \&c.\right) \\ m = 2m' + 1, & \sin mx = m\left(\sin x - \frac{m^2 - 1^2}{2\cdot 3}\sin^3 x + \&c.\right) \end{array}\right\} \quad (467)$$

all of which terminate, and give the equations of Arts. 77 and 78.

240. *To develop the sine and cosine of the multiple angle in a series of ascending powers of the tangent of the simple angle.*

We have

$$\cos m(2n\pi + x) + \sqrt{-1}\sin m(2n\pi + x) = (\cos x + \sqrt{-1}\sin x)^m$$
$$= \cos^m x\,(1 + \sqrt{-1}\tan x)^m$$

Expanding by the Binomial Theorem, and putting

$$T = 1 - \frac{m(m-1)}{1\cdot 2}\tan^2 x + \frac{m(m-1)(m-2)(m-3)}{1\cdot 2\cdot 3\cdot 4}\tan^4 x - \&c.$$

$$T' = m\tan x - \frac{m(m-1)(m-2)}{1\cdot 2\cdot 3}\tan^3 x + \&c.$$

we have

$$\cos m(2n\pi + x) + \sqrt{-1}\sin m(2n\pi + x) = \cos^m x\,(T + \sqrt{-1}\,T')$$

But the imaginary and real quantities are not yet distinctly separated in the second member, for m being fractional $\cos^m x$ has a number of imaginary values. If we designate its real value by $\cos^m x$, all its values are included in the expression

$$\cos^m x\,(1)^m = \cos^m x\,(\cos 2mn'\pi + \sqrt{-1}\sin 2mn'\pi)$$

which, substituted above for $\cos^m x$ gives

$$\cos m(2n\pi + x) + \sqrt{-1}\sin m(2n\pi + x) = \cos^m x\,(\cos 2mn'\pi\,.\,T - \sin 2mn'\pi\,.\,T')$$
$$+ \sqrt{-1}\cos^m x\,(\sin 2mn'\pi\,.\,T + \cos 2mn'\pi\,.\,T')$$

Comparing the real and imaginary terms, we now have

$$\cos m(2n\pi + x) = \cos^m x\,(\cos 2mn'\pi\,.\,T - \sin 2mn\pi\,.\,T')$$
$$\sin m(2n\pi + x) = \cos^m x\,(\sin 2mn'\pi\,.\,T + \cos 2mn'\pi\,.\,T')$$

and it is shown as in the preceding problems that $n = n'$, whence

$$\cos m(2n\pi + x) = \cos^m x\,(\cos 2mn\pi\,.\,T - \sin 2mn\pi\,.\,T') \quad (468)$$

$$\sin m(2n\pi + x) = \cos^m x\,(\sin 2mn\pi\,.\,T + \cos 2mn\pi\,.\,T') \quad (469)$$

in which m being a fraction $= \frac{p}{q}$, n is any number of the series, 0, 1, 2, . . . $q - 1$;

and $\cos^m x$ denotes only the real value of $\sqrt[q]{(\cos x)^p}$.

241. By the division of (469) by (468)

$$\tan m\,(2\,n\pi + x) = \frac{\tan 2\,mn\pi\,.\,T + T'}{T - \tan 2\,mn\pi\,.\,T'} \tag{470}$$

242. *When m is an integer*, both the series T and T' terminate, and in all cases $\cos 2\,mn\,\pi = 1$, $\sin 2\,mn\,\pi = 0$; and (468), (469) and (470) give

$$\cos mx = \cos^m x\,.\,T \tag{471}$$

$$\sin mx = \cos^m x\,.\,T' \tag{472}$$

$$\tan mx = \frac{T'}{T} \tag{473}$$

which last expression embraces all the equations of Art. 79.*

243. Before the memoir of Poinsot, developments were given for the multiple arcs in series of *descending* powers of the sine or cosine of the simple arc; but he has shown that these developments are impossible, except when m is integral, and in this case the series are the same as the preceding, with the terms written in inverse order.

244. *To develop any power of the cosine of the simple angle in a series of sines or cosines of the multiple angles, the cosine of the simple angle being positive.*

If $y = \cos x + \sqrt{-1}\,\sin x$, we have, by (434) and the Binomial Theorem,

$$(2\cos x)^m = (y + y^{-1})^m = y^m + my^{m-2} + \frac{m\,(m-1)}{2}\,y^{m-4} + \&c.$$

and by Moivre's Formula,

$$y^m = \cos m\,(2\,n\pi + x) + \sqrt{-1}\,\sin m\,(2\,n\pi + x)$$

$$my^{m-2} = m\cos(m-2)\,(2\,n\pi + x) + m\sqrt{-1}\,\sin(m-2)\,(2\,n\pi + x)$$

$$\frac{m(m-1)}{2}\,y^{m-4} = \frac{m\,(m-1)}{2}\cos(m-4)\,(2\,n\pi + x) + \frac{m\,(m-1)}{2}\sqrt{-1}\,\sin(m-4)(2n\pi + x)$$

&c. &c.

Therefore, if we put

$$P_{2n\pi+x} = \cos m\,(2\,n\pi + x) + m\cos(m-2)\,(2\,n\pi + x) + \&c.$$

$$P'_{2n\pi+x} = \sin m\,(2\,n\pi + x) + m\sin(m-2)\,(2\,n\pi + x) + \&c.$$

we have

$$(2\cos x)^m = P_{2n\pi+x} + \sqrt{-1}\,P'_{2n\pi+x} \tag{a}$$

Now m being a fraction $(2\cos x)^m$ has imaginary values, but *when* $\cos x$ *is positive*, it will have at least one real positive value, and then $(2\cos x)^m$ being understood to denote only this real value, all the values are included in the formula

$$(2\cos x)^m \times (1)^m = (2\cos x)^m\,(\cos 2\,mn'\pi + \sqrt{-1}\,\sin 2\,mn'\pi)$$

---

* Although the formulæ for multiple angles require, in general, the combination of two series when m is not an integer, yet there are certain cases, even when m is a fraction, in which one or the other of the series will disappear. See the memoir of Poinsot, cited at the beginning of this chapter.

Therefore we have

$$(2 \cos x)^m (\cos 2 mn' \pi + \sqrt{-1} \sin 2 mn' \pi) = P_{2n\pi+x} + \sqrt{-1}\, P'_{2n\pi+x}$$

Comparing the real and imaginary terms,

$$(2 \cos x)^m \cos 2mn'\pi = P_{2n\pi+x}$$
$$(2 \cos x)^m \sin 2 mn'\pi = P'_{2n\pi+x}$$

and to find the corresponding values of n and n', let $x = 0$, then $(2 \cos x)^m = 2^m$, and the series become

$$P_{2n\pi} = \cos 2 mn\pi \left(1 + m + \frac{m(m-1)}{2} + \&c.\right)$$
$$= \cos 2 mn\pi\, (1+1)^m$$
$$= 2^m \cos 2 mn\pi$$

and in the same way

$$P'_{2n\pi} = 2^m \sin 2 mn\pi$$

Therefore our formulæ become

$$2^m \cos 2 mn'\pi = 2^m \cos 2 mn\pi$$
$$2^m \sin 2 mn'\pi = 2^m \sin 2 mn\pi$$

and as in former cases, it is shown that $n = n'$, so that we have finally

$$(2 \cos x)^m = \frac{P_{2n\pi+x}}{\cos 2 mn\pi} \qquad (474)$$

$$(2 \cos x)^m = \frac{P'_{2n\pi+x}}{\sin 2 mn\pi} \qquad (475)$$

From this it appears that the real and positive value of $(2 \cos x)^m$ may be expressed either by a series of cosines or by one of sines of the multiple angles, and by comparing (474) and (475), we have the following constant relation between these series.

$$\frac{P'_{2n\pi+x}}{P_{2n\pi+x}} = \frac{\sin 2 mn\pi}{\cos 2 mn\pi}$$

245. If $n = 0$, (474) gives

$$(2 \cos x)^m = P_x = \cos mx + m \cos (m-2)\, x + \frac{m(m-1)}{2} \cos (m-4)\, x + \&c. \quad (476)$$

which may be employed as the general development of the real value of $(2 \cos x)^m$, when $x < \frac{\pi}{2}$.

246. The same supposition of $n = 0$, gives $\sin 2 mn\pi = 0$, and (475) gives therefore,

$$0 = P'_x = \sin mx + m \sin (m-2)\, x + \frac{m(m-1)}{2} \sin (m-4)\, x + \&c. \qquad (477)$$

a remarkable property of this series of sines of multiple arcs, which holds for all values of m, provided $x < \frac{\pi}{2}$.

247. *To develop any power of the cosine of the simple angle in a series of sines or cosines of the multiple angles, the cosine of the simple angle being negative.*

If the denominator of m is even, there is no real value of $(2 \cos x)^m$ when $\cos x$ is negative; but we may put

$$(2 \cos x)^m = (-2 \cos x)^m (-1)^m$$
$$= (-2 \cos x)^m [\cos m (2n' + 1) \pi + \sqrt{-1} \sin m (2n' + 1) \pi]$$

which, substituted in equation (a) of Art. 244, gives

$$(-2 \cos x)^m \cos m (2n' + 1) \pi = P_{2n\pi + x}$$
$$(-2 \cos x)^m \sin m (2n' + 1) \pi = P'_{2n\pi + x}$$

Making $x = \pi$, $\cos x = -1$, $(-2 \cos x)^m = 2^m$, and the series become, by the process shown in Art. 244,

$$P_{(2n+1)\pi} = 2^m \cos m (2n + 1) \pi$$
$$P'_{(2n+1)\pi} = 2^m \sin m (2n + 1) \pi$$

and we have

$$2^m \cos m (2n' + 1) \pi = 2^m \cos m (2n + 1) \pi$$
$$2^m \sin m (2n' + 1) \pi = 2^m \sin m (2n + 1) \pi$$

whence, as before, $n = n'$, and our formulæ are

$$(-2 \cos x)^m = \frac{P_{2n\pi + x}}{\cos m (2n + 1) \pi} \quad (478)$$

$$(-2 \cos x)^m = \frac{P'_{2n\pi + x}}{\sin m (2n + 1) \pi} \quad (479)$$

by which it appears that the real value of $(-2 \cos x)^m$ is also expressed either by a series of cosines or of sines of multiple arcs, which series have the constant relation

$$\frac{P'_{2n\pi + x}}{P_{2n\pi + x}} = \frac{\sin m (2n + 1) \pi}{\cos m (2n + 1) \pi}$$

248. If $n = 0$, (478) and (479) give

$$(-2 \cos x)^m = \frac{P_x}{\cos m\pi} = \frac{1}{\cos m\pi} (\cos mx + m \cos (m - 2) x + \&c.) \quad (480)$$

$$(-2 \cos x)^m = \frac{P'_x}{\sin m\pi} = \frac{1}{\sin m\pi} (\sin mx + m \sin (m - 2) x + \&c.) \quad (481)$$

In this case $\sin m\pi$ is not zero, unless m is an integer, so that the series P'_x does not become zero when $x > \frac{\pi}{2}$, and both (480) and (481) may be employed as the true developments of $(-2 \cos x)^m$.

249. *When m is an integer*, the series (476) and (480) always terminate at the $(m + 1)$th term; and, since in (480) $\cos m\pi = \pm 1$, according as m is even or odd, and $(-2 \cos x)^m = \pm (2 \cos x)^m$ in the same cases, both (476) and (480) become

$$(2 \cos x)^m = \cos mx + m \cos (m - 2) x + \frac{m (m - 1)}{2} \cos (m - 4) x + \&c. \quad (482)$$

But the series (481) becomes zero, so that (482) is the only series by which $(2 \cos x)^m$ can be developed in functions of the multiple arcs, when m is integral.

250. *To develop any power of the sine of the simple angle, in a series of sines or cosines of the multiple angles.*

If $y = \cos x + \sqrt{-1} \sin x$, we have, by (435) and the Binomial Theorem,

$$(\sqrt{-1})^m (2 \sin x)^m = (y - y^{-1})^m$$

$$= y^m - m y^{m-2} + \frac{m(m-1)}{2} y^{m-4} - \&c.$$

in which y^m, y^{m-2}, &c. have the same values as in Art. 244, but the signs of the coefficients are alternately + and —, so that if we put

$$Q_{2n\pi+x} = \cos m(2n\pi + x) - m \cos(m-2)(2n\pi + x) + \&c.$$
$$Q'_{2n\pi+x} = \sin m(2n\pi + x) - m \sin(m-2)(2n\pi + x) + \&c.$$

we have

$$(\sqrt{-1})^m (2 \sin x)^m = Q_{2n\pi+x} + \sqrt{-1}\, Q'_{2n\pi+x}$$

Substituting the value of $(\sqrt{-1})^m$ by (446), and comparing the real and imaginary terms, we find

$$(2 \sin x)^m \cos \frac{m(4n'+1)\pi}{2} = Q_{2n\pi+x}$$

$$(2 \sin x)^m \sin \frac{m(4n'+1)\pi}{2} = Q'_{2n\pi+x}$$

and if we make $x = \frac{\pi}{2}$, we shall find by the process frequently employed above, that $n = n'$; whence

$$(2 \sin x)^m = \frac{Q_{2n\pi+x}}{\cos \frac{1}{2} m (4n+1)\pi} \tag{483}$$

$$(2 \sin x)^m = \frac{Q'_{2n\pi+x}}{\sin \frac{1}{2} m (4n+1)\pi} \tag{484}$$

so that the real value of $(2 \sin x)^m$ may be developed in either the cosines or sines of the multiples. The two series have the constant relation

$$\frac{Q'_{2n\pi+x}}{Q_{2n\pi+x}} = \frac{\sin \frac{1}{2} m (4n+1)\pi}{\cos \frac{1}{2} m (4n+1)\pi}$$

251. If $n = 0$ in (483) and (484),

$$(2 \sin x)^m = \frac{Q_x}{\cos \frac{1}{2} m\pi} = \frac{1}{\cos \frac{1}{2} m\pi} (\cos mx - m \cos(m-2)x + \&c.) \tag{485}$$

$$(2 \sin x)^m = \frac{Q'_x}{\sin \frac{1}{2} m\pi} = \frac{1}{\sin \frac{1}{2} m\pi} (\sin mx - m \sin(m-2)x + \&c.) \tag{486}$$

both of which series are applicable when m is fractional.

252. *When m is an integer*, one or the other of the series (485), (486), will always be zero, according to the form of m, and there will be but one series to express $(2 \sin x)^m$.

If $m = 4m'$, $(2 \sin x)^m = \cos mx - m \cos(m-2)x + \&c.$ (487)

$m = 4m' + 1$, $(2 \sin x)^m = \sin mx - m \sin(m-2)x + \&c.$ (488)

$m = 4m' + 2$, $(2 \sin x)^m = -(\cos mx - m \cos(m-2)x + \&c.)$ (489)

$m = 4m' + 3$, $(2 \sin x)^m = -(\sin mx - m \sin(m-2)x + \&c.)$ (490)

253. The series (485) and (486) become zero when m is an integer, as follows:

If $\quad m = 2m', \qquad 0 = \sin mx - m \sin (m-2)x + \&c. \qquad (491)$

$m = 2m' + 1, \quad 0 = \cos mx - m \cos (m-2)x + \&c. \qquad (492)$

The reason why these series are zero is obvious, since they terminate at the $(m+1)$th term, the terms equally distant from the first and last are equal with opposite signs, and the middle term of (491) is zero.

254. *Given the equation*

$$\tan x = p \tan y \qquad (493)$$

to express $x \pm y$ in a series of multiples of y.

Substituting the values of $\tan x$ and $\tan y$ given by (431)

$$\frac{e^{2x\sqrt{-1}} - 1}{e^{2x\sqrt{-1}} + 1} = p \cdot \frac{e^{2y\sqrt{-1}} - 1}{e^{2y\sqrt{-1}} + 1}$$

whence

$$e^{2x\sqrt{-1}} = \frac{(p+1)\,e^{2y\sqrt{-1}} - (p-1)}{p + 1 - (p-1)\,e^{2y\sqrt{-1}}}$$

or putting

$$q = \frac{p-1}{p+1} \qquad (494)$$

$$e^{2x\sqrt{-1}} = \frac{e^{2y\sqrt{-1}} - q}{1 - q\,e^{2y\sqrt{-1}}} = e^{2y\sqrt{-1}}\left(\frac{1 - q\,e^{-2y\sqrt{-1}}}{1 - q\,e^{2y\sqrt{-1}}}\right) \qquad (a)$$

$$e^{2(x-y)\sqrt{-1}} = \frac{1 - q\,e^{-2y\sqrt{-1}}}{1 - q\,e^{2y\sqrt{-1}}}$$

g the Naperian logarithms of both members,

$$2(x-y)\sqrt{-1} = \log\left(1 - q\,e^{-2y\sqrt{-1}}\right) - \log\left(1 - q\,e^{2y\sqrt{-1}}\right)$$

eloping the second member by the formula

$$\log(1-n) = -n - \tfrac{1}{2}n^2 - \tfrac{1}{3}n^3 - \&c.$$

we have

$$2(x-y)\sqrt{-1} = -q\,e^{-2y\sqrt{-1}} - \tfrac{1}{2}q^2\,e^{-4y\sqrt{-1}} - \tfrac{1}{3}q^3\,e^{-6y\sqrt{-1}} - \&c.$$
$$+\,q\,e^{2y\sqrt{-1}} + \tfrac{1}{2}q^2\,e^{4y\sqrt{-1}} + \tfrac{1}{3}q^3\,e^{6y\sqrt{-1}} + \&c.$$

Substituting in the second member by (430),

$$x - y = q \sin 2y + \tfrac{1}{2}q^2 \sin 4y + \tfrac{1}{3}q^3 \sin 6y + \&c. \qquad (b)$$

The equation (a) might have been put under the form

$$e^{2(x+y)\sqrt{-1}} = \frac{1 - \dfrac{1}{q}\,e^{2y\sqrt{-1}}}{1 - \dfrac{1}{q}\,e^{-2y\sqrt{-1}}}$$

from which, by taking the logarithms and substituting as before,

$$x + y = -\frac{\sin 2y}{q} - \frac{\sin 4y}{2q^2} - \frac{\sin 6y}{3q^3} - \&c. \qquad (c)$$

In this investigation, we have, in effect, used Moivre's formula, in its limited or less general form; but the requisite generality may be given to our results, by observing, that (493) would hold if we were to substitute $\tan x = \tan(n'\pi + x)$, $\tan y = \tan(n''\pi + y)$, and therefore we may substitute for the first member of (b),

$n'\pi + x - (n''\pi + y) = x - y - (n'' - n')\pi = x - y - n\pi$, n being (like n' and n'') an arbitrary integer or zero. Hence, the required general development of $x - y$ in series is

$$x - y = n\pi + q \sin 2y + \tfrac{1}{2} q^2 \sin 4y + \tfrac{1}{3} q^3 \sin 6y + \&c. \tag{495}$$

In like manner, since $\tan x = \tan(x - n'\pi)$, $\tan y = \tan(y - n''\pi)$, we may substitute in the first member of (c), $x - n'\pi + y - n''\pi = x + y - n\pi$, and the general development of $x + y$ in series is

$$x + y = n\pi - \frac{\sin 2y}{q} - \frac{\sin 4y}{2q^2} - \frac{\sin 6y}{3q^3} + \&c. \tag{496}$$

In these formulæ x and y are supposed to be expressed in arc, and to obtain $x \mp y$ in seconds, the terms of the series must be divided by $\sin 1''$.

255. The preceding problem is particularly useful in finding x when p and y are given, and x is nearly equal to y; in which case p is nearly equal to unity, either q or $\frac{1}{q}$ is a small fraction, and one of the series (495), (496) converges rapidly.

Examples.

1. Given $y = 50°$ and $p = 1{\cdot}00065$, to find x from (493).

Taking only the first term of the series (495), and assuming $n = 0$,

$$2y = 100° \qquad \log \sin 2y \quad 9{\cdot}99335$$

$$q = \frac{{\cdot}00065}{2{\cdot}00065} \qquad \log q \quad 6{\cdot}51174$$

$$\text{ar co log} \sin 1'' \quad 5{\cdot}31443$$

$$x - y = 65''{\cdot}995 \qquad \log(x - y) \quad 1{\cdot}81952$$

$$x = 50°\ 1'\ 5''{\cdot}995$$

2. Given $y = 50°$ and $p = -1{\cdot}00065$, to find x from (493). In this case

$$q = \frac{2{\cdot}00065}{{\cdot}00065}$$

and the computation by (496), if we assume $n = 0$, is

$$2y = 100° \qquad \log \sin 2y \quad 9{\cdot}99335$$

$$-\frac{1}{q} = -\frac{{\cdot}00065}{2{\cdot}00065} \qquad \log\left(-\frac{1}{q}\right) - 6{\cdot}51174$$

$$\text{ar co log} \sin 1'' \quad 5{\cdot}31443$$

$$x + y = -65''{\cdot}995 \qquad \log(x + y) - 1{\cdot}81952$$

$$x = -y - 65''{\cdot}995 = -50°\ 1'\ 5''{\cdot}995$$

or, if $n = 1$, $x = 180° - 50°\ 1'\ 5''.995 = 129°\ 58'\ 54''.005$.

In general, (493) is to be solved by (495) when p is positive, and by (496) when p is negative.

256. *Given the equation*

$$\sin(\alpha + z) = m \sin z \tag{497}$$

to express z in a series of multiples of α.

We deduce as in Art. 168,

$$\tan(z + \tfrac{1}{2}\alpha) = \frac{m + 1}{m - 1} \tan \tfrac{1}{2}\alpha$$

which is reduced to (493) by putting

$$x = z + \tfrac{1}{2}\alpha \qquad y = \tfrac{1}{2}\alpha \qquad p = \frac{m+1}{m-1}$$

whence

$$q = \frac{p-1}{p+1} = \frac{1}{m}$$

and (495) becomes

$$z = n\pi + \frac{\sin\alpha}{m} + \frac{\sin 2\alpha}{2m^2} + \frac{\sin 3\alpha}{3m^3} + \&c. \tag{498}$$

which is to be employed when $m > 1$; and (496) becomes

$$z + \alpha = n\pi - m\sin\alpha - \tfrac{1}{2}m^2\sin 2\alpha - \tfrac{1}{3}m^3\sin 3\alpha - \&c. \tag{499}$$

which is to be employed when $m < 1$, n being any integer or zero.

257. *Given the equation*

$$\tan z = \frac{m\sin\alpha}{1 + m\cos\alpha} \tag{500}$$

to express z in a series of multiples of α.

This equation in the form

$$\frac{\sin z}{\cos z} = \frac{m\sin\alpha}{1 + m\cos\alpha}$$

gives

$$\sin z + m\sin z\cos\alpha = m\cos z\sin\alpha$$

$$\sin z = m\sin(\alpha - z)$$

$$\tan(z - \tfrac{1}{2}\alpha) = \frac{m-1}{m+1}\tan\tfrac{1}{2}\alpha$$

which is reduced to (493) by substituting

$$x = z - \tfrac{1}{2}\alpha \qquad y = \tfrac{1}{2}\alpha \qquad p = \frac{m-1}{m+1}$$

whence

$$q = \frac{p-1}{p+1} = -\frac{1}{m}$$

and the series (495) and (496) become

$$z - \alpha = n\pi - \frac{\sin\alpha}{m} + \frac{\sin 2\alpha}{2m^2} - \frac{\sin 3\alpha}{3m^3} + \&c. \tag{501}$$

$$z = n\pi + m\sin\alpha - \tfrac{1}{2}m^2\sin 2\alpha + \tfrac{1}{3}m^3\sin 3\alpha - \&c. \tag{502}$$

258. *Given the equation*

$$\tan z = \frac{m\sin\alpha}{1 - m\cos\alpha} \tag{503}$$

to express z in a series of multiples of α.

The equation (500) becomes (503) by changing the signs of both m and α; the same changes in (501) and (502) give

$$z + \alpha = n\pi - \frac{\sin\alpha}{m} - \frac{\sin 2\alpha}{2m^2} - \frac{\sin 3\alpha}{3m^3} - \&c. \tag{504}$$

$$z = n\pi + m\sin\alpha + \tfrac{1}{2}m^2\sin 2\alpha + \tfrac{1}{3}m^3\sin 3\alpha + \&c. \tag{505}$$

259. *In a plane triangle A B C, given a, b and C, to find A or B by a series of multiples of C.*

By (262)

$$\tan A = \frac{\frac{a}{b} \sin C}{1 - \frac{a}{b} \cos C}$$

which, compared with (503), gives, by (505),

$$A = \frac{a}{b} \sin C + \frac{a^2}{b^2} \cdot \frac{\sin 2\,C}{2} + \frac{a^3}{b^3} \cdot \frac{\sin 3\,C}{3} + \&c. \tag{506}$$

n being necessarily $= 0$ in this case. B is found by the same series, interchanging a and b.

260. *In a plane triangle, A B C, given a, b and C, to find c by a series of multiples of C.*

We have

$$c^2 = a^2 + b^2 - 2\,ab \cos C \tag{507}$$

$$\frac{c^2}{a^2} = \frac{b^2}{a^2} - \frac{2\,b}{a} \cos C + 1$$

by (451)

$$= \left[\frac{b}{a} - (\cos C + \sqrt{-1} \sin C)\right]$$

$$\times \left[\frac{b}{a} - (\cos C - \sqrt{-1} \sin C)\right]$$

$$\frac{c^2}{b^2} = \left[1 - \frac{a}{b} (\cos C + \sqrt{-1} \sin C)\right] \times \left[1 - \frac{a}{b} (\cos C - \sqrt{-1} \sin C)\right]$$

Taking the common logarithms, employing in the second member the formula

$$\log (1 - n) = - M (n + \tfrac{1}{2} n^2 + \tfrac{1}{3} n^3 + \&c.)$$

and applying Moivre's Formula (440) in expressing the powers of $\cos C \pm \sqrt{-1} \sin C$, we have

$2 \log c - 2 \log b =$

$$- M \left[\frac{a}{b} (\cos C + \sqrt{-1} \sin C) + \frac{a^2}{2\,b^2} (\cos 2\,C + \sqrt{-1} \sin 2\,C) + \&c.\right]$$

$$- M \left[\frac{a}{b} (\cos C - \sqrt{-1} \sin C) + \frac{a^2}{2\,b^2} (\cos 2\,C - \sqrt{-1} \sin 2\,C) + \&c.\right]$$

$$\log c = \log b - M \left(\frac{a}{b} \cos C + \frac{a^2}{b^2} \cdot \frac{\cos 2\,C}{2} + \frac{a^3}{b^3} \cdot \frac{\cos 3C}{3} + \&c.\right) \tag{508}$$

This series was first given by Legendre. The series (495) and (496), upon which are based those of the subsequent articles, (Arts. 256, 257, 258 and 259), are due to Lagrange.

PART II.

SPHERICAL TRIGONOMETRY.

CHAPTER I.

GENERAL FORMULÆ.

1. SPHERICAL TRIGONOMETRY treats of the methods of computing the unknown from the known parts of a spherical triangle.

It is shown in geometry,* that a spherical triangle may, in general, be *constructed* when any three of its six parts are given, (not excepting the case where the three angles are given). We are now to investigate the methods by which, in the same cases, the unknown parts may be *computed.*

We shall at first confine our attention to such triangles only as are treated of in geometry, namely, those whose sides are each less than a semicircumference, and whose angles are each less than two right angles; that is, those in which every part is less than 180°.

2. It is shown in geometry, that if a solid angle is formed at the center of a sphere by three planes, the three arcs in which these planes intersect the surface of the sphere form a spherical triangle. Now the real objects of investigation in spherical trigonometry are the mutual relations of the angles of inclination of the faces and edges of a solid angle; but, for convenience, the spherical triangle which forms the base of the solid angle is substituted for it. The sides of the triangle being proportional to the angles of inclination of the edges of the solid angle, are taken to represent those angles; and the angles which those sides form with each other are regarded

---

* The student is here supposed to be acquainted with Spherical Geometry, at least so much of it as is to be found in Legendre's treatise, or in that of Prof. Peirce, of Harvard University.

as identical with the angles of inclination of the faces of the solid angle. But, since varying the radius of the sphere would not, in any respect, change the solid angle, or the values of the angles which enter into it, the mutual relations in question ought to be deduced without any reference to the magnitude of the radius of the sphere. In fact, we shall deduce our fundamental formulæ from a direct consideration of the solid angle itself.

3. *In a spherical triangle, the sines of the sides are proportional to the sines of the opposite angles.*

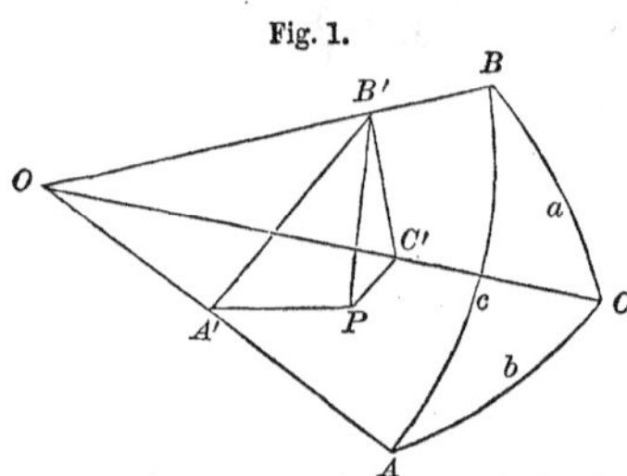

Let ABC, Fig. 1, be a spherical triangle, O the center of the sphere. The angles of the triangle are the inclinations of the planes AOB, AOC and BOC, to each other, and will be designated by A, B and C; their opposite sides respectively will be designated by a, b and c, as in plane triangles. The trigonometric functions of these sides will be the same as those of the angles BOC, AOC, AOB, which they subtend at the center of the sphere. (Pl. Trig. Art. 20.)

From any point B' in OB, let fall $B'P$ perpendicular to the plane AOC; and through $B'P$ let the planes $B'PA'$, $B'PC'$ be drawn perpendicular to OA and OC, intersecting the plane OAC in the lines PA', PC', and the planes AOB, BOC in the lines $A'B'$, $B'C'$. The plane triangles $A'PB'$, $B'PC'$ are right angled at P; and $OA'B'$, $OC'B'$ are right angled at A' and C'. The angle $B'A'P$, being formed by two lines perpendicular to OA, is the measure of the inclination of the planes AOB, AOC, or of the angle A; and $B'C'P$ is the measure of the angle C.

We have therefore, by Pl. Trig. Art. 15,

$$\sin A = \sin B'A'P = \frac{B'P}{B'A'}$$

$$\sin C = \sin B'C'P = \frac{B'P}{B'C'}$$

whence

$$\frac{\sin A}{\sin C} = \frac{B'P}{B'A'} \times \frac{B'C'}{B'P} = \frac{B'C'}{B'A'} \qquad (m)$$

Again, $$\sin a = \sin B'OC' = \frac{B'C'}{B'O}$$

$$\sin c = \sin B'OA' = \frac{B'A'}{B'O}$$

whence $$\frac{\sin a}{\sin c} = \frac{B'C'}{B'O} \times \frac{B'O}{B'A'} = \frac{B'C'}{B'A'} \qquad (n)$$

Comparing (m) and (n),

$$\frac{\sin a}{\sin c} = \frac{\sin A}{\sin C} \qquad (1)$$

which in the form of a proportion is

$$\sin a : \sin c = \sin A : \sin C$$

which is the theorem that was to be proved.

4. In Fig. 1, A, a, C and c, are each less than 90°, but the construction would not vary if any of these parts were greater than 90°, except that the points A' and C' might be found in the lines AO, CO, *produced* through O; and one or more of the right triangles $A'B'P$, &c., would contain the supplements of A, a, C, or c instead of these quantities themselves. But the sine of an angle and of its supplement being the same, the preceding demonstration would still be valid, so that the theorem is applicable to any spherical triangle.

Indeed, according to Pl. Trig. Art. 49, this result follows from the nature of the trigonometric functions themselves, and the demonstration of the preceding theorem might therefore be considered as general, without requiring a special examination of the various positions of the lines of the diagram.

5. *In a spherical triangle, the cosine of any side is equal to the product of the cosines of the other two sides,* plus *the continued product of the sines of those sides and the cosine of the included angle.*

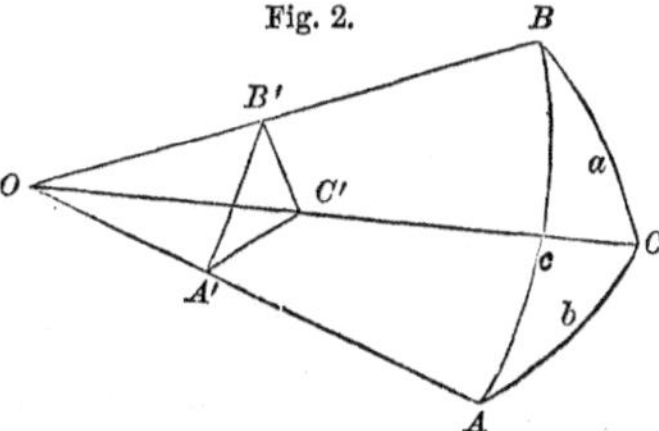

Fig. 2.

Let the plane $B'A'C'$, Fig. 2, be drawn perp. to OA, intersecting the planes AOB, BOC and AOC, in the lines $A'B'$, $B'C'$ and $A'C'$. Then the angle $B'A'C' = A$, and $B'OC' = a$, and by Pl. Trig. Art. 119, in the triangles $A'B'C'$, $OB'C'$, we have

$$B'C'^2 = A'B'^2 + A'C'^2 - 2\,A'B' \,.\, A'C' \cos A$$

$$B'C'^2 = OB'^2 + OC'^2 - 2\,OB' \,.\, OC' \cos a$$

Subtracting the first of these equations from the second, and observing that in the right triangles $OA'B'$, $OA'C'$,

$$OB'^2 - A'B'^2 = OA'^2, \qquad OC'^2 - A'C'^2 = OA'^2$$

we have

$$0 = 2\,OA'^2 + 2\,A'B'\,.\,A'C' \cos A - 2\,OB'\,.\,OC' \cos a$$

whence $$\cos a = \frac{OA'\,.\,OA'}{OB'\,.\,OC'} + \frac{A'B'\,.\,A'C'}{OB'\,.\,OC'} \cos A$$

Substituting the trigonometric functions derived from the right triangles $OA'B'$, $OA'C'$,

$$\cos a = \cos b \cos c + \sin b \sin c \cos A \tag{2}$$

which is the theorem to be proved. It may be regarded as the fundamental theorem, for the preceding (1) can be deduced from it, but as the process is somewhat circuitous, we have preferred deducing the two theorems from independent constructions.

6. In the construction of Fig. 2, both b and c are supposed less than 90°, while no restriction is placed upon A and a; but the equation (2) is no less applicable to all the other cases if the principle of Pl. Trig. Art. 49 be granted. As that principle may not be sufficiently evident to the student unacquainted with analytical geometry, we shall verify it in this case, as follows.*

Fig. 3.

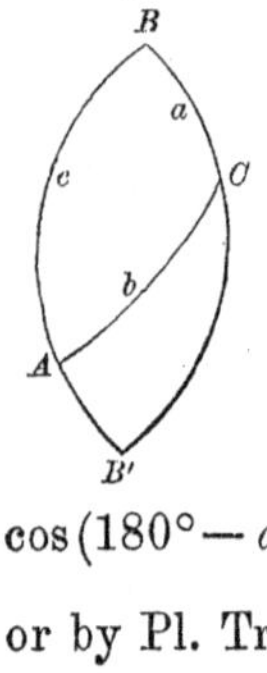

1st. In the triangle ABC, (Fig. 3), let $b < 90°$ and $c > 90°$. Produce BA, BC to meet in B', forming the lune BB'; then $AB' = 180° - c$, and b are both $< 90°$, and the preceding demonstration would apply to the triangle $AB'C$. Therefore, applying (2) to $AB'C$, we have

$$\cos(180° - a) = \cos b \cos(180° - c) + \sin b \sin(180° - c) \cos(180° - A)$$

or by Pl. Trig. (64),

$$-\cos a = -\cos b \cos c - \sin b \sin c \cos A$$

and changing all the signs

$$\cos a = \cos b \cos c + \sin b \sin c \cos A$$

the same result that would have been found by applying (2) directly to ABC.

* Hymer's Spherical Trigonometry. Cambridge, 1841.

2d. In the triangle ABC, Fig. 4, let $b > 90°$, $c > 90°$; produce AB and AC to meet in A'; then $A'B$ and $A'C$ being both less than 90°, the formula (2) is applicable to $A'BC$. Therefore

Fig. 4.

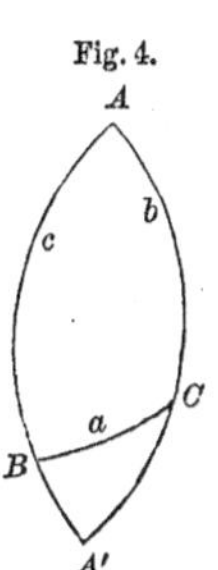

$$\begin{aligned}\cos a &= \cos(180° - b)\cos(180° - c) \\ &\qquad + \sin(180° - b)\sin(180° - c)\cos A \\ &= (-\cos b)(-\cos c) + \sin b \sin c \cos A \\ &= \cos b \cos c + \sin b \sin c \cos A\end{aligned}$$

the same result as before.

7. The theorems expressed by (1) and (2) being applied successively to the several parts of the triangle, give the two following groups:

$$\left.\begin{aligned}\sin a \sin B &= \sin b \sin A \\ \sin b \sin C &= \sin c \sin B \\ \sin c \sin A &= \sin a \sin C\end{aligned}\right\} \quad (3)$$

$$\left.\begin{aligned}\cos a &= \cos b \cos c + \sin b \sin c \cos A \\ \cos b &= \cos c \cos a + \sin c \sin a \cos B \\ \cos c &= \cos a \cos b + \sin a \sin b \cos C\end{aligned}\right\} \quad (4)$$

8. Let $A'B'C'$, Fig. 5, be the polar triangle of ABC, and designate its angles and sides by A', B', C', a', b' and c'. Then, by geometry,

Fig. 5.

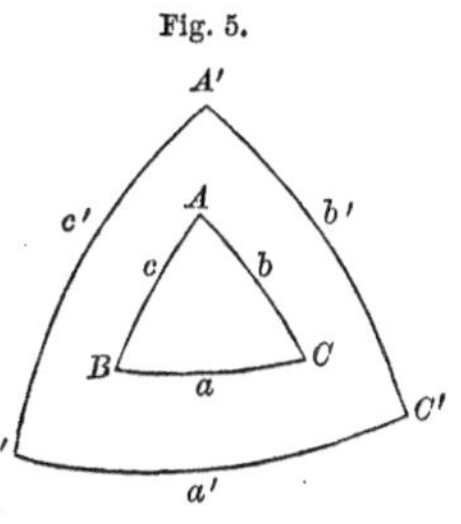

$$\begin{aligned}A' &= 180° - a, & a' &= 180° - A \\ B' &= 180° - b, & b' &= 180° - B \\ C' &= 180° - c, & c' &= 180° - C\end{aligned}$$

and applying the first equation of (4) to $A'B'C'$,

$$\cos a' = \cos b' \cos c' + \sin b' \sin c' \cos A'$$

or by Pl. Trig. (64),

$$- \cos A = (-\cos B)(-\cos C) + \sin B \sin C(-\cos a)$$

$$- \cos A = \cos B \cos C - \sin B \sin C \cos a$$

Changing the signs of this, we have the first of the following group:

$$\left.\begin{aligned}\cos A &= -\cos B \cos C + \sin B \sin C \cos a \\ \cos B &= -\cos C \cos A + \sin C \sin A \cos b \\ \cos C &= -\cos A \cos B + \sin A \sin B \cos c\end{aligned}\right\} \quad (5)$$

It is thus that, by means of the polar triangle, any formula of a spherical triangle may be immediately transformed into another, in which angles take the place of sides, and sides of angles.

9. Several other important fundamental groups of formulæ are obtained from the preceding with the greatest ease.

The first of (4) multiplied by $\cos c$ is

$$\cos a \cos c = \cos b \cos^2 c + \sin b \sin c \cos c \cos A$$

and the second of (4) is the same as

$$\cos a \cos c + \sin a \sin c \cos B = \cos b$$

the difference of which is

$$\sin a \sin c \cos B = (1 - \cos^2 c) \cos b - \sin b \sin c \cos c \cos A$$

Since $1 - \cos^2 c = \sin^2 c$, this may be divided by $\sin c$, and gives

whence
$$\left.\begin{aligned} \sin a \cos B &= \sin c \cos b - \cos c \sin b \cos A \\ \sin b \cos C &= \sin a \cos c - \cos a \sin c \cos B \\ \sin c \cos A &= \sin b \cos a - \cos b \sin a \cos C \end{aligned}\right\} \quad (6)$$

If we interchange B and C, and therefore also b and c, the group becomes

$$\left.\begin{aligned} \sin a \cos C &= \sin b \cos c - \cos b \sin c \cos A \\ \sin b \cos A &= \sin c \cos a - \cos c \sin a \cos B \\ \sin c \cos B &= \sin a \cos b - \cos a \sin b \cos C \end{aligned}\right\} \quad (7)$$

10. If (6) and (7) are applied to the polar triangle, they give, after changing the signs of all the terms,

$$\left.\begin{aligned} \sin A \cos b &= \sin C \cos B + \cos C \sin B \cos a \\ \sin B \cos c &= \sin A \cos C + \cos A \sin C \cos b \\ \sin C \cos a &= \sin B \cos A + \cos B \sin A \cos c \end{aligned}\right\} \quad (8)$$

and

$$\left.\begin{aligned} \sin A \cos c &= \sin B \cos C + \cos B \sin C \cos a \\ \sin B \cos a &= \sin C \cos A + \cos C \sin A \cos b \\ \sin C \cos b &= \sin A \cos B + \cos A \sin B \cos c \end{aligned}\right\} \quad (9)$$

11. Dividing the first of (6) by the following derived from (3),

$$\frac{\sin a \sin B}{\sin A} = \sin b$$

we find the first of the following group

$$\left.\begin{aligned} \sin A \cot B &= \sin c \cot b - \cos c \cos A \\ \sin B \cot C &= \sin a \cot c - \cos a \cos B \\ \sin C \cot A &= \sin b \cot a - \cos b \cos C \end{aligned}\right\} \quad (10)$$

and in the same way from (7), or by interchanging the letters B and C, b and c in (10), we find

$$\left.\begin{aligned} \sin A \cot C &= \sin b \cot c - \cos b \cos A \\ \sin B \cot A &= \sin c \cot a - \cos c \cos B \\ \sin C \cot B &= \sin a \cot b - \cos a \cos C \end{aligned}\right\} \quad (11)$$

If (10) are applied to the polar triangle, we find (11), so that no new relations are elicited.

12. The preceding formulæ are sufficient to furnish a theoretical solution for every case of spherical triangles, but some transformations are required to facilitate their application in practice.

In the first of (4) substitute, by Pl. Trig. (139),

$$\cos A = 1 - 2 \sin^2 \tfrac{1}{2} A$$

we find, by Pl. Trig. (39),

$$\cos a = \cos (b - c) - 2 \sin b \sin c \sin^2 \tfrac{1}{2} A \quad (12)$$

and we have similar expressions for $\cos b$ and $\cos c$.

If we substitute in (4), by Pl. Trig. (138),

$$\cos A = -1 + 2 \cos^2 \tfrac{1}{2} A$$

we find, by Pl. Trig. (38),

$$\cos a = \cos (b + c) + 2 \sin b \sin c \cos^2 \tfrac{1}{2} A \quad (13)$$

and, of course, similar expressions for $\cos b$ and $\cos c$.

13. Substituting in (5)

$$\cos a = 1 - 2 \sin^2 \tfrac{1}{2} a = -1 + 2 \cos^2 \tfrac{1}{2} a$$

we find by the same process

$$\cos A = -\cos (B + C) - 2 \sin B \sin C \sin^2 \tfrac{1}{2} a \quad (14)$$

$$\cos A = -\cos (B - C) + 2 \sin B \sin C \cos^2 \tfrac{1}{2} a \quad (15)$$

which might have been obtained by applying (12) and (13) to the polar triangle.

14. If in (12) we substitute $\cos a = 1 - 2\sin^2 \frac{1}{2} a$, $\cos (b - c) = 1 - 2\sin^2 \frac{1}{2}(b - c)$, we obtain the first of the following equations; and the others are obtained by a similar process from (12), (13), (14) and (15).

$$\sin^2 \tfrac{1}{2} a = \sin^2 \tfrac{1}{2} (b - c) + \sin b \sin c \sin^2 \tfrac{1}{2} A \qquad (16)$$
$$\sin^2 \tfrac{1}{2} a = \sin^2 \tfrac{1}{2} (b + c) - \sin b \sin c \cos^2 \tfrac{1}{2} A \qquad (17)$$
$$\cos^2 \tfrac{1}{2} a = \cos^2 \tfrac{1}{2} (b - c) - \sin b \sin c \sin^2 \tfrac{1}{2} A \qquad (18)$$
$$\cos^2 \tfrac{1}{2} a = \cos^2 \tfrac{1}{2} (b + c) + \sin b \sin c \cos^2 \tfrac{1}{2} A \qquad (19)$$
$$\sin^2 \tfrac{1}{2} A = \cos^2 \tfrac{1}{2} (B + C) + \sin B \sin C \sin^2 \tfrac{1}{2} a \qquad (20)$$
$$\sin^2 \tfrac{1}{2} A = \cos^2 \tfrac{1}{2} (B - C) - \sin B \sin C \cos^2 \tfrac{1}{2} a \qquad (21)$$
$$\cos^2 \tfrac{1}{2} A = \sin^2 \tfrac{1}{2} (B + C) - \sin B \sin C \sin^2 \tfrac{1}{2} a \qquad (22)$$
$$\cos^2 \tfrac{1}{2} A = \sin^2 \tfrac{1}{2} (B - C) + \sin B \sin C \cos^2 \tfrac{1}{2} a \qquad (23)$$

15. By Pl. Trig. we have

$$1 = \cos^2 \tfrac{1}{2} A + \sin^2 \tfrac{1}{2} A$$
$$\cos A = \cos^2 \tfrac{1}{2} A - \sin^2 \tfrac{1}{2} A$$

whence

$$\cos b \cos c = \cos b \cos c \cos^2 \tfrac{1}{2} A + \cos b \cos c \sin^2 \tfrac{1}{2} A$$
$$\sin b \sin c \cos A = \sin b \sin c \cos^2 \tfrac{1}{2} A - \sin b \sin c \sin^2 \tfrac{1}{2} A$$

the sum of which is, by (4),

$$\cos a = \cos (b - c) \cos^2 \tfrac{1}{2} A + \cos (b + c) \sin^2 \tfrac{1}{2} A \qquad (24)$$

and substituting $1 - 2\sin^2 \frac{1}{2} a$, &c., for $\cos a$, &c.

$$\sin^2 \tfrac{1}{2} a = \sin^2 \tfrac{1}{2} (b - c) \cos^2 \tfrac{1}{2} A + \sin^2 \tfrac{1}{2} (b + c) \sin^2 \tfrac{1}{2} A \qquad (25)$$
$$\cos^2 \tfrac{1}{2} a = \cos^2 \tfrac{1}{2} (b - c) \cos^2 \tfrac{1}{2} A + \cos^2 \tfrac{1}{2} (b + c) \sin^2 \tfrac{1}{2} A \qquad (26)$$

In the same manner we deduce from (5)

$$\cos A = -\cos (B - C) \sin^2 \tfrac{1}{2} a - \cos (B + C) \cos^2 \tfrac{1}{2} a \qquad (27)$$
$$\sin^2 \tfrac{1}{2} A = \cos^2 \tfrac{1}{2} (B - C) \sin^2 \tfrac{1}{2} a + \cos^2 \tfrac{1}{2} (B + C) \cos^2 \tfrac{1}{2} a \qquad (28)$$
$$\cos^2 \tfrac{1}{2} A = \sin^2 \tfrac{1}{2} (B - C) \sin^2 \tfrac{1}{2} a + \sin^2 \tfrac{1}{2} (B + C) \cos^2 \tfrac{1}{2} a \qquad (29)$$

It is hardly necessary to add that each of the equations (12 to 29) gives a group of three, by applying it successively to the three sides or three angles of the triangle.

16. From (12) we find

$$\sin^2 \tfrac{1}{2} A = \frac{\cos (b - c) - \cos a}{2 \sin b \sin c}$$

If, in Pl. Trig. (108), we put

$$x = a, \qquad y = b - c$$

whence

$$\tfrac{1}{2}(x + y) = \tfrac{1}{2}(a + b - c), \qquad \tfrac{1}{2}(x - y) = \tfrac{1}{2}(a - b + c)$$

we find

$$\cos (b - c) - \cos a = 2 \sin \tfrac{1}{2}(a - b + c) \sin \tfrac{1}{2}(a + b - c)$$

which, substituted in the above equation, gives

$$\sin^2 \tfrac{1}{2} A = \frac{\sin \tfrac{1}{2}(a - b + c) \sin \tfrac{1}{2}(a + b - c)}{\sin b \sin c} \qquad (30)$$

Let s denote the half sum of the sides, that is, let

$$a + b + c = 2s, \qquad \tfrac{1}{2}(a + b + c) = s$$

then

$$a - b + c = a + b + c - 2b = 2s - 2b = 2(s - b)$$
$$a + b - c = a + b + c - 2c = 2s - 2c = 2(s - c)$$

which substituted in (30) give

$$\left.\begin{aligned} \sin^2 \tfrac{1}{2} A &= \frac{\sin(s - b)\sin(s - c)}{\sin b \sin c} \\ \text{whence also} \qquad \sin^2 \tfrac{1}{2} B &= \frac{\sin(s - c)\sin(s - a)}{\sin c \sin a} \\ \sin^2 \tfrac{1}{2} C &= \frac{\sin(s - a)\sin(s - b)}{\sin a \sin b} \end{aligned}\right\} \quad (31)$$

17. From (13) we find

$$\cos^2 \tfrac{1}{2} A = \frac{\cos a - \cos(b + c)}{2 \sin b \sin c}$$

and from Pl. Trig. (108), by making

$$x = b + c, \qquad y = a$$

$$\tfrac{1}{2}(x + y) = \tfrac{1}{2}(a + b + c), \qquad \tfrac{1}{2}(x - y) = \tfrac{1}{2}(b + c - a)$$

we find

$$\cos a - \cos(b + c) = 2 \sin \tfrac{1}{2}(a + b + c) \sin \tfrac{1}{2}(b + c - a)$$

which, substituted above, gives

$$\cos^2 \tfrac{1}{2} A = \frac{\sin \tfrac{1}{2}(a + b + c) \sin \tfrac{1}{2}(b + c - a)}{\sin b \sin c} \quad (32)$$

Introducing, as in the preceding article, $s = \tfrac{1}{2}(a + b + c)$,

$$\left.\begin{aligned} \cos^2 \tfrac{1}{2} A &= \frac{\sin s \sin(s - a)}{\sin b \sin c} \\ \cos^2 \tfrac{1}{2} B &= \frac{\sin s \sin(s - b)}{\sin c \sin a} \\ \cos^2 \tfrac{1}{2} C &= \frac{\sin s \sin(s - c)}{\sin a \sin b} \end{aligned}\right\} \quad (33)$$

18. The quotient of (31) divided by (33) gives

$$\left.\begin{aligned}\tan^2\tfrac{1}{2}A &= \frac{\sin(s-b)\sin(s-c)}{\sin s\sin(s-a)}\\ \tan^2\tfrac{1}{2}B &= \frac{\sin(s-c)\sin(s-a)}{\sin s\sin(s-b)}\\ \tan^2\tfrac{1}{2}C &= \frac{\sin(s-a)\sin(s-b)}{\sin s\sin(s-c)}\end{aligned}\right\}\quad(34)$$

19. From (14) we find

$$\sin^2\tfrac{1}{2}a = -\frac{\cos A + \cos(B+C)}{2\sin B\sin C}$$

from which, by Pl. Trig. (107), we deduce

$$\sin^2\tfrac{1}{2}a = \frac{-\cos\tfrac{1}{2}(A+B+C)\cos\tfrac{1}{2}(B+C-A)}{\sin B\sin C}\quad(35)$$

and if we put

$$\tfrac{1}{2}(A+B+C) = S$$

$$\left.\begin{aligned}\sin^2\tfrac{1}{2}a &= \frac{-\cos S\cos(S-A)}{\sin B\sin C}\\ \sin^2\tfrac{1}{2}b &= \frac{-\cos S\cos(S-B)}{\sin C\sin A}\\ \sin^2\tfrac{1}{2}c &= \frac{-\cos S\cos(S-C)}{\sin A\sin B}\end{aligned}\right\}\quad(36)$$

The first member of each of these equations being a square, the second member must be essentially positive, although its algebraic sign is negative; in fact, since by geometry $2S > 180°$, $S > 90°$, $\cos S$ is negative, and $-\cos S$ is positive.

20. From (15) we find

$$\cos^2\tfrac{1}{2}a = \frac{\cos A + \cos(B-C)}{2\sin B\sin C}$$

from which we deduce, by a process similar to the preceding,

$$\cos^2\tfrac{1}{2}a = \frac{\cos\tfrac{1}{2}(A-B+C)\cos\tfrac{1}{2}(A+B-C)}{\sin B\sin C}\quad(37)$$

$$\left.\begin{aligned}\cos^2 \tfrac{1}{2} a &= \frac{\cos(S-B)\cos(S-C)}{\sin B \sin C}\\ \cos^2 \tfrac{1}{2} b &= \frac{\cos(S-C)\cos(S-A)}{\sin C \sin A}\\ \cos^2 \tfrac{1}{2} c &= \frac{\cos(S-A)\cos(S-B)}{\sin A \sin B}\end{aligned}\right\} \quad (38)$$

21. From (36) and (38)

$$\left.\begin{aligned}\tan^2 \tfrac{1}{2} a &= \frac{-\cos S \cos(S-A)}{\cos(S-B)\cos(S-C)}\\ \tan^2 \tfrac{1}{2} b &= \frac{-\cos S \cos(S-B)}{\cos(S-C)\cos(S-A)}\\ \tan^2 \tfrac{1}{2} c &= \frac{-\cos S \cos(S-C)}{\cos(S-A)\cos(S-B)}\end{aligned}\right\} \quad (39)$$

We might have deduced (36), (38), (39), by applying (31), (33), (34) to the polar triangle.

22. *Napier's Analogies.* Dividing the 1st of (34) by the 2d, we find

$$\frac{\tan \tfrac{1}{2} A}{\tan \tfrac{1}{2} B} = \frac{\sin(s-b)}{\sin(s-a)}$$

Regarding this as a proportion, we have, by composition and division,

$$\frac{\tan \tfrac{1}{2} A + \tan \tfrac{1}{2} B}{\tan \tfrac{1}{2} A - \tan \tfrac{1}{2} B} = \frac{\sin(s-b) + \sin(s-a)}{\sin(s-b) - \sin(s-a)} \quad (m)$$

In Pl. Trig. (109), if we put $x = s - b$, $y = s - a$, whence

$$x + y = 2s - a - b = c$$
$$x - y = a - b$$

we have

$$\frac{\sin(s-b) + \sin(s-a)}{\sin(s-b) - \sin(s-a)} = \frac{\tan \tfrac{1}{2} c}{\tan \tfrac{1}{2}(a-b)}$$

and by Pl. Trig. (126),

$$\frac{\tan \tfrac{1}{2} A + \tan \tfrac{1}{2} B}{\tan \tfrac{1}{2} A - \tan \tfrac{1}{2} B} = \frac{\sin \tfrac{1}{2}(A+B)}{\sin \tfrac{1}{2}(A-B)}$$

Therefore (m) becomes

$$\left.\begin{aligned} \frac{\sin\frac{1}{2}(A+B)}{\sin\frac{1}{2}(A-B)} &= \frac{\tan\frac{1}{2}c}{\tan\frac{1}{2}(a-b)} \\ \text{or}\quad \sin\tfrac{1}{2}(A+B):\sin\tfrac{1}{2}(A-B) &= \tan\tfrac{1}{2}c:\tan\tfrac{1}{2}(a-b) \end{aligned}\right\} \quad (40)$$

which is the first of Napier's Analogies.

23. Again, the product of the 1st and 2d of (34) gives

$$\tan\tfrac{1}{2}A\tan\tfrac{1}{2}B = \frac{\sin(s-c)}{\sin s}$$

or

$$1:\tan\tfrac{1}{2}A\tan\tfrac{1}{2}B = \sin s:\sin(s-c)$$

whence, by composition and division,

$$\frac{1-\tan\frac{1}{2}A\tan\frac{1}{2}B}{1+\tan\frac{1}{2}A\tan\frac{1}{2}B} = \frac{\sin s - \sin(s-c)}{\sin s + \sin(s-c)} \quad (n)$$

By Pl. Trig. (109), if $x = s$, $y = s - c$, we have

$$\frac{\sin s - \sin(s-c)}{\sin s + \sin(s-c)} = \frac{\tan\frac{1}{2}c}{\tan\frac{1}{2}(a+b)}$$

and by Pl. Trig. (127),

$$\frac{1-\tan\frac{1}{2}A\tan\frac{1}{2}B}{1+\tan\frac{1}{2}A\tan\frac{1}{2}B} = \frac{\cos\frac{1}{2}(A+B)}{\cos\frac{1}{2}(A-B)}$$

Therefore (n) becomes

$$\left.\begin{aligned} \frac{\cos\frac{1}{2}(A+B)}{\cos\frac{1}{2}(A-B)} &= \frac{\tan\frac{1}{2}c}{\tan\frac{1}{2}(a+b)} \\ \text{or}\quad \cos\tfrac{1}{2}(A+B):\cos\tfrac{1}{2}(A-B) &= \tan\tfrac{1}{2}c:\tan\tfrac{1}{2}(a+b) \end{aligned}\right\} \quad (41)$$

which is the second of Napier's Analogies.

24. If (40) and (41) are applied to the polar triangle, we shall find

$$\left.\begin{aligned} \frac{\sin\frac{1}{2}(a+b)}{\sin\frac{1}{2}(a-b)} &= \frac{\cot\frac{1}{2}C}{\tan\frac{1}{2}(A-B)} \\ \text{or}\quad \sin\tfrac{1}{2}(a+b):\sin\tfrac{1}{2}(a-b) &= \cot\tfrac{1}{2}C:\tan\tfrac{1}{2}(A-B) \end{aligned}\right\} \quad (42)$$

$$\left.\begin{aligned} \frac{\cos\frac{1}{2}(a+b)}{\cos\frac{1}{2}(a-b)} &= \frac{\cot\frac{1}{2}C}{\tan\frac{1}{2}(A+B)} \\ \text{or}\quad \cos\tfrac{1}{2}(a+b):\cos\tfrac{1}{2}(a-b) &= \cot\tfrac{1}{2}C:\tan\tfrac{1}{2}(A+B) \end{aligned}\right\} \quad (43)$$

which are the third and fourth of Napier's Analogies.

25. *Gauss's Theorem.* If

$$p = \cos \tfrac{1}{2} c \sin \tfrac{1}{2} (A + B) \qquad P = \cos \tfrac{1}{2} C \cos \tfrac{1}{2} (a - b)$$
$$q = \cos \tfrac{1}{2} c \cos \tfrac{1}{2} (A + B) \qquad Q = \sin \tfrac{1}{2} C \cos \tfrac{1}{2} (a + b)$$
$$r = \sin \tfrac{1}{2} c \sin \tfrac{1}{2} (A - B) \qquad R = \cos \tfrac{1}{2} C \sin \tfrac{1}{2} (a - b)$$
$$s = \sin \tfrac{1}{2} c \cos \tfrac{1}{2} (A - B) \qquad S = \sin \tfrac{1}{2} C \sin \tfrac{1}{2} (a + b)$$

then the products $p \times q$, $p \times r$, $p \times s$, $q \times r$, $q \times s$, $r \times s$, *are respectively equal to the products* $P \times Q$, $P \times R$, $P \times S$, $Q \times R$, $Q \times S$, $R \times S$.

First. From (3) we have

$$\sin c (\sin A \pm \sin B) = \sin C (\sin a \pm \sin b)$$

which, by Pl. Trig. (105), (106) and (135), are reduced to

$$\sin \tfrac{1}{2} c \cos \tfrac{1}{2} c \sin \tfrac{1}{2} (A + B) \cos \tfrac{1}{2} (A - B) = \sin \tfrac{1}{2} C \cos \tfrac{1}{2} C \sin \tfrac{1}{2} (a + b) \cos \tfrac{1}{2} (a - b)$$
$$\sin \tfrac{1}{2} c \cos \tfrac{1}{2} c \cos \tfrac{1}{2} (A + B) \sin \tfrac{1}{2} (A - B) = \sin \tfrac{1}{2} C \cos \tfrac{1}{2} C \cos \tfrac{1}{2} (a + b) \sin \tfrac{1}{2} (a - b)$$

or

$$ps = PS \qquad \text{and} \quad qr = QR$$

Second. From (6) and (7)

$$\sin c (\cos B \pm \cos A) = (1 \mp \cos C) \sin (a \pm b)$$

which, by Pl. Trig. are reduced to

$$\sin \tfrac{1}{2} c \cos \tfrac{1}{2} c \cos \tfrac{1}{2} (A + B) \cos \tfrac{1}{2} (A - B) = \sin \tfrac{1}{2} C \sin \tfrac{1}{2} C \sin \tfrac{1}{2} (a + b) \cos \tfrac{1}{2} (a + b)$$
$$\sin \tfrac{1}{2} c \cos \tfrac{1}{2} c \sin \tfrac{1}{2} (A + B) \sin \tfrac{1}{2} (A - B) = \cos \tfrac{1}{2} C \cos \tfrac{1}{2} C \sin \tfrac{1}{2} (a - b) \cos \tfrac{1}{2} (a - b)$$

or

$$qs = QS \qquad \text{and} \quad pr = PR$$

Third. From (8) and (9)

$$(1 \pm \cos c) \sin (A \pm B) = \sin C (\cos b \pm \cos a)$$

which, by Pl. Trig. are reduced to

$$\cos \tfrac{1}{2} c \cos \tfrac{1}{2} c \sin \tfrac{1}{2} (A + B) \cos \tfrac{1}{2} (A + B) = \sin \tfrac{1}{2} C \cos \tfrac{1}{2} C \cos \tfrac{1}{2} (a + b) \cos \tfrac{1}{2} (a - b)$$
$$\sin \tfrac{1}{2} c \sin \tfrac{1}{2} c \sin \tfrac{1}{2} (A - B) \cos \tfrac{1}{2} (A - B) = \sin \tfrac{1}{2} C \cos \tfrac{1}{2} C \sin \tfrac{1}{2} (a + b) \sin \tfrac{1}{2} (a - b)$$

or

$$pq = PQ \qquad \text{and} \quad rs = RS$$

26. *The notation of the preceding article being still employed, the quantities* p^2, q^2, r^2, s^2, *are respectively equal to* P^2, Q^2, R^2, S^2.

We have

$$pq \times pr = PQ \times PR$$

and

$$qr = QR$$

the quotient of which is

$$p^2 = P^2 \quad \text{whence} \quad p = \pm P$$

and in the same way

$$q^2 = Q^2 \qquad q = \pm Q$$
$$r^2 = R^2 \qquad r = \pm R$$
$$s^2 = S^2 \qquad s = \pm S$$

27. *In these last equations, the positive sign must be used in all the second members, or the negative sign in all of them.* For if we take

$$p = + P$$

the equations

$$pq = PQ, \qquad pr = PR, \qquad ps = PS$$

being divided by this, give

$$q = + Q, \qquad r = + R, \qquad s = + S$$

and if we take

$$p = - P$$

the same equations, divided by this, give

$$q = - Q, \qquad r = - R, \qquad s = - S$$

We have therefore the following, which are generally cited as *Gauss's Equations.*

$$\left.\begin{aligned} \cos \tfrac{1}{2} c \sin \tfrac{1}{2} (A + B) &= \cos \tfrac{1}{2} C \cos \tfrac{1}{2} (a - b) \\ \cos \tfrac{1}{2} c \cos \tfrac{1}{2} (A + B) &= \sin \tfrac{1}{2} C \cos \tfrac{1}{2} (a + b) \\ \sin \tfrac{1}{2} c \sin \tfrac{1}{2} (A - B) &= \cos \tfrac{1}{2} C \sin \tfrac{1}{2} (a - b) \\ \sin \tfrac{1}{2} c \cos \tfrac{1}{2} (A - B) &= \sin \tfrac{1}{2} C \sin \tfrac{1}{2} (a + b) \end{aligned}\right\} \quad (44)$$

or

$$\left.\begin{aligned} \cos \tfrac{1}{2} c \sin \tfrac{1}{2} (A + B) &= - \cos \tfrac{1}{2} C \cos \tfrac{1}{2} (a - b) \\ \cos \tfrac{1}{2} c \cos \tfrac{1}{2} (A + B) &= - \sin \tfrac{1}{2} C \cos \tfrac{1}{2} (a + b) \\ \sin \tfrac{1}{2} c \sin \tfrac{1}{2} (A - B) &= - \cos \tfrac{1}{2} C \sin \tfrac{1}{2} (a - b) \\ \sin \tfrac{1}{2} c \cos \tfrac{1}{2} (A - B) &= - \sin \tfrac{1}{2} C \sin \tfrac{1}{2} (a + b) \end{aligned}\right\} \quad (45)$$

If, however, we consider only those triangles whose parts are all less than 180°, the first of these groups, (44), is alone applicable, for we must then have $p = + P$; since $\cos \tfrac{1}{2} c$, $\sin \tfrac{1}{2} (A + B)$, $\cos \tfrac{1}{2} C$, $\cos \tfrac{1}{2} (a - b)$ are then all positive quantities. The use of (45) will be seen in the chapter on the solution of the general spherical triangle.

Napier's Analogies, (40), (41), (42) and (43) can be deduced directly from (44).

Additional Formulæ.

28. We shall here add some formulæ which, though not so frequently used as the preceding, are either remarkable for their elegance and symmetry, or of importance in certain inquiries of astronomy and geodesy.

29. The product of (30) and (32) gives

$$\sin^2 A = \frac{4 \sin s \sin (s - a) \sin (s - b) \sin (s - c)}{\sin^2 b \sin^2 c} \qquad (46)$$

Put

$$n^2 = \sin s \sin (s - a) \sin (s - b) \sin (s - c) \qquad (47)$$

then

$$\sin A = \frac{2n}{\sin b \sin c}$$

and in the same manner

$$\sin B = \frac{2n}{\sin a \sin c} \qquad (48)$$

the quotient of which is

$$\frac{\sin A}{\sin B} = \frac{\sin a}{\sin b}$$

which is our first theorem, Art. 3. As (48) was obtained from (30) and (32), and these from (4) without the aid of (3), we may consider the whole fabric of spherical trigonometry as resting upon the fundamental formulæ (4).

30. We have also from (35) and (37)

$$\sin^2 a = \frac{-4 \cos S \cos (S - A) \cos (S - B) \cos (S - C)}{\sin^2 B \sin^2 C} \tag{49}$$

and if

$$N^2 = -\cos S \cos (S - A) \cos (S - B) \cos (S - C) \tag{50}$$

$$\sin a = \frac{2N}{\sin B \sin C} \tag{51}$$

From (48) and (51),

$$\frac{n}{N} = \frac{\sin a}{\sin A} = \frac{\sin b}{\sin B} = \frac{\sin c}{\sin C} \tag{52}$$

31. If we develop (47) and (50) by Pl. Trig. (173) and (174)

$$4n^2 = 1 - \cos^2 a - \cos^2 b - \cos^2 c + 2 \cos a \cos b \cos c \tag{53}$$

$$4N^2 = 1 - \cos^2 A - \cos^2 B - \cos^2 C - 2 \cos A \cos B \cos C \tag{54}$$

32. The following simple results are easily deduced from the equations (31 to 38).

$$\left.\begin{aligned}
\frac{\cos \frac{1}{2} A \cos \frac{1}{2} B}{\sin \frac{1}{2} C} &= \frac{\sin s}{\sin c} \\
\frac{\cos \frac{1}{2} A \sin \frac{1}{2} B}{\cos \frac{1}{2} C} &= \frac{\sin (s - a)}{\sin c} \\
\frac{\sin \frac{1}{2} A \cos \frac{1}{2} B}{\cos \frac{1}{2} C} &= \frac{\sin (s - b)}{\sin c} \\
\frac{\sin \frac{1}{2} A \sin \frac{1}{2} B}{\sin \frac{1}{2} C} &= \frac{\sin (s - c)}{\sin c}
\end{aligned}\right\} \tag{55}$$

$$\left.\begin{aligned}
\frac{\sin \frac{1}{2} a \sin \frac{1}{2} b}{\cos \frac{1}{2} c} &= \frac{-\cos S}{\sin C} \\
\frac{\sin \frac{1}{2} a \cos \frac{1}{2} b}{\sin \frac{1}{2} c} &= \frac{\cos (S - A)}{\sin C} \\
\frac{\cos \frac{1}{2} a \sin \frac{1}{2} b}{\sin \frac{1}{2} c} &= \frac{\cos (S - B)}{\sin C} \\
\frac{\cos \frac{1}{2} a \cos \frac{1}{2} b}{\cos \frac{1}{2} c} &= \frac{\cos (S - C)}{\sin C}
\end{aligned}\right\} \tag{56}$$

33. By means of (55) and (56) we can deduce expressions for the functions of s, $s - a$, &c., in terms of the angles, or of S, $S - A$, &c., in terms of the sides. We have, from (51),

$$\sin c = \frac{2N}{\sin A \sin B} = \frac{N}{2 \sin \frac{1}{2} A \cos \frac{1}{2} A \sin \frac{1}{2} B \cos \frac{1}{2} B}$$

which, substituted in (55), gives

$$\sin s = \frac{N}{2 \sin \frac{1}{2} A \sin \frac{1}{2} B \sin \frac{1}{2} C} \tag{57}$$

$$\sin (s - c) = \frac{N}{2 \cos \frac{1}{2} A \cos \frac{1}{2} B \sin \frac{1}{2} C} \tag{58}$$

whence, by interchanging the letters, we have also $\sin (s - a)$ and $\sin (s - b)$.

Again, we have

$$\sin(s-c) = \sin s \cos c - \cos s \sin c$$

whence

$$\cos s = \frac{\sin s \cos c - \sin(s-c)}{\sin c}$$

which, by (55), is reduced to

$$\cos s = \frac{\cos\frac{1}{2}A \cos\frac{1}{2}B \cos c - \sin\frac{1}{2}A \sin\frac{1}{2}B}{\sin\frac{1}{2}C} \qquad (59)$$

and from the equation

$$\cos(s-c) = \cos s \cos c + \sin s \sin c$$

we find, by substituting (55) and (59),

$$\cos(s-c) = \frac{-\sin\frac{1}{2}A \sin\frac{1}{2}B \cos c + \cos\frac{1}{2}A \cos\frac{1}{2}B}{\sin\frac{1}{2}C} \qquad (60)$$

34. To eliminate c from the second members of (59) and (60), we have, by (5),

$$\cos c = \frac{\cos C + \cos A \cos B}{\sin A \sin B}$$

whence

$$\cos\tfrac{1}{2}A \cos\tfrac{1}{2}B \cos c = \frac{\cos C + \cos A \cos B}{4 \sin\frac{1}{2}A \sin\frac{1}{2}B}$$

$$\sin\tfrac{1}{2}A \sin\tfrac{1}{2}B \cos c = \frac{\cos C + \cos A \cos B}{4 \cos\frac{1}{2}A \cos\frac{1}{2}B}$$

which, substituted in (59) and (60), give

$$\cos s = \frac{\cos A + \cos B + \cos C - 1}{4 \sin\frac{1}{2}A \sin\frac{1}{2}B \sin\frac{1}{2}C} \qquad (61)$$

$$\cos(s-c) = \frac{\cos A + \cos B - \cos C + 1}{4 \cos\frac{1}{2}A \cos\frac{1}{2}B \sin\frac{1}{2}C} \qquad (62)$$

$$\cos s = \frac{1 - \sin^2\frac{1}{2}A - \sin^2\frac{1}{2}B - \sin^2\frac{1}{2}C}{2 \sin\frac{1}{2}A \sin\frac{1}{2}B \sin\frac{1}{2}C} \qquad (63)$$

$$\cos(s-c) = \frac{\cos^2\frac{1}{2}A + \cos^2\frac{1}{2}B + \sin^2\frac{1}{2}C - 1}{2 \cos\frac{1}{2}A \cos\frac{1}{2}B \sin\frac{1}{2}C} \qquad (64)$$

From the preceding, we easily deduce

$$\tan s = \frac{2N}{\cos A + \cos B + \cos C - 1} \qquad (65)$$

$$\tan(s-c) = \frac{2N}{\cos A + \cos B - \cos C + 1} \qquad (66)$$

$$\cot s = \frac{\cos c - \tan\frac{1}{2}A \tan\frac{1}{2}B}{\sin c} \qquad (67)$$

$$\cot(s-c) = \frac{\cot\frac{1}{2}A \cot\frac{1}{2}B - \cos c}{\sin c} \qquad (68)$$

35. The equations (57 to 68) applied to the polar triangle, give

$$-\cos S = \frac{n}{2\cos\frac{1}{2}a\cos\frac{1}{2}b\cos\frac{1}{2}c} \tag{69}$$

$$\cos(S - C) = \frac{n}{2\sin\frac{1}{2}a\sin\frac{1}{2}b\cos\frac{1}{2}c} \tag{70}$$

$$\sin S = \frac{\sin\frac{1}{2}a\sin\frac{1}{2}b\cos C + \cos\frac{1}{2}a\cos\frac{1}{2}b}{\cos\frac{1}{2}c} \tag{71}$$

$$\sin(S - C) = \frac{\cos\frac{1}{2}a\cos\frac{1}{2}b\cos C + \sin\frac{1}{2}a\sin\frac{1}{2}b}{\cos\frac{1}{2}c} \tag{72}$$

$$\sin S = \frac{\cos a + \cos b + \cos c + 1}{4\cos\frac{1}{2}a\cos\frac{1}{2}b\cos\frac{1}{2}c}$$

$$= \frac{\cos^2\frac{1}{2}a + \cos^2\frac{1}{2}b + \cos^2\frac{1}{2}c - 1}{2\cos\frac{1}{2}a\cos\frac{1}{2}b\cos\frac{1}{2}c} \tag{73}$$

$$\sin(S - C) = \frac{1 - \cos a - \cos b + \cos c}{4\sin\frac{1}{2}a\sin\frac{1}{2}b\cos\frac{1}{2}c}$$

$$= \frac{\sin^2\frac{1}{2}a + \sin^2\frac{1}{2}b + \cos^2\frac{1}{2}c - 1}{2\sin\frac{1}{2}a\sin\frac{1}{2}b\cos\frac{1}{2}c} \tag{74}$$

$$-\cot S = \frac{2n}{\cos a + \cos b + \cos c + 1} \tag{75}$$

$$\cot(S - C) = \frac{2n}{1 - \cos a - \cos b + \cos c} \tag{76}$$

$$-\tan S = \frac{\cos C + \cot\frac{1}{2}a\cot\frac{1}{2}b}{\sin C} \tag{77}$$

$$\tan(S - C) = \frac{\cos C + \tan\frac{1}{2}a\tan\frac{1}{2}b}{\sin C} \tag{78}$$

36. From (73) we find

$$1 - \sin S = \frac{2\cos\frac{1}{2}a\cos\frac{1}{2}b\cos\frac{1}{2}c - \cos^2\frac{1}{2}a - \cos^2\frac{1}{2}b - \cos^2\frac{1}{2}c + 1}{2\cos\frac{1}{2}a\cos\frac{1}{2}b\cos\frac{1}{2}c}$$

$$1 + \sin S = \frac{2\cos\frac{1}{2}a\cos\frac{1}{2}b\cos\frac{1}{2}c + \cos^2\frac{1}{2}a + \cos^2\frac{1}{2}b + \cos^2\frac{1}{2}c - 1}{2\cos\frac{1}{2}a\cos\frac{1}{2}b\cos\frac{1}{2}c}$$

the numerators of which may be reduced by Pl. Trig. (173) and (174), by making $x = \frac{1}{2}a$, $y = \frac{1}{2}b$, $z = \frac{1}{2}c$, whence $v = \frac{1}{4}(a + b + c) = \frac{1}{2}s$, $v - x = \frac{1}{2}(s - a)$, &c.: therefore,

$$1 - \sin S = \frac{2\sin\frac{1}{2}s\sin\frac{1}{2}(s - a)\sin\frac{1}{2}(s - b)\sin\frac{1}{2}(s - c)}{\cos\frac{1}{2}a\cos\frac{1}{2}b\cos\frac{1}{2}c}$$

$$1 + \sin S = \frac{2\cos\frac{1}{2}s\cos\frac{1}{2}(s - a)\cos\frac{1}{2}(s - b)\cos\frac{1}{2}(s - c)}{\cos\frac{1}{2}a\cos\frac{1}{2}b\cos\frac{1}{2}c}$$

The product of these equations reproduces (69); their quotient is, by Pl. Trig. (154),

$$\tan^2(45° - \tfrac{1}{2}S) = \tan\tfrac{1}{2}s\tan\tfrac{1}{2}(s - a)\tan\tfrac{1}{2}(s - b)\tan\tfrac{1}{2}(s - c) \tag{79}$$

37. *To deduce the formulæ of plane triangles from those of spherical triangles.*

The analogy of many of the preceding formulæ with those of plane triangles is sufficiently obvious. We can, in fact, deduce the plane formulæ from those of this chapter, by regarding the plane triangle as *described upon a sphere whose radius is infinite, the triangle being an infinitely small portion of the sphere.* The quantities a, b and c, must, in this case, express the absolute lengths of the sides; and the angles which they subtend at the center of the sphere, expressed in arc, will be $\frac{a}{r}, \frac{b}{r}, \frac{c}{r}$, r being the radius of the sphere. When r is very large, $\frac{a}{r}, \frac{b}{r}, \frac{c}{r}$, are very small, and we may express the values of $\sin\frac{a}{r}$, $\cos\frac{a}{r}$, &c. approximately, by one or two terms of their expansions in series, Pl. Trig. (405) and (406), and if their values be substituted in our spherical formulæ, we shall obtain *approximate* relations between the sides and angles of the triangle. If we then make r infinite we shall obtain *exact* relations between the sides and angles of a plane triangle.

Thus we have

$$\frac{\sin A}{\sin B} = \frac{\sin\frac{a}{r}}{\sin\frac{b}{r}} = \frac{\frac{a}{r} - \frac{a^3}{2.3\,r^3} + \&c.}{\frac{b}{r} - \frac{b^3}{2.3\,r^3} + \&c.} = \frac{a - \frac{a^3}{2.3\,r^2} + \&c.}{b - \frac{b^3}{2.3\,r^2} + \&c.}$$

and making r infinite, we find the formula of Pl. Trig.

$$\frac{\sin A}{\sin B} = \frac{a}{b}$$

In the same manner

$$\cos A = \frac{\cos\frac{a}{r} - \cos\frac{b}{r}\cos\frac{c}{r}}{\sin\frac{b}{r}\sin\frac{c}{r}} = \frac{1 - \frac{a^2}{2r^2} + \&c. - \left(1 - \frac{b^2}{2r^2} - \frac{c^2}{2r^2} + \frac{b^2c^2}{4r^4} - \&c.\right)}{\left(\frac{b}{r} - \frac{b^3}{2.3\,r^3} + \&c.\right)\left(\frac{c}{r} - \frac{c^3}{2.3\,r^3} + \&c.\right)}$$

$$= \frac{b^2 + c^2 - a^2 - \frac{b^2c^2}{2r^2} + \&c.}{2\,bc - \frac{b^3c}{3r^2} - \&c.}$$

and making r infinite, we have the formula of Pl. Trig.

$$\cos A = \frac{b^2 + c^2 - a^2}{2\,bc}$$

Formulæ that involve only the sines or tangents of the sides may be reduced immediately to the plane formulæ by substituting a, b, &c., for $\sin a$, $\tan a$, &c. Thus, (31 to 34) give the corresponding formulæ of Pl. Trig. by omitting the symbol *sin.*; and (40), (41), by omitting the symbol *tan.* when these symbols are prefixed to sides.

CHAPTER II.

SOLUTION OF SPHERICAL RIGHT TRIANGLES.

38. WHEN one of the angles of a spherical triangle is a right angle, the general formulæ of the preceding chapter assume forms that are remarkably analogous to the relations established for the solution of plane right triangles, and equally simple in their application.

39. Let $C = 90°$, Fig. 6. From (3) we have

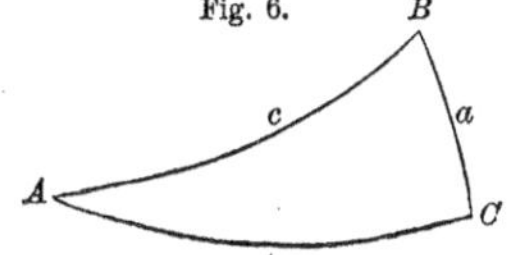

Fig. 6.

$$\sin A = \frac{\sin a}{\sin c} \sin C$$

but since $C = 90°$, $\sin C = 1$; therefore,

$$\sin A = \frac{\sin a}{\sin c}$$

and, in the same manner,

$$\sin B = \frac{\sin b}{\sin c} \qquad (80)$$

that is, *the sine of either oblique angle of a spherical right triangle is equal to the quotient of the sine of the opposite side divided by the sine of the hypotenuse.* Compare Pl. Trig. (1).

40. From (7), we find

$$\cos A = \frac{\sin b \cos c - \sin a \cos C}{\cos b \sin c}$$

but if $C = 90°$, $\cos C = 0$; therefore,

$$\cos A = \frac{\sin b \cos c}{\cos b \sin c} = \tan b \cot c$$

or

$$\cos A = \frac{\tan b}{\tan c} \qquad \cos B = \frac{\tan a}{\tan c} \qquad (81)$$

that is, *the cosine of either angle is equal to the tangent of the adjacent side, divided by the tangent of the hypotenuse.* Compare Pl. Trig. (1).

41. From (10), we have,

$$\cot A = \frac{\sin b \cot a - \cos b \cos C}{\sin C}$$

which, when $C = 90°$, becomes

$$\cot A = \sin b \cot a = \frac{\sin b}{\tan a}$$

or, taking the reciprocals,

$$\tan A = \frac{\tan a}{\sin b} \qquad \tan B = \frac{\tan b}{\sin a} \tag{82}$$

that is, *the tangent of either angle is equal to the tangent of the opposite side, divided by the sine of the adjacent side.* Compare Pl. Trig. (1).

42. From (8), we find,

$$\sin A = \frac{\sin C \cos B + \cos C \sin B \cos a}{\cos b}$$

and if $C = 90°$,

$$\sin A = \frac{\cos B}{\cos b} \qquad \sin B = \frac{\cos A}{\cos a} \tag{83}$$

that is, *the cosine of either angle, divided by the cosine of its opposite side, is equal to the sine of the other angle.* In Pl. Trig. we have $\sin A = \cos B$.

43. From (4), we have,

$$\cos c = \cos a \cos b + \sin a \sin b \cos C$$

or, when $C = 90°$,

$$\cos c = \cos a \cos b \tag{84}$$

that is, *the cosine of the hypotenuse is equal to the product of the cosines of the two sides.* In Pl. Trig. $c^2 = a^2 + b^2$.

44. From (5),

$$\cos c = \frac{\cos C + \cos A \cos B}{\sin A \sin B}$$

or, when $C = 90°$,

$$\cos c = \frac{\cos A \cos B}{\sin A \sin B} = \cot A \cot B \tag{85}$$

that is, *the cosine of the hypotenuse is equal to the product of the cotangents of the two angles.* In Pl. Trig., $1 = \cot A \cot B$.

45. No difficulty will be found in remembering the preceding formulæ for spherical right triangles, if they are associated with the corresponding ones for plane triangles: thus,

In plane right triangles.		*In spherical right triangles.*	
$\sin A = \frac{a}{c}$	$\sin B = \frac{b}{c}$	$\sin A = \frac{\sin a}{\sin c}$	$\sin B = \frac{\sin b}{\sin c}$
$\cos A = \frac{b}{c}$	$\cos B = \frac{a}{c}$	$\cos A = \frac{\tan b}{\tan c}$	$\cos B = \frac{\tan a}{\tan c}$
$\tan A = \frac{a}{b}$	$\tan B = \frac{b}{a}$	$\tan A = \frac{\tan a}{\sin b}$	$\tan B = \frac{\tan b}{\sin a}$
$\sin A = \cos B,$	$\sin B = \cos A$	$\sin A = \frac{\cos B}{\cos b}$	$\sin B = \frac{\cos A}{\cos a}$
$c^2 = a^2 + b^2$		$\cos c = \cos a \cos b$	
$1 = \cot A \cot B$		$\cos c = \cot A \cot B$	

46. *Napier's Rules.* By putting these ten equations under a different form, Napier contrived to express them all in two rules, which, though artificial, are very generally employed as aids to the memory.

In these rules, the complements of the hypotenuse and of the two oblique angles are employed instead of the hypotenuse and the angles themselves. The right angle not entering into the formulæ, they express the relations of five parts, but in the rules the five parts considered are a, b, co. c, co. A and co. B. Any one of these parts being called a *middle part,* the two immediately adjacent may be called *adjacent parts,* and the remaining two, *opposite parts.* The right angle not being considered, the two sides including it are regarded as adjacent. The rules are:

I. *The sine of the middle part is equal to the product of the tangents of the adjacent parts.*

II. *The sine of the middle part is equal to the product of the cosines of the opposite parts.*

The correctness of these rules will be shown by taking each of the five parts as middle part, and comparing the equations thus found with those already demonstrated.

1st. Let co. c be the middle part; then co. A and co. B are the adjacent parts, a and b the opposite parts, and the rules give

$$\sin(\text{co. } c) = \tan(\text{co. } A)\tan(\text{co. } B) \quad \text{or} \quad \cos c = \cot A \cot B$$
$$\sin(\text{co. } c) = \cos a \cos b \qquad\qquad \cos c = \cos a \cos b$$

which are (85) and (84).

2d. Let co. A be the middle part; then co. c and b are the adjacent parts, co. B and a the opposite parts, and the rules give

$$\sin(\text{co. } A) = \tan(\text{co. } c)\tan b \quad \text{or} \quad \cos A = \cot c \tan b$$
$$\sin(\text{co. } A) = \cos(\text{co. } B)\cos a \qquad\qquad \cos A = \sin B \cos a$$

In the same manner, if co. B is taken as the middle part,

$$\sin(\text{co. } B) = \tan(\text{co. } c)\tan a \qquad \text{or} \qquad \cos B = \cot c \tan a$$
$$\sin(\text{co. } B) = \cos(\text{co. } A)\cos b \qquad\qquad \cos B = \sin A \cos b$$

and these four equations are the same as (81) and (83).

3d. Let a be the middle part; then co. B and b are the adjacent parts, co. A and co. c the opposite parts, and the rules give,

$$\sin a = \tan(\text{co. } B)\tan b \qquad \text{or} \qquad \sin a = \cot B \tan b$$
$$\sin a = \cos(\text{co. } A)\cos(\text{co. } c) \qquad\qquad \sin a = \sin A \sin c$$

In the same manner, if b is taken as the middle part,

$$\sin b = \tan(\text{co. } A)\tan a \qquad \text{or} \qquad \sin b = \cot A \tan a$$
$$\sin b = \cos(\text{co. } B)\cos(\text{co. } c) \qquad\qquad \sin b = \sin B \sin c$$

and these four equations are the same as (80) and (82).

It appears, therefore, that these rules include all the ten equations previously proved; and they include no others, since we have taken each part successively as the middle part.

In the application of these rules, it is unnecessary to use the notation co. A, co. B, co. c, since we may write down at once sin A for cos (co. A), &c.*

47. In order to solve a spherical right triangle, two parts must be given, and from the equations of Art. 45, that equation must be selected which expresses the relation between these two parts and the required part.

When Napier's Rules are employed, it is only necessary to determine which of the three parts—the two given and the one required—is to be taken as the middle part. "These three parts are either all adjacent to each other, in which case the middle one is taken as the middle part, and the other two are adjacent parts; or one is separated from the other two, and then the part which stands by itself is the middle part, and the other two are opposite parts."†

48. In order to distinguish the functions of parts less than 90° from those greater than 90°, it will be necessary carefully to observe their algebraic signs, according to Pl. Trig. Art. 40. But when a required part is determined by its sine, since the sine of an angle and of its supplement are the same, there will be two angles, both of which may be regarded as solutions, except when this ambiguity is removed by either of the following principles.

---

* If we employ as the five parts, the hypotenuse, the two angles, and the complements of the two sides including the right angle, these parts will be the complements of those used in Napier's Rules, and we shall have

MAUDUIT'S RULES.—I. *The cosine of the middle part is equal to the product of the cotangents of the adjacent parts.*

II. *The cosine of the middle part is equal to the product of the sines of the opposite parts.*

With a little attention at the commencement, however, and by observing the analogy exhibited in Art. 45, the student will find that he will have little use for either of these artificial rules.

† Peirce's Spherical Trigonometry.

49. *In a right spherical triangle, an angle and its opposite side are always in the same quadrant, that is, either both less or both greater than* 90°. For, by (83),

$$\sin A = \frac{\cos B}{\cos b}$$

in which, since $\sin A$ is always positive, ($A < 180°$), $\cos B$ and $\cos b$ must have the same sign; that is, B and b must be either both less or both greater than 90°.

50. *When the two sides including the right angle are in the same quadrant, the hypotenuse is less than* 90°, *and when the two sides are in different quadrants, the hypotenuse is greater than* 90°. For, by (84),

$$\cos c = \cos a \cos b$$

in which, if a and b are in the same quadrant, $\cos a$ and $\cos b$ have like signs, and $\cos c$ is positive, that is, $c < 90°$; but if a and b are in different quadrants, $\cos a$ and $\cos b$ have different signs, and $\cos c$ is negative, that is, $c > 90°$.

We proceed now to the solution of the several cases.

51. CASE I. *Given the hypotenuse and one angle*, or c and A, Fig. 6.

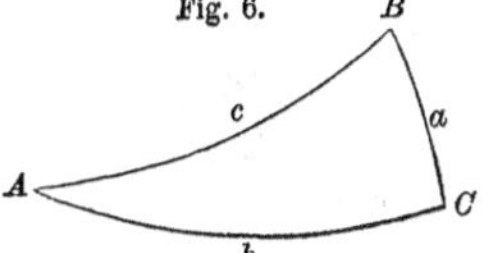

To find a. The relation among the three parts, c, A, and a, (as in Pl. Trig. with the same data), is given by the sine of A; and by Art. 45,

$$\sin A = \frac{\sin a}{\sin c}$$

from which we find*

$$\sin a = \sin c \sin A \tag{86}$$

There will be two values of a corresponding to the same sine, but, by Art. 49, the true value is that which is in the same quadrant as A.

To find b. The relation among the three parts, c, A, and b, (as in Pl. Trig. with the same data), is given by the cosine of A, or,

$$\cos A = \frac{\tan b}{\tan c}$$

from which†

$$\tan b = \tan c \cos A \tag{87}$$

---

* This equation would be found by Napier's Rules, taking a as the middle part.

† We find the same result by Napier's Rules, taking co. A as the middle part.

To find B. We have, by (85),*

$$\cos c = \cot A \cot B$$

from which
$$\cot B = \frac{\cos c}{\cot A} = \cos c \tan A \qquad (88)$$

The quadrants in which b and B are to be taken, will be determined by means of the signs of $\tan b$ and $\cot B$, according to Pl. Trig. Art. 40.

Check. To guard against numerical errors, it is often expedient to compute the same quantity by two different and independent methods. In many cases, however, we may test the accuracy of several operations by a single formula, which may be called the *check*. In the present instance, when the three parts, a, b, and B, have been found, we should have, by (82), the relation

$$\sin a = \tan b \cot B$$

so that if the work is correct, we shall find

$$\log \sin a = \log \tan b + \log \cot B$$

EXAMPLES.

1. Given $c = 110° 46' 20''$, $A = 80° 10' 30''$, to solve the triangle.

By (86).	By (87).	By (88).
c, log sin 9·9708106	log tan − 0·4210061	log cos − 9·5498045
A, log sin 9·9935833	log cos + 9·2320794	log tan + 0·7615038
log sin a 9·9643939	log tan b − 9·6530855	log cot B − 0.3113083
		log tan b − 9·6530855
	Check.	log sin a + 9·9643938

Ans. $a = 67° 6' 52''.7$, $b = 155° 46' 42''.7$, $B = 153° 58' 24''.5$

2. Given $c = 120°$, $A = 120°$; solve the triangle.

Ans. $a = 131° 24' 34''.7$ $b = 40° 53' 36''.2$ $B = 49° 6' 23''.8$

52. If A = 90°, we must also have, by (85), $c = 90°$, and then

$$\tan b = \frac{0}{0} \qquad \tan B = \frac{0}{0}$$

so that b and B are both indeterminate; that is, there is an indefinite number of triangles which satisfy the given values of c and A; but since

$$\cos B = \cos b \sin A = \cos b$$

we always have $B = b$; and since

$$\sin a = \sin c \sin A = 1$$

we have $a = 90°$, and all the parts of the triangle are equal to 90°, except b and B.

If only c is given $= 90°$, all the parts of the triangle are equal to 90°, except A and a; and we have $A = a$.

---

* Or by Napier's Rules, taking co. c as the middle part.

53. Case II. *Given the hypotenuse and a side,* or c and a.

To find A. We have by (80),

$$\sin A = \frac{\sin a}{\sin c} = \operatorname{cosec} c \sin a \tag{89}$$

To find B. By (81),

$$\cos B = \frac{\tan a}{\tan c} = \cot c \tan a \tag{90}$$

To find b. By (84),

$$\cos c = \cos a \cos b$$

from which

$$\cos b = \frac{\cos c}{\cos a} = \cos c \sec a \tag{91}$$

Check. We have between A, B, and b, the relation

$$\cos B = \sin A \cos b$$

EXAMPLES.

1. Given $c = 140°$, $a = 20°$; solve the triangle.

By (89).		By (90).		By (91).	
c, log cosec	0·1919325	log cot	− 0·0761865	log cos	− 9·8842540
a, log sin	9·5340517	log tan	+ 9·5610659	log sec	+ 0·0270142
log sin A	9·7259842	log cos B	− 9·6372524	log cos b	− 9·9112682
				log sin A	+ 9·7259842
			Check.	log cos B	− 9·6372524

Ans. $A = 32° \; 8' 48''·1$
$B = 115° \, 42' \, 23''·8$
$b = 144° \, 36' \, 28''·4$

2. Given $c = 101° \, 16' \, 16''·7$, $b = 115° \, 42' \, 38''·5$; find A.

Ans. $A = 65° \, 32' \, 56''·4$

54. When $a = c$ and consequently both $= 90°$, $\sin A = 1$, $A = 90°$, and

$$\cos B = \frac{0}{0} \qquad \cos b = \frac{0}{0} \qquad \cos B = \cos b$$

so that $B = b$, but both are indeterminate as in Art. 52.

55. Case III. *Given one angle and its opposite side*, or A and a. We shall have

$$\sin A = \frac{\sin a}{\sin c} \qquad \text{whence} \qquad \sin c = \operatorname{cosec} A \sin a \tag{92}$$

$$\tan A = \frac{\tan a}{\sin b} \qquad \sin b = \cot A \tan a \tag{93}$$

$$\sin B = \frac{\cos A}{\cos a} \qquad \sin B = \cos A \sec a \tag{94}$$

Check. $\sin b = \sin c \sin B$

In this case, there are always two solutions, all the required parts being determined by their sines, and the ambiguity not being removed by either Art. 49 or Art. 50. This also appears from Fig. 7.

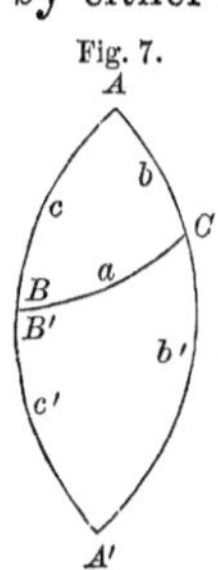

Fig. 7.

If AB and AC be produced to meet in A', ABA' and ACA' are semicircumferences and $A = A'$; the triangles ABC and $A'BC$ both contain the given parts A and a, but c', b' and B' are respectively the supplements of c, b and B. It must not be inferred that in every case all the required parts are less than 90° in one triangle, and greater than 90° in the other; but the proper values for each triangle must be selected by Arts. 49 and 50.

Examples.

1. Given $A = 100°$, $a = 112°$; solve the triangle.

Ans. $\left.\begin{aligned} c &= 70° \, 18' \, 10''.2 \\ b &= 154° \, 7' \, 26''.5 \\ B &= 152° \, 23' \, 1''.3 \end{aligned}\right\}$ or $\left\{\begin{aligned} c &= 109° \, 41' \, 49''.8 \\ b &= 25° \, 52' \, 33''.5 \\ B &= 27° \, 36' \, 58''.7 \end{aligned}\right.$

2. Given $A = 80°$, $a = 68°$; solve the triangle.

Ans. $\left.\begin{aligned} c &= 70° \, 18' \, 10''.2 \\ b &= 25° \, 52' \, 33''.5 \\ B &= 27° \, 36' \, 58''.7 \end{aligned}\right\}$ or $\left\{\begin{aligned} c &= 109° \, 41' \, 49''.8 \\ b &= 154° \, 7' \, 26''.5 \\ B &= 152° \, 23' \, 1''.3 \end{aligned}\right.$

3. Given $B = 150°$, $b = 160°$; solve the triangle.

Ans. $\left.\begin{aligned} c &= 136° \, 50' \, 23''.3 \\ a &= 39° \, 4' \, 50''.7 \\ A &= 67° \, 9' \, 42''.7 \end{aligned}\right\}$ or $\left\{\begin{aligned} c &= 43° \, 9' \, 36''.7 \\ a &= 140° \, 55' \, 9''.3 \\ A &= 112° \, 50' \, 17''.3 \end{aligned}\right.$

56. CASE IV. *Given one angle and its adjacent side,* or A and b. We shall find the required parts by the equations

$$\cos B = \sin A \cos b \tag{95}$$

$$\tan a = \tan A \sin b \tag{96}$$

$$\cot c = \cos A \cot b \tag{97}$$

Check. $\cos B = \tan a \cot c$

EXAMPLES.

1. Given $A = 80° 10' 30''$, $b = 155° 46' 42''.7$; solve the triangle.

Ans. $B = 153° 58' 24''.5$
$a = 67° 6' 52''.6$
$c = 110° 46' 20''.0$

2. Given $B = 152° 23' 1''.3$, $a = 112° 0' 0''$; solve the triangle.

Ans. $A = 100°$
$b = 154° 7' 26''.5$
$c = 70° 18' 10''.2$

57. CASE V. *Given the two sides,* a and b.
We find the required parts by the equations

$$\cos c = \cos a \cos b \tag{98}$$

$$\cot A = \cot a \sin b \tag{99}$$

$$\cot B = \sin a \cot b \tag{100}$$

Check. $\cos c = \cot A \cot B$

EXAMPLE.

Given $a = 116°$, $b = 16°$; solve the triangle.

Ans. $c = 114° 55' 20''.4$
$A = 97° 39' 24''.4$
$B = 17° 41' 39''.9$

58. CASE VI. *Given the two angles,* A and B.
The required parts are found by the formulæ

$$\cos c = \cot A \cot B \tag{101}$$

$$\cos a = \cos A \operatorname{cosec} B \tag{102}$$

$$\cos b = \operatorname{cosec} A \cos B \tag{103}$$

Check. $\cos c = \cos a \cos b$

Example.

Given $A = 60° \, 47' \, 24''.3$, $B = 57° \, 16' \, 20''.2$; solve the triangle.

Ans. $c = 68° \, 56' \, 28''.9$
$a = 54° \, 32' \, 32''.1$
$b = 51° \, 43' \, 36''.1$

Additional Formulæ for the Solution of Spherical Right Triangles.

59. As in plane trigonometry, cases occur in which particular solutions of greater accuracy than the ordinary ones are required. (Pl. Trig. Art. 112.)

60. From (89) we find

$$\frac{1 - \sin A}{1 + \sin A} = \frac{\sin c - \sin a}{\sin c + \sin a}$$

which by Pl. Trig. (154) and (109) is reduced to

$$\tan^2 (45° - \tfrac{1}{2} A) = \frac{\tan \frac{1}{2}(c - a)}{\tan \frac{1}{2}(c + a)} \tag{104}$$

which will give a more accurate result than (89), when A is nearly 90°.

61. From (91) we find

$$\frac{1 - \cos b}{1 + \cos b} = \frac{\cos a - \cos c}{\cos a + \cos c}$$

or

$$\tan^2 \tfrac{1}{2} b = \tan \tfrac{1}{2}(c + a) \tan \tfrac{1}{2}(c - a) \tag{105}$$

which may be employed instead of (91) when b is small, or nearly 180°.

62. From (90) we find

$$\frac{1 - \cos B}{1 + \cos B} = \frac{\tan c - \tan a}{\tan c + \tan a}$$

$$\tan^2 \tfrac{1}{2} B = \frac{\sin (c - a)}{\sin (c + a)} \tag{106}$$

which may be employed instead of (90) when B is small, or nearly 180°.

63. By similar transformations the formulæ (101), (102) and (103) become

$$\tan^2 \tfrac{1}{2} c = \frac{-\cos (A + B)}{\cos (A - B)} \tag{107}$$

$$\tan^2 \tfrac{1}{2} a = \tan [\tfrac{1}{2}(A + B) - 45°] \tan [45° + \tfrac{1}{2}(A - B)] \tag{108}$$

$$\tan^2 \tfrac{1}{2} b = \tan [\tfrac{1}{2}(A + B) - 45°] \tan [45° - \tfrac{1}{2}(A - B)] \tag{109}$$

We have also, by (14),

$$\sin^2 \tfrac{1}{2} c = \frac{-\cos (A + B) - \cos C}{2 \sin A \sin B}$$

which, when $C = 90°$, becomes

$$\sin^2 \tfrac{1}{2} c = \frac{-\cos (A + B)}{2 \sin A \sin B} \tag{110}$$

and from (15), in the same manner,

$$\cos^2 \tfrac{1}{2} c = \frac{\cos (A - B)}{2 \sin A \sin B} \tag{111}$$

of which (110) may be used when c is small, and (111) when c is nearly 180°, instead of (101).

64. The equations (92), (93), and (94), of CASE III. give

$$\tan^2(45° - \tfrac{1}{2}c) = \frac{\tan\frac{1}{2}(A - a)}{\tan\frac{1}{2}(A + a)} \tag{112}$$

$$\tan^2(45° - \tfrac{1}{2}b) = \frac{\sin(A - a)}{\sin(A + a)} \tag{113}$$

$$\tan^2(45° - \tfrac{1}{2}B) = \tan\tfrac{1}{2}(A - a)\tan\tfrac{1}{2}(A + a) \tag{114}$$

The roots of these equations having the double sign, we may take the angles $45° - \frac{1}{2}c$, etc. either with the positive or negative sign, whence the two solutions of the problem, as in Art. 55.

65. Some of the solutions may be adapted for computation by the table of natural sines. Thus from (86), (95), and (98),

$$\sin a = \tfrac{1}{2}[\cos(c - A) - \cos(c + A)] \tag{115}$$

$$\cos B = \tfrac{1}{2}[\sin(b + A) - \sin(b - A)] \tag{116}$$

$$\cos c = \tfrac{1}{2}[\cos(a + b) + \cos(a - b)] \tag{117}$$

66. The following relations are occasionally useful:

From (83) we have

$$\frac{\cos a}{\cos b} = \frac{\sin A \cos A}{\sin B \cos B} = \frac{\sin 2A}{\sin 2B} \tag{118}$$

From (80) and (83),

$$\frac{\sin B}{\sin c} = \frac{\sin A \cos A}{\sin a \cos a} = \frac{\sin 2A}{\sin 2a}. \tag{119}$$

From (80) and (84),

$$\frac{\sin A}{\cos b} = \frac{\sin a \cos a}{\sin c \cos c} = \frac{\sin 2a}{\sin 2c} \tag{120}$$

67. Various relations may be deduced from the general formulæ of the preceding chapter by making $C = 90°$. The following are easily obtained:

$$\sin(c - a) = \cos c \tan b \tan\tfrac{1}{2}B = \cos a \sin b \tan\tfrac{1}{2}B$$

$$\sin(c + a) = \cos c \tan b \cot\tfrac{1}{2}B = \cos a \sin b \cot\tfrac{1}{2}B$$

$$\cos(c - a) = \cos b + \sin a \sin b \tan\tfrac{1}{2}B$$

$$\cos(c + a) = \cos b - \sin a \sin b \cot\tfrac{1}{2}B$$

$$\sin(a - b) = 2\sin c \sin\tfrac{1}{2}(A + B)\sin\tfrac{1}{2}(A - B)$$

$$\sin(a + b) = 2\sin c \cos\tfrac{1}{2}(A + B)\cos\tfrac{1}{2}(A - B)$$

$$\sin S = \frac{\cos\frac{1}{2}a\cos\frac{1}{2}b}{\cos\frac{1}{2}c} \qquad \cos S = -\frac{\sin\frac{1}{2}a\sin\frac{1}{2}b}{\cos\frac{1}{2}c}$$

$$\tan S = -\cot\tfrac{1}{2}a\cot\tfrac{1}{2}b \tag{121}$$

QUADRANTAL AND ISOSCELES TRIANGLES.

68. The polar triangle of the right triangle is a *quadrantal* triangle, one *side* (the side opposite the angle C) being equal to 90°. The solution of such triangles is as simple as that of right triangles, the formulæ for the purpose being obtained from the preceding, by the process of Art. 8. It is unnecessary to produce them here, as quadrantal triangles are generally avoided in practice, and when unavoidable are readily solved by means of the polar triangle.

An isosceles triangle is easily solved by dividing it into two right triangles by a perpendicular from the angle included by the equal sides.

CHAPTER III.

SOLUTION OF SPHERICAL OBLIQUE TRIANGLES.

69. In the solution of spherical oblique triangles, a required part may sometimes be found by its *sine*, in which case there will be two values of that part, answering to the conditions, unless the proper value can be determined by other considerations. In certain cases, the true value can be selected by applying one or more of the following principles, some of which are demonstrated in geometry. We still consider only those triangles each of whose parts is less than 180°.

I. *The greater side is opposite the greater angle, and conversely.*

II. *Each side is less than the sum of the other two.*

III. *The sum of the sides is less than* 360°.

IV. *The sum of the angles is greater than* 180°.

V. *Each angle is greater than the difference between* 180° *and the sum of the other two angles.*

For, by IV., $A + B + C > 180°$

whence, $A > 180° - (B + C)$

Fig. 8.

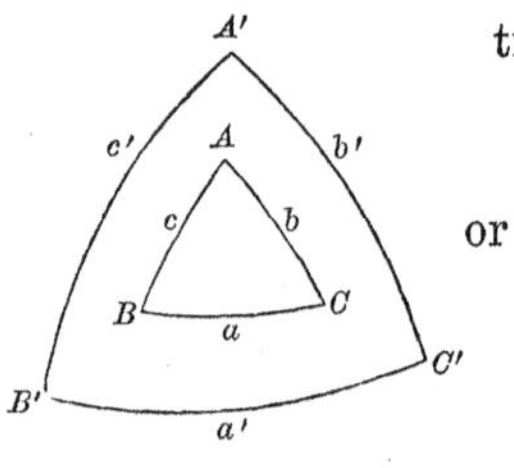

But if $B + C > 180°$, we have, in the polar triangle, $A'B'C'$, Fig. 8, by II.,

$$a' < b' + c'$$

or $$180° - A < 180° - B + 180° - C$$

$$-A < 180° - (B + C)$$

$$A > (B + C) - 180°$$

VI. *A side which differs more from* 90° *than another side, is in the same quadrant as its opposite angle.*

For, by (4), we have

$$\cos A = \frac{\cos a - \cos b \cos c}{\sin b \sin c}$$

in which the denominator is always positive. If, then, a differs more from 90° than b or than c, we have, (neglecting the signs for a moment),

$$\cos a > \cos b \text{ or } > \cos c$$

and still more

$$\cos a > \cos b \cos c$$

Hence $\cos a$ being *numerically* greater than $\cos b \cos c$, the sign of the whole numerator, and therefore the sign of $\cos A$, is the same as that of $\cos a$; that is, A and a are in the same quadrant.

VII. *An angle which differs more from 90° than another angle, is in the same quadrant as its opposite side.* For, by (5),

$$\cos a = \frac{\cos A + \cos B \cos C}{\sin B \sin C}$$

in which, if A differs more from 90° than B, or than C, $\cos A$ determines the sign of the whole fraction, and therefore the sign of $\cos a$.

VIII. *In every spherical triangle there are at least two sides which are in the same quadrants as their opposite angles respectively.* This follows from VI. and VII.

IX. *The sum of two sides is greater than, equal to, or less than, 180°, according as the sum of the two opposite angles is greater than, equal to, or less than, 180°.* In other words, the half sum of two sides is in the same quadrant as the half sum of the opposite angles. For, by (41),

$$\tan \tfrac{1}{2}(a+b) \cos \tfrac{1}{2}(A+B) = \tan \tfrac{1}{2} c \cos \tfrac{1}{2}(A-B)$$

the second member of which is always positive, so that $\tan \frac{1}{2}(a+b)$ and $\cos \frac{1}{2}(A+B)$ must have the same sign.

70. CASE I. *Given two sides and the included angle*, or b, c and A. (Fig. 9.)

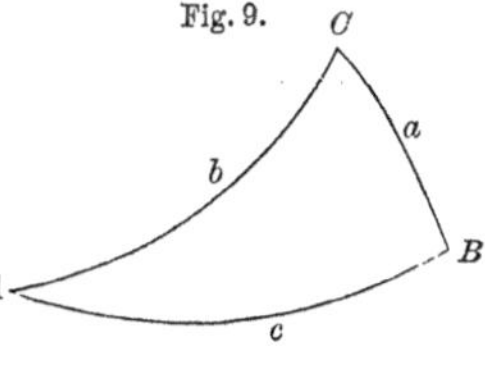

First Solution; when the third side and one of the remaining angles are required.

To find a. The relation between the given parts b, c, A and the required part a is expressed by the first equation of (4),

$$\cos a = \cos c \cos b + \sin c \sin b \cos A \qquad \text{(M)}$$

by which a may be found by computing separately the two terms of the second member and adding their values to form the natural cosine of a; but we should thus be required to use, besides the table of log. sines, also the table of logarithms of numbers, and the table of natural sines and cosines. To adapt it for logarithmic computation by the table of log. sines exclusively, we employ the process of

Pl. Trig., Arts. 174, 175. Thus, let k be a number and ϕ an auxiliary angle such that

$$\left.\begin{aligned} k \sin \phi &= \sin b \cos A \\ k \cos \phi &= \cos b \end{aligned}\right\} \qquad (m)$$

then (M) becomes

$$\begin{aligned} \cos a &= k\,(\cos c \cos \phi + \sin c \sin \phi) \\ &= k \cos (c - \phi) \end{aligned} \qquad (m')$$

so that k and ϕ being found from (m) we may find a by (m'). But we may eliminate k by dividing the first equation of (m) by the second, and substituting in m' the value of $k = \dfrac{\cos b}{\cos \phi}$, whence we have, for finding a,

$$\left.\begin{aligned} \tan \phi &= \tan b \cos A \\ \cos a &= \frac{\cos (c - \phi) \cos b}{\cos \phi} \end{aligned}\right\} \qquad (122)$$

which are the formulæ commonly employed.*

To find B. The relation between b, c, A and B, is, by the first equation of (10),

$$\cot B = \frac{\sin c \cot b - \cos c \cos A}{\sin A} \qquad (N)$$

This may be adapted for logarithms by the process above employed, but to assimilate it to (M) we multiply the numerator and denominator of the second member by $\sin b$, whence

$$\cot B = \frac{\sin c \cos b - \cos c \sin b \cos A}{\sin b \sin A}$$

which by (m) becomes

$$\cot B = \frac{k \sin (c - \phi)}{\sin b \sin A} \qquad (n)$$

or substituting the value of $k = \dfrac{\sin b \cos A}{\sin \phi}$, the formulæ for finding B are

$$\left.\begin{aligned} \tan \phi &= \tan b \cos A \\ \cot B &= \frac{\sin (c - \phi) \cot A}{\sin \phi} \end{aligned}\right\} \qquad (123)$$

* We might have assumed $k \sin \phi = \cos b$, $k \cos \phi = \sin b \cos A$, which would have reduced (M) to $\cos a = k \sin (c + \phi)$. In this way all the solutions that follow may be varied.

In the use of these formulæ, as indeed of all that follow, the signs of all the functions must be carefully observed, according to Pl. Trig. Arts. 37 and 40.

We may take ϕ between 0 and 180°, less or greater than 90°, according as the sign of its tangent is positive or negative; or we may take it numerically less than 90° in all cases, but positive or negative according to the sign of its tangent, (Pl. Trig. Arts. 37 and 174).

Check. The quotient of (n) divided by (m') is

$$\frac{\cot B}{\cos a} = \frac{\tan (c - \phi)}{\sin b \sin A}$$

which multiplied by the following, from (3),

$$\sin a \sin B = \sin b \sin A$$

gives

$$\tan a \cos B = \tan (c - \phi) \tag{124}$$

by which the values of a and B, found by (122) and (123), may be verified.

71. If a and C were required, the solution would evidently be similar, only interchanging b and c, B and C. By the fundamental formulæ we should have

$$\left.\begin{aligned} \cos a &= \cos b \cos c + \sin b \sin c \cos A \\ \cot C &= \frac{\sin b \cos c - \cos b \sin c \cos A}{\sin c \sin A} \end{aligned}\right\} \tag{O}$$

and denoting the auxiliary angle in this case by χ, the logarithmic solution would be

$$\left.\begin{aligned} \tan \chi &= \tan c \cos A \\ \cos a &= \frac{\cos (b - \chi) \cos c}{\cos \chi} \\ \cot C &= \frac{\sin (b - \chi) \cot A}{\sin \chi} \\ \textit{Check.} \quad \tan a \cos C &= \tan (b - \chi) \end{aligned}\right\} \tag{125}$$

EXAMPLES.

1. Given $b = 120° 30' 30''$, $c = 70° 20' 20''$, $A = 50° 10' 10''$; find a and B.

By (122).

$b =$	$120° 30' 30''$	$\log \tan b$ —	0·2297071
$A =$	$50° 10' 10''$	$\log \cos A$ +	9·8065322
$\phi =$	$132° 36' 44''.2$*	$\log \tan \phi$ —	0·0362393
$c =$	$70° 20' 20''.0$		
$c - \phi = -$	$62° 16' 24''.2$		

By (122).	By (123).	By (124).
$\log \cos (c-\phi) +$ 9·6676893	$\log \sin (c-\phi) -$ 9·9470304	$\log \tan (c-\phi) -$ 0·2793410
ar co $\log \cos \phi -$ 0·1693898	ar co $\log \sin \phi +$ 0·1331505	$\log \tan a +$ 0·4291648
$\log \cos b -$ 9·7055761	$\log \cot A +$ 9·9212038	$\log \cos B -$ 9·8501762
$\log \cos a +$ 9·5426552	$\log \cot B -$ 0·0013847	*Check.* — 0·2793410
$a = 69° 34' 55''·9$	$B = 135° 5' 28''·8$	

2. Given $b = 120° 30' 30''$, $c = 70° 20' 20''$, $A = 50° 10' 10''$; find a and C.

Ans. $a = 69° 34' 55''·9$
$C = 50° 30' 8''·4$

3. Given $b = 99° 40' 48''$, $c = 100° 49' 30''$, $A = 65° 33' 10''$; find a and B.

Ans. $a = 64° 23' 15''.0$
$B = 95° 38' 4''.0$

4. Given $b = 99° 40' 48''$, $c = 100° 49' 30''$, $A = 65° 33' 10''$; find a and C.

Ans. $a = 64° 23' 15''.0$
$C = 97° 26' 29''.1$

5. Given $b = 98° 2' 20''$, $c = 80° 35' 40''$, $A = 10° 16' 30''$; find a and C.

Ans. $a = 20° 13' 30''·1$
$C = 30° 35' 56''·7$

72. If B, C and a were all required, we might find a and C by (125), and then B by Art. 3, which gives

$$\sin a : \sin b = \sin A : \sin B$$

or

$$\sin B = \frac{\sin b \sin A}{\sin a}$$

---

* We may also take $\phi = -47° 23' 15''·8$, whence $c - \phi = 117° 43' 35''·8$, which will give the same values of a and B as found in the text.

Of the two values of B less than 180° given by this formula, the proper one may generally be selected by the principles of Art. 69. There are cases, however, in which all the conditions there given are satisfied by both values of B,* and on this account it is preferable, in general, to combine (123) and (125), or to employ the following solution, when the three unknown parts are all to be found.

73. CASE I. Given b, c and A. *Second Solution;* when the two remaining angles are required, or when the three unknown parts are all required.

We have, by Napier's Analogies, (42) and (43),

$$\sin \tfrac{1}{2}(b+c) : \sin \tfrac{1}{2}(b-c) = \cot \tfrac{1}{2} A : \tan \tfrac{1}{2}(B-C)$$

$$\cos \tfrac{1}{2}(b+c) : \cos \tfrac{1}{2}(b-c) = \cot \tfrac{1}{2} A : \tan \tfrac{1}{2}(B+C)$$

whence

$$\tan \tfrac{1}{2}(B-C) = \frac{\sin \frac{1}{2}(b-c)}{\sin \frac{1}{2}(b+c)} \cot \tfrac{1}{2} A \qquad (126)$$

$$\tan \tfrac{1}{2}(B+C) = \frac{\cos \frac{1}{2}(b-c)}{\cos \frac{1}{2}(b+c)} \cot \tfrac{1}{2} A \qquad (127)$$

which determine $\frac{1}{2}(B-C)$ and $\frac{1}{2}(B+C)$; then the half difference added to the half sum gives the greater angle, and the half difference subtracted from the half sum gives the less angle.

If $c > b$, we may write $c-b$, $C-B$, in the place of $b-c$, $B-C$.

We may now find a by either of Napier's Analogies, (40), (41), which give†

$$\tan \tfrac{1}{2} a = \frac{\sin \frac{1}{2}(B+C)}{\sin \frac{1}{2}(B-C)} \tan \tfrac{1}{2}(b-c) \qquad (128)$$

$$\tan \tfrac{1}{2} a = \frac{\cos \frac{1}{2}(B+C)}{\cos \frac{1}{2}(B-C)} \tan \tfrac{1}{2}(b+c) \qquad (129)$$

---

* By Art. 69, VI., if b differs more from 90° than c, B is in the same quadrant as b, and all ambiguity is removed. If c differs more from 90° than b, we may find a and B by (122) and (123), and then C by the formula

$$\sin C = \frac{\sin c \sin A}{\sin a}$$

C being taken in the same quadrant as c.

† We may also find a from any one of Gauss's Equations (44), which become, in the present case,

$$\cos \tfrac{1}{2} a \sin \tfrac{1}{2}(B+C) = \cos \tfrac{1}{2} A \cos \tfrac{1}{2}(b-c)$$
$$\cos \tfrac{1}{2} a \cos \tfrac{1}{2}(B+C) = \sin \tfrac{1}{2} A \cos \tfrac{1}{2}(b+c)$$
$$\sin \tfrac{1}{2} a \sin \tfrac{1}{2}(B-C) = \cos \tfrac{1}{2} A \sin \tfrac{1}{2}(b-c)$$
$$\sin \tfrac{1}{2} a \cos \tfrac{1}{2}(B-C) = \sin \tfrac{1}{2} A \sin \tfrac{1}{2}(b+c)$$

EXAMPLES.

1. Given $b = 120° \, 30' \, 30''$, $c = 70° \, 20' \, 20''$, $A = 50° \, 10' \, 10''$; find B, C and a. We have

$$\tfrac{1}{2}(b+c) = 95° \, 25' \, 25''$$
$$\tfrac{1}{2}(b-c) = 25° \;\; 5' \;\; 5''$$
$$\tfrac{1}{2}A = 25° \;\; 5' \;\; 5''$$

By (126).

$$\begin{aligned} \text{ar co log sin } \tfrac{1}{2}(b+c) &+ 0.0019477 \\ \log\sin \tfrac{1}{2}(b-c) &+ 9.6273228 \\ \log\cot \tfrac{1}{2}A &+ 0.3296529 \\ \hline \log\tan \tfrac{1}{2}(B-C) &+ 9.9589234 \\ \tfrac{1}{2}(B-C) &= 42° \, 17' \, 40''.0 \\ B &= 135° \;\; 5' \, 28''.6 \end{aligned}$$

By (127).

$$\begin{aligned} \text{ar co log cos } \tfrac{1}{2}(b+c) &- 1.0244829 \\ \log\cos \tfrac{1}{2}(b-c) &+ 9.9569757 \\ \log\cot \tfrac{1}{2}A &+ 0.3296529 \\ \hline \log\tan \tfrac{1}{2}(B+C) &- 1.3111115 \\ \tfrac{1}{2}(B+C) &= 92° \, 47' \, 48''.6 \\ C &= 50° \, 30' \;\; 8''.6 \end{aligned}$$

By (128).

$$\begin{aligned} \text{ar co log sin } \tfrac{1}{2}(B-C) &+ 0.1720232 \\ \log\sin \tfrac{1}{2}(B+C) &+ 9.9994824 \\ \log\tan \tfrac{1}{2}(b-c) &+ 9.6703471 \\ \hline \log\tan \tfrac{1}{2}a &+ 9.8418527 \\ \tfrac{1}{2}a &= 34° \, 47' \, 28''.1 \end{aligned}$$

By (129).

$$\begin{aligned} \text{ar co log cos } \tfrac{1}{2}(B-C) &+ 0.1309465 \\ \log\cos \tfrac{1}{2}(B+C) &- 8.6883710 \\ \log\tan \tfrac{1}{2}(b+c) &- 1.0225352 \\ \hline \log\tan \tfrac{1}{2}a &+ 9.8418527 \end{aligned}$$

Ans. $B = 135° \;\; 5' \, 28''.6$
$C = \;\; 50° \, 30' \;\; 8''.6$
$a = \;\; 69° \, 34' \, 56''.2$

2. Given $b = 99° \, 40' \, 48''$, $c = 100° \, 49' \, 30''$, $A = 65° \, 33' \, 10''$; find B, C and a.

Ans. $B = 95° \, 38' \;\; 4''.0$
$C = 97° \, 26' \, 29''.1$
$a = 64° \, 23' \, 15''.1$

74. It may be remarked with regard to (128) and (129) that, when b and c (and consequently B and C) are nearly equal, a small error in the previous determination of the small angle $\frac{1}{2}(B - C)$ may produce a large one in $\log\sin\frac{1}{2}(B-C)$, and consequently in $\log\tan\frac{1}{2}a$ found by (128). In that case, therefore, (129) must be preferred.

In like manner, if $\frac{1}{2}(b+c)$, and consequently $\frac{1}{2}(B+C)$, are nearly equal to 90°, (129) will become inaccurate, and then (128) is to be preferred.

Formula (128) would fail entirely if $B = C$, and formula (129) would fail if $\frac{1}{2}(B+C) = 90°$, since the second members in these cases would assume the indeterminate form $\frac{0}{0}$.

75. CASE I. Given b, c and A. *Third Solution. When the third side is alone required*, the computation by (122) is in most cases as convenient as any other; but there are various other methods

derived from the formulæ of the preceding chapter, which have been employed with advantage in particular applications. Among the most convenient are the following, from (12) and (13):

$$\cos a = \cos (b - c) - 2 \sin b \sin c \sin^2 \tfrac{1}{2} A \qquad (130)$$

$$\cos a = \cos (b + c) + 2 \sin b \sin c \cos^2 \tfrac{1}{2} A \qquad (131)$$

The computation of these requires the use of natural cosines and numbers, the signs of which must be carefully observed.

EXAMPLE.

Given $b = 99° 40' 48''$, $c = 100° 49' 30''$, $A = 65° 33' 10''$; find a.

By (130).*

$\tfrac{1}{2} A = 32° 46' 35''$ $\quad \log \sin^2 \tfrac{1}{2} A = 2 \log \sin \tfrac{1}{2} A$ 9.4669752
$b - c = -1° \; 8' 42''$ $\quad \log \sin b$ 9.9922023
$\log \sin c$ 9.9937722
$\log 2$ 0.3010300

$-2 \sin b \sin c \sin^2 \tfrac{1}{2} A = -0.5675181$ $\quad \log$ 9.7539797
$\text{nat} \cos (b - c) = +0.9998003$

$\text{nat} \cos a = +0.4322822$ $\quad a = 64° 23' 15''$

By (131).

$\tfrac{1}{2} A = 32° 46' 35''$ $\quad \log \cos^2 \tfrac{1}{2} A = 2 \log \cos \tfrac{1}{2} A$ 9.8493748
$b + c = 200° 30' 18''$ $\quad \log \sin b$ 9.9922023
$\log \sin c$ 9.9937722
$\log 2$ 0.3010300

$+2 \sin b \sin c \cos^2 \tfrac{1}{2} A = +1.3689240$ $\quad \log$ 0.1363793
$\text{nat} \cos (b + c) = -0.9366416$

$\text{nat} \cos a = +0.4322824$ $\quad a = 64° 23' 15''$

76. In Art. 14, we have deduced several formulæ by which $\tfrac{1}{2} a$ may be computed. We may adapt (17) and (18) for logarithmic computation, as follows:

$$\left.\begin{aligned} \sin^2 \phi &= \sin b \sin c \cos^2 \tfrac{1}{2} A \\ \sin^2 \tfrac{1}{2} a &= \sin^2 \tfrac{1}{2} (b + c) - \sin^2 \phi \\ &= \sin [\tfrac{1}{2} (b + c) + \phi] \sin [\tfrac{1}{2} (b + c) - \phi] \end{aligned}\right\} \qquad (132)$$

$$\left.\begin{aligned} \sin^2 \phi &= \sin b \sin c \sin^2 \tfrac{1}{2} A \\ \cos^2 \tfrac{1}{2} a &= \cos^2 \tfrac{1}{2} (b - c) - \sin^2 \phi \\ &= \cos [\tfrac{1}{2} (b - c) + \phi] \cos [\tfrac{1}{2} (b - c) - \phi] \end{aligned}\right\} \qquad (133)$$

of which (132) is to be preferred when $\tfrac{1}{2} a < 45°$, and (133) when $\tfrac{1}{2} a > 45°$.

* The computation of (130) is facilitated by the use of a special table (given in many treatises on navigation), from which, with the argument A, is taken the logarithm of $2 \sin^2 \tfrac{1}{2} A = \text{versin } A$. [Pl. Trig. (4) and (139)].

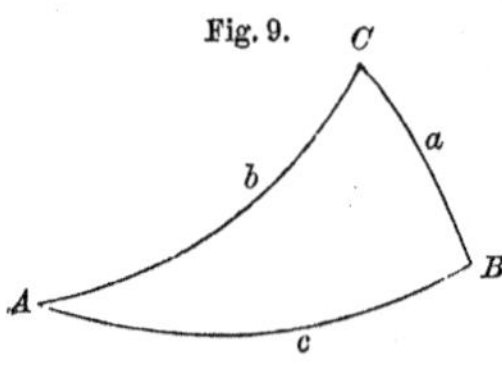

77. CASE II. *Given two angles and the included side*, or A, C and b. (Fig. 9).

First Solution; when the third angle and one of the remaining sides are required.

To find B. The relation between A, C, b and B, is, by (5),

$$\cos B = -\cos C \cos A + \sin C \sin A \cos b \qquad (\text{M})$$

which is adapted for logarithms by the method employed in the preceding case. Thus, let

$$\left.\begin{aligned} h \sin \vartheta &= \cos A \\ h \cos \vartheta &= \sin A \cos b \end{aligned}\right\} \qquad (m)$$

then (M) becomes

$$\begin{aligned} \cos B &= h\,(\sin C \cos \vartheta - \cos C \sin \vartheta) \\ &= h \sin (C - \vartheta) \end{aligned} \qquad (m')$$

or, eliminating $h = \dfrac{\cos A}{\sin \vartheta}$, the formulæ for finding B are

$$\left.\begin{aligned} \cot \vartheta &= \tan A \cos b \\ \cos B &= \frac{\sin (C - \vartheta) \cos A}{\sin \vartheta} \end{aligned}\right\} \qquad (134)$$

To find a. From the third equation of (10), we find,

$$\cot a = \frac{\sin C \cot A + \cos C \cos b}{\sin b}$$

$$= \frac{\sin C \cos A + \cos C \sin A \cos b}{\sin A \sin b} \qquad (\text{N})$$

which, by (m), becomes

$$\cot a = \frac{h \cos (C - \vartheta)}{\sin A \sin b} \qquad (n)$$

or, eliminating $h = \dfrac{\sin A \cos b}{\cos \vartheta}$, we have, for finding a,

$$\left.\begin{aligned} \cot \vartheta &= \tan A \cos b \\ \cot a &= \frac{\cos (C - \vartheta) \cot b}{\cos \vartheta} \end{aligned}\right\} \qquad (135)$$

As in the preceding case, we may either take ϑ always between 0 and 180°, less or greater than 90° according as its tangent is positive or negative; or we may take ϑ numerically less than 90° in all cases, positive or negative, according to the sign of its tangent. (Pl. Trig. Art. 174.)

Check. The quotient of (n) by (m') is

$$\frac{\cot a}{\cos B} = \frac{\cot (C - \vartheta)}{\sin A \sin b}$$

which, multiplied by

$$\sin B \sin a = \sin A \sin b$$

gives

$$\tan B \cos a = \cot (C - \vartheta) \tag{136}$$

by which the values of B and a, found by (134) and (135), may be verified.

78. If B and c were required, the solution would be similar, only interchanging a and c, A and C. By the fundamental formulæ, we should have,

$$\left.\begin{aligned} \cos B &= -\cos A \cos C + \sin A \sin C \cos b \\ \cot c &= \frac{\sin A \cos C + \cos A \sin C \cos b}{\sin C \sin b} \end{aligned}\right\} \quad (o)$$

and denoting the auxiliary angle by ζ, the logarithmic solution would be

$$\left.\begin{aligned} \cot \zeta &= \tan C \cos b \\ \cos B &= \frac{\sin (A - \zeta) \cos C}{\sin \zeta} \\ \cot c &= \frac{\cos (A - \zeta) \cot b}{\cos \zeta} \\ \textit{Check.} \quad \tan B \cos c &= \cot (A - \zeta) \end{aligned}\right\} \quad (137)$$

EXAMPLES.

1. Given $A = 135° \, 5' \, 28''.8$, $C = 50° \, 30' \, 8''.4$, $b = 69° \, 34' \, 55''.9$; find B and a.

By (134).

$A =$	135° 5′ 28″·8	log tan A — 9·9986154
$b =$	69° 34′ 55″·9	log cos b + 9·5426553
$\vartheta =$	109° 10′ 31″·0*	log cot ϑ — 9·5412707
$C =$	50° 30′ 8″·4	
$C - \vartheta =$	— 58° 40′ 22″·6	

* We may also take $\vartheta = -70° \, 49' \, 29''.0$, whence $C - \vartheta = 121° \, 19' \, 37''.4$, which will evidently give the same results as those obtained in the text.

By (134).	By (135).	By (136).
$\log\sin(C-\vartheta) - 9{\cdot}9315664$	$\log\cos(C-\vartheta) + 9{\cdot}7159386$	$\log\cot(C-\vartheta) - 9{\cdot}7843722$
$\text{ar co}\log\sin\vartheta + 0{\cdot}0247897$	$\text{ar co}\log\cos\vartheta - 0{\cdot}4835187$	$\log\tan B + 0{\cdot}0787962$
$\log\cos A - 9{\cdot}8501762$	$\log\cot b + 9{\cdot}7508352$	$\log\cos a - 9{\cdot}7055757$
$\log\cos B + 9{\cdot}8065323$	$\log\cot a - 9{\cdot}7702925$	*Check.* $-9{\cdot}7843719$
$B = 50^\circ\, 10'\, 10''{\cdot}0$	$a = 120^\circ\, 30'\, 29''{\cdot}9$	

Ans. $B = 50^\circ\, 10'\, 10''{\cdot}0$
$a = 120^\circ\, 30'\, 29''{\cdot}9$

2. Given $A = 135^\circ\ 5'\, 28''{\cdot}8$, $C = 50^\circ\, 30'\, 8''{\cdot}4$, $b = 69^\circ\, 34'\, 55''{\cdot}9$; find B and c.

Ans. $B = 50^\circ\, 10'\, 10''{\cdot}0$
$c = 70^\circ\, 20'\, 20''{\cdot}0$

3. Given $A = 65^\circ\, 33'\, 10''$, $C = 95^\circ\, 38'\, 4''$, $b = 100^\circ\, 49'\, 30''$; find B and a.

Ans. $B = 97^\circ\, 26'\, 29''$
$a = 64^\circ\, 23'\, 15''$

4. Given $A = 97^\circ\, 26'\, 29''$, $C = 95^\circ\, 38'\, 4''$, $b = 64^\circ\, 23'\, 15''$; find B and a.

Ans. $B = 65^\circ\, 33'\, 10''$
$a = 100^\circ\, 49'\, 30'$

79. If a, c and B were all required, we might find B and c by (137), and then a by Art. 3, which gives,

$$\sin B : \sin A = \sin b : \sin a$$

$$\sin a = \frac{\sin A \sin b}{\sin B} \tag{138}$$

Of the two values of a given by this equation, the proper one is to be selected, if possible, by the principles of Art. 69.* But as cases occur in which all the conditions there given are satisfied by both values of a, it is preferable, in general, to combine (135) and (137), or to employ the following solution when the three unknown parts are all to be found.

---

* By Art. 69, VII., when A differs more from 90° than C, a must be taken in the same quadrant with A, and all ambiguity is removed. If, then, by A we always denote that angle which differs more from 90° than the other given angle, we may always solve this case by means of (137) and (138), without meeting with any difficulty in determining the quadrant in which a is to be taken.

80. Case II. Given A, C and b. *Second Solution;* when the two remaining sides, or when the three unknown parts are all required.

We have, by Napier's Analogies, (40) and (41),

$$\sin \tfrac{1}{2}(A+C) : \sin \tfrac{1}{2}(A-C) = \tan \tfrac{1}{2} b : \tan \tfrac{1}{2}(a-c)$$

$$\cos \tfrac{1}{2}(A+C) : \cos \tfrac{1}{2}(A-C) = \tan \tfrac{1}{2} b : \tan \tfrac{1}{2}(a+c)$$

whence

$$\left.\begin{aligned} \tan \tfrac{1}{2}(a-c) &= \frac{\sin \frac{1}{2}(A-C)}{\sin \frac{1}{2}(A+C)} \tan \tfrac{1}{2} b \\ \tan \tfrac{1}{2}(a+c) &= \frac{\cos \frac{1}{2}(A-C)}{\cos \frac{1}{2}(A+C)} \tan \tfrac{1}{2} b \end{aligned}\right\} \quad (139)$$

which determine $\frac{1}{2}(a-c)$ and $\frac{1}{2}(a+c)$; then the half difference added to the half sum gives the greater side, and the half difference subtracted from the half sum gives the less side. If $C > A$, we may write $C-A$, $c-a$ in the place of $A-C$, $a-c$.

We may now find B by either of Napier's Analogies, (42) and (43), which give*

$$\cot \tfrac{1}{2} B = \frac{\sin \frac{1}{2}(a+c)}{\sin \frac{1}{2}(a-c)} \tan \tfrac{1}{2}(A-C) \qquad (140)$$

$$\cot \tfrac{1}{2} B = \frac{\cos \frac{1}{2}(a+c)}{\cos \frac{1}{2}(a-c)} \tan \tfrac{1}{2}(A+C) \qquad (141)$$

Examples.

1. Given $A = 135° \, 5' \, 28''.6$, $C = 50° \, 30' \, 8''.6$, $b = 69° \, 34' \, 56''.2$; find a, c and B.

We have
$$\tfrac{1}{2}(A+C) = 92° \, 47' \, 48''.6$$
$$\tfrac{1}{2}(A-C) = 42° \, 17' \, 40''.0$$
$$\tfrac{1}{2} b = 34° \, 47' \, 28''.1$$

Then, by (139),

ar co log $\sin\frac{1}{2}(A+C)$	$+0.0005176$	ar co log $\cos\frac{1}{2}(A+C)$	-1.3116286
log $\sin\frac{1}{2}(A-C)$	$+9.8279768$	log $\cos\frac{1}{2}(A-C)$	$+9.8690535$
log $\tan\frac{1}{2}b$	$+9.8418527$	log $\tan\frac{1}{2}b$	$+9.8418527$
log $\tan\frac{1}{2}(a-c)$	$+9.6703471$	log $\tan\frac{1}{2}(a+c)$	-1.0225348
$\frac{1}{2}(a-c) =$	$25° \, 5' \, 5''.0$	$\frac{1}{2}(a+c) =$	$95° \, 25' \, 25''.0$
$a =$	$120° \, 30' \, 30''.0$	$c =$	$70° \, 20' \, 20''.0$

* We may also find B by any one of Gauss's Equations, (44), interchanging B and C, b and c.

By (140).		By (141).	
ar co log $\sin\frac{1}{2}(a-c)$	$+\ 0.3726772$	ar co log $\cos\frac{1}{2}(a-c)$	$+\ 0.0430243$
log $\sin\frac{1}{2}(a+c)$	$+\ 9.9980523$	log $\cos\frac{1}{2}(a+c)$	$-\ 8.9755171$
log $\tan\frac{1}{2}(A-C)$	$+\ 9.9589234$	log $\tan\frac{1}{2}(A+C)$	$-\ 1.3111110$
log $\cot\frac{1}{2}B$	$+\ 0.3296529$	*log $\cot\frac{1}{2}B$	$+\ 0.3296524$
$\frac{1}{2}B = 25°\ 5'\ 5''.0$			

Ans. $a = 120°\ 30'\ 30''$
$c = 70°\ 20'\ 20''$
$B = 50°\ 10'\ 10''$

2. Given $A = 95°\ 38'\ 4''$, $C = 97°\ 26'\ 29''$, $b = 64°\ 23'\ 15''$; find a, c and B.

Ans. $a = 99°\ 40'\ 48''$
$c = 100°\ 49'\ 30''$
$B = 65°\ 33'\ 10''$

81. Case II. Given A, C and b. *Third Solution. When the third angle B is alone required*, the computation by (134) is in most cases as convenient as any other, but there are other methods (corresponding to those given in Art. 75 for finding a) which may occasionally be serviceable. By (14) and (15) we have

$$\cos B = -\cos(A+C) - 2\sin A \sin C \sin^2\tfrac{1}{2}b \qquad (142)$$

$$\cos B = -\cos(A-C) + 2\sin A \sin C \cos^2\tfrac{1}{2}b \qquad (143)$$

the computation of which is similar to that of (130) and (131).

Example.

Given $A = 95°\ 38'\ 4''$, $C = 97°\ 26'\ 29''$, $b = 64°\ 23'\ 15''$; find B.

By (142).

$\frac{1}{2}b = 32°\ 11'\ 37''.5$	$\log\sin^2\frac{1}{2}b = 2\log\sin\frac{1}{2}b$	9.4531022
$A + C = 193°\ 4'\ 33''$	$\log\sin A$	9.9978967
	$\log\sin C$	9.9963268
	$\log 2$	0.3010300
$-2\sin A\sin C\sin^2\frac{1}{2}b = -0.5602162$	log	9.7483557
$-\text{nat}\cos(A+C) = +0.9740715$		
$\text{nat}\cos B = +0.4138553$		$B = 65°\ 33'\ 9''.9$

* For the reasons given in Art. 74, (141) is, in this example, not so accurate as (140).

82. In Art. 14, several formulæ are given, by which $\frac{1}{2}B$ may be computed. By (21) and (22) we have

$$\sin^2 \tfrac{1}{2} B = \cos^2 \tfrac{1}{2}(A - C) - \sin A \sin C \cos^2 \tfrac{1}{2} b$$
$$\cos^2 \tfrac{1}{2} B = \sin^2 \tfrac{1}{2}(A + C) - \sin A \sin C \sin^2 \tfrac{1}{2} b$$

which may be adapted for logarithms, thus:

$$\left.\begin{aligned} \sin^2 \phi &= \sin A \sin C \cos^2 \tfrac{1}{2} b \\ \sin^2 \tfrac{1}{2} B &= \cos^2 \tfrac{1}{2}(A - C) - \sin^2 \phi \\ &= \cos [\tfrac{1}{2}(A - C) + \phi] \cos [\tfrac{1}{2}(A - C) - \phi] \end{aligned}\right\} \quad (144)$$

$$\left.\begin{aligned} \sin^2 \phi &= \sin A \sin C \sin^2 \tfrac{1}{2} b \\ \cos^2 \tfrac{1}{2} B &= \sin^2 \tfrac{1}{2}(A + C) - \sin^2 \phi \\ &= \sin [\tfrac{1}{2}(A + C) + \phi] \sin [\tfrac{1}{2}(A + C) - \phi] \end{aligned}\right\} \quad (145)$$

of which (144) is to be preferred when $\frac{1}{2}B < 45°$, and (145) when $\frac{1}{2}B > 45°$.

83. Case II. might have been reduced to Case I. by means of the polar triangle, Art. 8; for there will be known in the polar triangle, two sides and an angle opposite one of them, being the supplements of the given angles and side of the proposed triangle. The polar triangle being solved, therefore, by Case I., and its two remaining angles and third side found, the supplements of these parts would be the two sides and third angle required in the proposed triangle. It is easily seen, also, that all the formulæ above given for this case might have been obtained by these considerations.

84. Case III. *Given two sides and an angle opposite one of them;* or a, b, and A. Fig. 9.

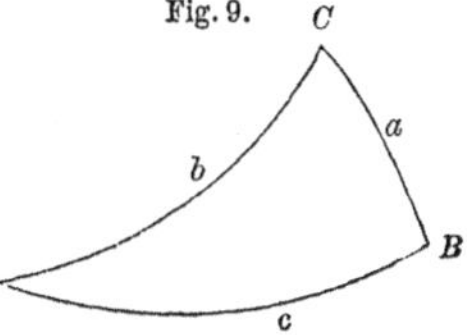

Fig. 9.

First Solution, in which each required part is deduced directly from fundamental formulæ independently of the other two parts.

To find c. We have, by (4),*

$$\cos c \cos b + \sin c \sin b \cos A = \cos a \quad (\text{M})$$

to solve which, let

$$\left.\begin{aligned} k \sin \phi &= \sin b \cos A \\ k \cos \phi &= \cos b \end{aligned}\right\} \quad (m)$$

then (M) becomes

$$k \cos (c - \phi) = \cos a$$

or putting $c - \phi = \phi'$,

$$\left.\begin{aligned} k \cos \phi' &= \cos a \\ c &= \phi + \phi' \end{aligned}\right\} \quad (m')$$

* This formula has been already employed and adapted for logarithms in Case I; but, for the sake of clearness, it is repeated. The student will remark that a simple transformation of (122) gives (146). It will also be observed that the given angle and the given side adjacent to it, in each of the first four cases, are denoted by A and b, in order that the auxiliaries ϕ and ϑ may have the same values throughout.

The auxiliary ϕ will be fully determined by (m), being taken between 0 and 180°, and always positive (Pl. Trig. Art. 174); but, as the cosine of an angle is also the cosine of the negative of that angle [Pl. Trig. (56)], we may take ϕ' in (m') either with the positive or the negative sign, so that $c = \phi \pm \phi'$. There will thus be two values of c answering to the same data, both of which will be admissible, except when $\phi + \phi'$ exceeds 180°, in which case the only solution is $c = \phi - \phi'$; and except when ϕ' exceeds ϕ (which would make c negative), in which case the only solution is $c = \phi + \phi'$.

Therefore, eliminating k, we have for finding c,

$$\left.\begin{aligned} \tan \phi &= \tan b \cos A \\ \cos \phi' &= \frac{\cos \phi \cos a}{\cos b} \\ c &= \phi \pm \phi' \end{aligned}\right\} \quad (146)$$

To find C. We have by (10),

$$\cos C \cos b + \sin C \cot A = \sin b \cot a$$

or, multiplying by $\sin A$,

$$\cos C \sin A \cos b + \sin C \cos A = \sin A \sin b \cot a \qquad (\text{N})$$

to solve which, let

$$\left.\begin{aligned} h \sin \vartheta &= \cos A \\ h \cos \vartheta &= \sin A \cos b \end{aligned}\right\} \quad (n)$$

then (N) becomes

$$h \cos (C - \vartheta) = \sin A \sin b \cot a$$

or putting $C - \vartheta = \vartheta'$,

$$\left.\begin{aligned} h \cos \vartheta' &= \sin A \sin b \cot a \\ C &= \vartheta + \vartheta' \end{aligned}\right\} \quad (n')$$

Here ϑ will be fully determined, while ϑ' found by its cosine may be either positive or negative, so that we shall have in general two values of $C = \vartheta \pm \vartheta'$, corresponding respectively to the two values of c; but, as before, values greater than 180°, and negative values, being excluded, there will in certain cases be but one solution.

Eliminating $h = \dfrac{\sin A \cos b}{\cos \vartheta}$, we have, then, for finding C,

$$\left.\begin{aligned} \cot \vartheta &= \tan A \cos b \\ \cos \vartheta' &= \cos \vartheta \tan b \cot a \\ C &= \vartheta \pm \vartheta' \end{aligned}\right\} \tag{147}$$

To find B. We have several methods: 1st, directly by (3),

$$\sin B = \frac{\sin A \sin b}{\sin a} \tag{148}$$

which gives two values of B, supplements of each other, corresponding respectively to the two values of c and C. We shall presently see how to determine which are the corresponding values of c, C and B.

2d. In (123), ϕ has the same value as in (146), and therefore putting in (123), $c - \phi = \phi'$, we have

$$\cot B = \frac{\sin \phi' \cot A}{\sin \phi} \tag{149}$$

which gives two values of B by the positive and negative values of ϕ'.

3d. By (124),

$$\cos B = \tan \phi' \cot a \tag{150}$$

which also gives two values of B by the positive and negative values of ϕ'.

4th. In (134), ϑ has the same value as in (147), and therefore putting in (134), $C - \vartheta = \vartheta'$,

$$\cos B = \frac{\sin \vartheta' \cos A}{\sin \vartheta} \tag{151}$$

which gives two values of B, as before.

5th. By (136),

$$\cot B = \tan \vartheta' \cos a \tag{152}$$

which gives two values of B, as before.

The formula (149) shows that when ϕ' is positive, $\cot B$ and $\cot A$ have the same sign, that is, B and A are in the same quadrant; and that, when ϕ' is negative, $\cot B$ and $\cot A$ have different signs, that is, B and A are in different quadrants. A like result follows from (151), with reference to ϑ'. Hence, *that value of B which is in the same quadrant as A, belongs to the triangle in which $c = \phi + \phi'$, $C = \vartheta + \vartheta'$; and that value of B which is in a different quadrant from A, belongs to the triangle in which $c = \phi - \phi'$, $C = \vartheta - \vartheta'$.* This precept enables us to employ (148) without ambiguity. In the use of (149), (150), (151) and (152), it is only necessary carefully to observe the signs of the several terms.

Checks. Of the various formulæ above given for finding B, one or more may be employed for the purpose of verification. When c and C have been found, the most simple check is the following, from (3),

$$\frac{\sin C}{\sin c} = \frac{\sin A}{\sin a} \tag{153}$$

which, indeed, might have been employed to find C, after c was found, and reciprocally, but for the ambiguity attaching to the sines.

85. According to Art. 69, VI., if b differs more from 90° than a, B must be in the same quadrant as b, and, since but one of the two values of B can satisfy this condition, there will be but one solution. In that case c and C will each be found to have but one admissible value.

86. The problem will be altogether impossible, when a differs more from 90° than b, and is yet not in the same quadrant with A. In such case, we should find that $\phi + \phi' > 180°$, and $\phi - \phi' < 0$; $\vartheta + \vartheta' > 180°$, $\vartheta - \vartheta' < 0$.

The problem will also be impossible, when $\sin A \sin b > \sin a$, since, by (148), we shall then have $\sin B > 1$.

Examples.

1. Given $a = 40° 16'$, $b = 47° 44'$, $A = 52° 30'$; find B.

By (148).

$a = 40° 16'$	ar co log sin a	0·1895350
$b = 47° 44'$	log sin b	9·8692449
$A = 52° 30'$	log sin A	9·8994667
$B = 65° 16' 35''$	log sin B	9·9582466
or $B = 114° 43' 25''$		

2. With the same data, find c and B.

By (146.)

$a = 40° 16'$		log cos a +9·8825499
$b = 47° 44'$	log tan b +0·0414996	ar co log cos b +0·1722547
$A = 52° 30'$	log cos A +9·7844471	
$\phi = 33° 48' 51''·4$	log tan ϕ +9·8259467	log cos ϕ +9·9195204
$\phi' = \pm 19° 30' 29''·0$		log cos ϕ' +9·9743250
$c_1 = 53° 19' 20''·4$		
$c_2 = 14° 18' 22''·4$		

	By (149).	Check. (150).
$\phi = 33°48'51''.4$	ar co log $\sin\phi + 0.2545328$	$\log\cot a + 0.0720848$
$\phi' = \pm\ 19°30'29''.0$	$\log\sin\phi' \pm 9.5236676$	$\log\tan\phi' \pm 9.5493427$
$A = 52°30'\ 0''$	$\log\cot A + 9.8849805$	± 9.6214275
$B_1 = 65°16'34''.9$ $B_2 = 114°43'25''.1$	$\log\cot B \pm 9.6631809$	$\log\cos B \pm 9.6214275$

$$\textit{Ans. } \left.\begin{array}{l} c = 53°\ 19'\ 20''.4 \\ B = 65°\ 16'\ 34''.9 \end{array}\right\} \text{ or } \left\{\begin{array}{l} c = 14°\ 18'\ 22''.4 \\ B = 114°\ 43'\ 25''.1 \end{array}\right.$$

3. Given $a = 120°$, $b = 70°$, $A = 130°$; find C and B.

By (147).

$a = 120°$		$\log\cot a - 9.7614394$
$b = 70°$	$\log\cos b + 9.5340517$	$\log\tan b + 0.4389341$
$A = 130°$	$\log\tan A - 0.0761865$	
$\vartheta = 112°\ 10'\ 33''.6$	$\log\cot\vartheta - 9.6102382$	$\log\cos\vartheta - 9.5768627$
$\vartheta' = \pm\ 53°\ 13'\ 13''.8$		$\log\cos\vartheta' + 9.7772362$
$C_1 = 165°\ 23'\ 47''.4$		
$C_2 = 58°\ 57'\ 19''.8$		

	By (151).	Check. (152).
$\vartheta = 112°10'33''.6$	ar co log $\sin\vartheta + 0.0333755$	$\log\cos a - 9.6989700$
$\vartheta' = \pm\ 53°13'13''.8$	$\log\sin\vartheta' \pm 9.9036030$	$\log\tan\vartheta' \pm 0.1263669$
$A = 130°\ 0'\ 0''$	$\log\cos A - 9.8080675$	∓ 9.8253369
$B_1 = 123°46'37''.5$ $B_2 = 56°13'22''.5$	$\log\cos B \mp 9.7450460$	$\log\cot B \mp 9.8253369$

$$\textit{Ans. } \left.\begin{array}{l} C = 165°\ 23'\ 47''.4 \\ B = 123°\ 46'\ 37''.5 \end{array}\right\} \text{ or } \left\{\begin{array}{l} C = 58°\ 57'\ 19''.8 \\ B = 56°\ 13'\ 22''.5 \end{array}\right.$$

4. Given $a = 70°$, $b = 120°$, $A = 130°$; find C.

By (147).

$a = 70°$		$\log\cot a + 9.5610659$
$b = 120°$	$\log\cos b - 9.6989700$	$\log\tan b - 0.2385606$
$A = 130°$	$\log\tan A - 0.0761865$	
$\vartheta = 59°\ 12'\ 37''.0$	$\log\cot\vartheta + 9.7751565$	$\log\cos\vartheta + 9.7091756$
$\vartheta' = \pm 108°\ 49'\ 35''.1$		$\log\cos\vartheta' - 9.5088021$

$C = 168°\ 2'\ 12''.1$, taking ϑ' with the positive sign only, since its negative value would render C negative.

5. Given $a = 99° 40' 48''$, $b = 64° 23' 15''$, $A = 95° 38' 4''$; find c, C and B.

Ans. $c = 100° 49' 30''$
$C = 97° 26' 29''$
$B = 65° 33' 10''$

6. Given $a = 40° 5' 25''.6$, $b = 118° 22' 7''.3$, $A = 29° 42' 33''.8$; find c, C and B.

$$\textit{Ans.}\quad \left.\begin{aligned} c &= 153° 38' 42''.4 \\ C &= 160° \ \ 1' 24''.4 \\ B &= 42° 37' 17''.5 \end{aligned}\right\} \text{ or } \left\{\begin{aligned} c &= 90° \ \ 5' 41''.0 \\ C &= 50° 18' 55''.2 \\ B &= 137° 22' 42''.5 \end{aligned}\right.$$

7. Given $a = 69° 34' 56''$, $b = 120° 30' 30''$, $A = 50° 10' 10''$; find c and C.

Ans. $c = 70° 20' 20'$
$C = 50° 30' \ 8''.4$

8. Given $a = 120° 30' 30''$, $b = 69° 34' 56''$, $A = 50° 10' 10''$; find c and C.

Ans. Impossible.

9. Given $a = 40°$, $b = 60°$, $A = 50°$; solve the triangle.

Ans. Impossible.

87. Case III. Given a, b and A. *Second Solution.* We find B by the formula

$$\sin B = \frac{\sin A \sin b}{\sin a}$$

and then by Napier's Analogies, (41) and (43),

$$\left.\begin{aligned} \tan \tfrac{1}{2} c &= \frac{\cos \frac{1}{2}(A + B)}{\cos \frac{1}{2}(A - B)} \tan \tfrac{1}{2}(a + b) \\ \cot \tfrac{1}{2} C &= \frac{\cos \frac{1}{2}(a + b)}{\cos \frac{1}{2}(a - b)} \tan \tfrac{1}{2}(A + B) \end{aligned}\right\} \quad (154)$$

or by (40) and (42),

$$\left.\begin{aligned} \tan \tfrac{1}{2} c &= \frac{\sin \frac{1}{2}(A + B)}{\sin \frac{1}{2}(A - B)} \tan \tfrac{1}{2}(a - b) \\ \cot \tfrac{1}{2} C &= \frac{\sin \frac{1}{2}(a + b)}{\sin \frac{1}{2}(a - b)} \tan \tfrac{1}{2}(A - B) \end{aligned}\right\} \quad (155)$$

in which we employ successively the two values of B, and obtain two solutions, except when for one of these values the second members become negative, for $\frac{1}{2} c$ and $\frac{1}{2} C$ being less than 90°, their tangents must be positive.

We leave it to the student to apply these formulæ to the preceding examples.

88. *To determine by inspection of the data a, b and A, whether there are two solutions, or but one.*

1st. It has already been seen, Art. 85, that when b differs more from 90° than a, B must be in the same quadrant as b, and there can be but one solution. It remains to show,

2d. That when a differs more from 90° than b, there will necessarily be two solutions. We have, by the first of (4),

$$\sin c = \frac{\cos a - \cos b \cos c}{\sin b \cos A}$$

Two solutions exist so long as both values of c are positive, and less than 180°, that is, so long as $\sin c$ is positive. Now when a differs more from 90° than b, we have, (neglecting the signs for a moment),

$$\cos a > \cos b > \cos b \cos c$$

therefore the numerator of the above value of $\sin c$ has the sign of $\cos a$. But by Art. 69, VI., a and A are in the same quadrant, and $\cos a$ and $\cos A$ have the same sign; consequently also, the numerator and denominator have the same sign, and the value of the fraction, or of $\sin c$, is positive, as was to be proved.*

Hence, *there is but one solution when the side opposite the given angle differs less from 90° than the other given side, and two solutions when the side opposite the given angle differs more from 90° than the other given side.*

89. Case IV. *Given two angles and a side opposite one of them,* or A, B and b. (Fig. 9).

First Solution, in which each required part is deduced directly from the fundamental formulæ.

Fig. 9.

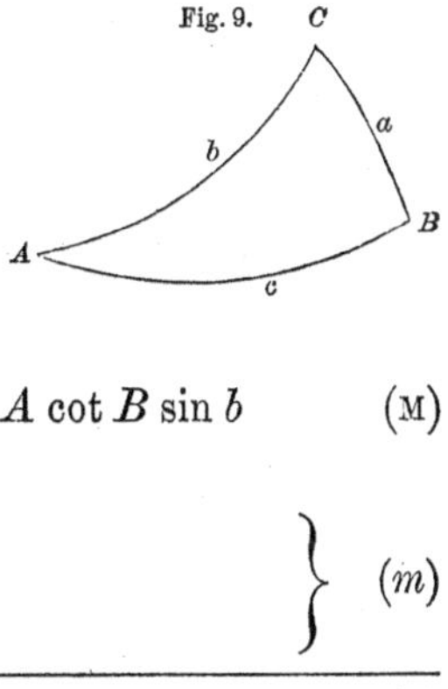

To find c. We have, by (10),

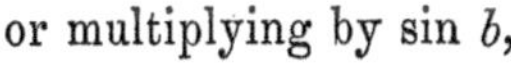

$$\sin c \cot b - \cos c \cos A = \sin A \cot B$$

or multiplying by $\sin b$,

$$\sin c \cos b - \cos c \sin b \cos A = \sin A \cot B \sin b \qquad \text{(M)}$$

to solve which we take

$$\left.\begin{aligned} k \sin \phi &= \sin b \cos A \\ k \cos \phi &= \cos b \end{aligned}\right\} \qquad (m)$$

---

* The same proposition may be otherwise proved thus. By the equations (m) and (m') Art. 84, we have

$$\cos \phi = \frac{\cos b}{k} \qquad \cos \phi' = \frac{\cos a}{k} \qquad k = \frac{\sin b}{\sin \phi} \cos A$$

from the third of which we see that k has the sign of $\cos A$; if then a differs more from 90° than b, that is, if $\cos a$ and $\cos A$ have the same sign, $\cos \phi'$ is positive, and $\phi' < 90°$. Also since, (neglecting signs), $\cos a > \cos b$, we have $\cos \phi' > \cos \phi$, or ϕ' differs more from 90° than ϕ. Hence $\phi' < \phi$ and $\phi' < 180° - \phi$, or $\phi - \phi' > 0$ and $\phi + \phi' < 180°$, or both values of c are between 0 and 180°.

then, putting $c - \phi = \phi'$, (M) becomes

$$\left.\begin{aligned} k \sin \phi' &= \sin A \cot B \sin b \\ c &= \phi + \phi' \end{aligned}\right\} \quad (m')$$

or eliminating k, we have, for finding c,

$$\left.\begin{aligned} \tan \phi &= \tan b \cos A \\ \sin \phi' &= \sin \phi \tan A \cot B \\ c &= \phi + \phi' \end{aligned}\right\} \quad (156)$$

Here ϕ' being determined by its sine, will have two values, supplements of each other, which being successively added to ϕ, give two values of c.

When the second member of the formula

$$\sin \phi' = \sin \phi \tan A \cot B.$$

is negative, $\sin \phi'$, and therefore ϕ' is negative, and the two supplemental values of ϕ' must be successively subtracted from ϕ. There will be two solutions, then, except when one of the values of c exceeds 180°, or when one of them is negative.

To find C. We have, by (5),

$$\sin C \sin A \cos b - \cos C \cos A = \cos B \quad (\text{N})$$

whence, if we put

$$\left.\begin{aligned} h \sin \vartheta &= \cos A \\ h \cos \vartheta &= \sin A \cos b \end{aligned}\right\} \quad (n)$$

and also $C - \vartheta = \vartheta'$, we have

$$\left.\begin{aligned} h \sin \vartheta' &= \cos B \\ C &= \vartheta + \vartheta' \end{aligned}\right\} \quad (n')$$

Eliminating h, we have

$$\left.\begin{aligned} \cot \vartheta &= \tan A \cos b \\ \sin \vartheta' &= \frac{\sin \vartheta \cos B}{\cos A} \\ C &= \vartheta + \vartheta' \end{aligned}\right\} \quad (157)$$

As ϑ' is also determined by its sine, it will have two supplemental values, which will both be added to or both subtracted from ϑ, (according to the sign of $\sin \vartheta'$,) thus giving two values of C, except when one of them exceeds 180°, or when one of them is negative.

To find a. We have several methods: 1st, directly by (3), which gives

$$\sin a = \frac{\sin b \sin A}{\sin B} \tag{158}$$

2d. By (146), where ϕ and ϕ' have the same values as in this case,

$$\cos a = \frac{\cos \phi' \cos b}{\cos \phi} \tag{159}$$

3d. By (150),

$$\cot a = \cot \phi' \cos B \tag{160}$$

4th. By (147), where ϑ and ϑ' have the same values as in this case,

$$\cot a = \frac{\cos \vartheta' \cot b}{\cos \vartheta} \tag{161}$$

5th. By (152),

$$\cos a = \cot \vartheta' \cot B \tag{162}$$

Each of the last four formulæ gives two supplemental values of a by the two values of ϕ' or ϑ', employed in the second members.

From (156) we have

$$\cos A = \tan \phi \cot b$$

which with (159) gives

$$\frac{\cos a}{\cos A} = \cos \phi' \times \frac{\sin b}{\sin \phi}$$

The sign of the second member of this equation depends upon that of $\cos \phi'$, since $\sin b$ and $\sin \phi$ are always positive. Hence when $\cos \phi'$ is positive, $\cos a$ and $\cos A$ must have like signs; and when $\cos \phi'$ is negative, $\cos a$ and $\cos A$ must have different signs. A like result follows from the first of (157) and (161) with reference to ϑ'. Hence, *that value of* a *which is in the same quadrant with* A *belongs to the triangle in which* $\phi' < 90°$, $\vartheta' < 90°$; *and that value of* a *which is in a different quadrant from* A *belongs to the triangle in which* $\phi' > 90°$, $\vartheta' > 90°$. This precept enables us to employ (158) without ambiguity. In the use of (159), (160), (161), and (162), it is only necessary to observe the algebraic signs of the several terms.

Checks. Of the various formulæ above given for finding a, one or more may be employed for the purpose of verification. When c and C have been found, however, the most simple check is

$$\frac{\sin C}{\sin c} = \frac{\sin B}{\sin b} \qquad (163)$$

which might have been employed for finding C after c was found, or reciprocally, but for the ambiguity attaching to the sines.

90. According to Art. 69, VII., if A differs more from 90° than B, a must be in the same quadrant with A. But since the two values of a are supplements of each other, only one of them can satisfy this condition, and there will then be but one solution. In such case c and C will each be found to have but one admissible value.

91. The problem will be impossible when B differs more from 90° than A, and yet is not in the same quadrant with b. In such case we should find both values of c (and both values of C) to be greater than 180°, or both negative.

The problem will also be impossible when $\sin b \sin A > \sin B$, since by (158) we shall then have $\sin a > 1$.

Examples.

1. Given $A = 132°\ 16'$, $B = 139°\ 44'$, $b = 127°\ 30'$; find a.

By (158).

$B = 139°\ 44'\ 0''$	ar co log sin B	0·1895350
$A = 132°\ 16'\ 0''$	log sin A	9·8692449
$b = 127°\ 30'\ 0''$	log sin b	9·8994667
$a = 65°\ 16'\ 35''·1$	log sin a	9·9582466
or $a = 114°\ 43'\ 24''·9$		

2. With the same data, find C and a.

By (157).

$B = 139°44'\ 0''$		log cos B — 9.8825499
$A = 132°16'\ 0''$	log tan A — 0·0414996	ar co log cos A — 0·1722547
$b = 127°30'\ 0''$	log cos b — 9·7844471	
$\vartheta = 56°11'\ 8''·6$	log cot ϑ + 9·8259467	log sin ϑ + 9·9195204
$\vartheta'_1 = +\ 70°29'31''·0$	. . . .	log sin ϑ' + 9·9743250
$\vartheta'_2 = +109°30'29''·0$		
$C_1 = 126°40'39''·6$		
$C_2 = 165°41'37''·6$		

	By (161).	*Check.* (162).
$\vartheta = 56°\,11'\,8''.6$	$\text{ar co log cos}\,\vartheta + 0·2545328$	$\log\cot B - 0·0720848$
$\vartheta'_1 = 70°\,29'31''.0$ $\vartheta'_2 = 109°\,30'29''.0$	$\log\cos\vartheta' \pm 9·5236676$	$\log\cot\vartheta' \pm 9·5493427$
$b = 127°\,30'\,0''$	$\log\cot b - 9·8849805$	∓ 9.6214275
$a_1 = 114°\,43'25''.1$ $a_2 = 65°\,16'34''.9$	$\log\cot a \mp 9·6631809$	$\log\cos a \mp 9·6214275$

Ans. $\left.\begin{array}{l} C = 126°\,40'\,39''·6 \\ a = 114°\,43'\,25''·1 \end{array}\right\}$ or $\left\{\begin{array}{l} C = 165°\,41'\,37''·6 \\ a = 65°\,16'\,34''·9 \end{array}\right.$

3. Given $A = 110°$, $B = 60°$, $b = 50°$; find c and a.

	By (156).	
$B = 60°$		$\log\cot B + 9·9614394$
$A = 110°$	$\log\cos A - 9·5340517$	$\log\tan A - 0·4389341$
$b = 50°$	$\log\tan b + 0·0761865$	
$\phi = 157°49'26''·4$	$\log\tan\phi - 9·6102382$	$\log\sin\phi + 9·5768627$
$\phi'_1 = -36°46'46''·2$ $\phi'_2 = -143°13'13''·8$	. . . .	$\log\sin\phi' - 9·7772362$
$c_1 = 121°\,2'40''·2$ $c_2 = 14°36'12''·6$		

	By (159).	*Check.* (160).
$\phi = 157°49'26''·4$	$\text{ar co log cos}\,\phi - 0·0333755$	$\log\cos B + 9·6989700$
$\phi'_1 = -36°46'46''·2$ $\phi'_2 = -143°13'13''·8$	$\log\cos\phi' \pm 9·9036030$	$\log\cot\phi' \mp 0·1263669$
$b = 50°\,0'\,0''$	$\log\cos b + 9·8080675$	$\mp 9·8253369$
$a_1 = 123°46'37''·5$ $a_2 = 56°13'22''·5$	$\log\cos a \mp 9·7450460$	$\log\cot a \mp 9·8253369$

Ans. $\left.\begin{array}{l} c = 121°\,2'\,40''·2 \\ a = 123°\,46'\,37''·5 \end{array}\right\}$ or $\left\{\begin{array}{l} c = 14°\,36'\,12''·6 \\ a = 56°\,13'\,22''·5 \end{array}\right.$

4. Given $A = 60°$, $B = 110°$, $b = 50°$; find c.

Ans. $c = 11°\,57'\,47''·9$

5. Given $A = 115°\,36'\,45''$, $B = 80°\,19'\,12''$, $b = 84°\,21'\,56''$; find a, c and C.

Ans. $a = 114°\,26'\,50''$
$c = 82°\,33'\,31''$
$C = 79°\,10'\,30''$

6. Given $A = 61°\,37'\,52''·7$, $B = 139°\,54'\,34''·4$, $b = 150°\,17'\,26''·2$; find a, c and C.

Ans. $\left.\begin{array}{l} a = 42°\,37'\,17''·5 \\ c = 129°\,41'\,4''·8 \\ C = 89°\,54'\,19''·0 \end{array}\right\}$ or $\left\{\begin{array}{l} a = 137°\,22'\,42''·5 \\ c = 19°\,58'\,35''·6 \\ C = 26°\,21'\,17''·6 \end{array}\right.$

7. Given $A = 70°$, $B = 120°$, $b = 80°$; solve the triangle.
Ans. Impossible.

8. Given $A = 60°$, $B = 40°$, $b = 50°$; solve the triangle.
Ans. Impossible.

92. Case IV. Given A, B and b. *Second Solution.* We find a by the formula

$$\sin a = \frac{\sin b \sin A}{\sin B}$$

and then by Napier's Analogies we find c and C, precisely as in Case III., Art. 87, employing successively, in (154) or (155), the two values of a given by the preceding equation. There will be but one solution, if one of these values renders the second members of (154) or (155) negative.

The student should apply this method to the preceding examples.

93. *To determine by inspection of the data A, B and b, whether there are two solutions or but one.*

1st. It has already been seen, Art. 90, that when A differs more from 90° than B, a must be in the same quadrant with A, and there can be but one solution. It remains to show that,

2d. When B differs more from 90° than A, there will necessarily be two solutions. We have, by (5),

$$\sin C = \frac{\cos B + \cos A \cos C}{\sin A \cos b}$$

Two solutions exist so long as both values of C are less than 180°, and both positive, that is, so long as $\sin C$ is positive. Now when B differs more from 90° than A, we have, (neglecting signs for a moment),

$$\cos B > \cos A > \cos A \cos C$$

therefore the numerator of the value of $\sin C$ has the sign of $\cos B$. But by Art. 69, VII., B and b are in the same quadrant, consequently the numerator and denominator have the same sign, and the value of the fraction, or of $\sin C$ is always positive, as was to be proved.*

Hence, *there is but one solution when the angle opposite the given side differs less from 90° than the other given angle; and two solutions when the angle opposite the given side differs more from 90° than the other given angle.*

94. Case IV. might have been reduced to Case III. by means of the polar triangle of Art. 8. For there will be known in the polar triangle two sides and an angle opposite one of them, being the supplements of the given angles and side of the proposed triangle. The polar triangle being solved, therefore, by Case III., and its two remaining angles and third side found, the supplements of these parts will be the required sides and third angle of the proposed triangle.

---

* It may be shown that both values of C will be admissible, by a process of reasoning similar to that employed in the note on page 197, applied to the equations of Art. 89.

95. Case V. *Given the three sides*, or a, b and c. (Fig. 9.) We have three methods for computing the half angles:

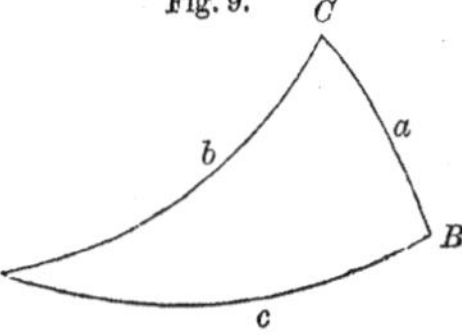

1st. By the sines, from (31), remembering that

$$s = \tfrac{1}{2}(a + b + c)$$

$$\left.\begin{aligned} \sin\tfrac{1}{2}A &= \sqrt{\left(\frac{\sin(s-b)\sin(s-c)}{\sin b \sin c}\right)} \\ \sin\tfrac{1}{2}B &= \sqrt{\left(\frac{\sin(s-c)\sin(s-a)}{\sin c \sin a}\right)} \\ \sin\tfrac{1}{2}C &= \sqrt{\left(\frac{\sin(s-a)\sin(s-b)}{\sin a \sin b}\right)} \end{aligned}\right\} \quad (164)$$

2d. By the cosines, from (33),

$$\left.\begin{aligned} \cos\tfrac{1}{2}A &= \sqrt{\left(\frac{\sin s \sin(s-a)}{\sin b \sin c}\right)} \\ \cos\tfrac{1}{2}B &= \sqrt{\left(\frac{\sin s \sin(s-b)}{\sin c \sin a}\right)} \\ \cos\tfrac{1}{2}C &= \sqrt{\left(\frac{\sin s \sin(s-c)}{\sin a \sin b}\right)} \end{aligned}\right\} \quad (165)$$

3d. By the tangents, from (34),

$$\left.\begin{aligned} \tan\tfrac{1}{2}A &= \sqrt{\left(\frac{\sin(s-b)\sin(s-c)}{\sin s \sin(s-a)}\right)} \\ \tan\tfrac{1}{2}B &= \sqrt{\left(\frac{\sin(s-c)\sin(s-a)}{\sin s \sin(s-b)}\right)} \\ \tan\tfrac{1}{2}C &= \sqrt{\left(\frac{\sin(s-a)\sin(s-b)}{\sin s \sin(s-c)}\right)} \end{aligned}\right\} \quad (166)$$

When only one of the angles is required, the simplest method will be by (165), but if the required angle is less than 90°, it will be found more accurately by (164), for then $\tfrac{1}{2}A < 45°$, and the sine varies more rapidly than the cosine. And, for a similar reason, if the angle is greater than 90°, we should prefer (165). By (166) we always have an accurate result, although the formula is not quite so simple.

When the three angles are required, (166) will require the least labor, since $\sin a$, $\sin b$, and $\sin c$, are not then required.

No ambiguity can arise in these solutions, since the half angles must be less than $90°$; they require therefore no attention to the algebraic signs.

EXAMPLES.

1. Given $a = 100°$, $b = 50°$, $c = 60°$; find A.

a	$= 100°$		
b	$= 50°$	log cosec	0·1157460
c	$= 60°$	log cosec	0·0624694
$2s$	$= 210°$		
s	$= 105°$	log sin	9·9849438
$s-a$	$= 5°$	log sin	8·9402960
			2)9·1034552
$\frac{1}{2}A$	$= 69°\ 7'\ 52''\cdot 7$	log cos	9·5517276
A	$= 138°\ 15'\ 45''\cdot 4$		

2. With the same data, find all the angles.

By (166).

$s = 105°$	l. cosec 0·0150562	l. cosec 0·0150562	l. cosec 0·0150562
$s-a = 5°$	l. cosec 1·0597040	l. sin 8·9402960	l. sin 8·9402960
$s-b = 55°$	l. sin 9·9133645	l. cosec 0·0866355	l. sin 9·9133645
$s-c = 45°$	l. sin 9·8494850	l. sin 9·8494850	l. cosec 0·1505150
	2)0·8376097	2)8·8914727	2)9·0192317
	l. tan 0·4188049	l. tan 9·4457364	l. tan 9·5096159
	$\frac{1}{2}A = 69°\ 7'\ 52''\cdot 7$	$\frac{1}{2}B = 15°\ 35'\ 37''\cdot 0$	$\frac{1}{2}C = 17°\ 54'\ 59''\cdot 1$
Ans.	$A = 138°\ 15'\ 45''\cdot 4$	$B = 31°\ 11'\ 14''\cdot 0$	$C = 35°\ 49'\ 58''\cdot 2$

3. Given $a = 10°$, $b = 7°$, $c = 4°$; find the angles.

Ans. $A = 128°\ 44'\ 45''\cdot 1$
$B = 33°\ 11'\ 12''\cdot 0$
$C = 18°\ 15'\ 31''\cdot 1$

96. The method by (166), may be put under the following convenient form. Let

$$P = \sqrt{\left(\frac{\sin(s-a)\sin(s-b)\sin(s-c)}{\sin s}\right)}$$

then

$$\tan\tfrac{1}{2}A = \frac{P}{\sin(s-a)},\quad \tan\tfrac{1}{2}B = \frac{P}{\sin(s-b)},\quad \tan\tfrac{1}{2}C = \frac{P}{\sin(s-c)} \qquad (167)$$

which are similar to the formulæ of Pl. Trig. Art. 146, and are computed in the same manner.

97. CASE V. Given a, b and c. *Second Solution.* If the whole angle is required directly,* we have

$$\cos A = \frac{\cos a - \cos b \cos c}{\sin b \sin c}$$

which may be adapted for logarithms by an auxiliary thus:

$$\left.\begin{aligned} \cos \phi &= \cos b \cos c \\ \cos A &= \frac{2 \sin \frac{1}{2}(\phi + a) \sin \frac{1}{2}(\phi - a)}{\sin b \sin c} \end{aligned}\right\} \quad (168)$$

Or thus,

$$\left.\begin{aligned} \cot \phi &= \frac{\cos b \cos c}{\sin a} \\ \cos A &= \frac{\sin (\phi - a)}{\sin b \sin c \sin \phi} \end{aligned}\right\} \quad (169)$$

98. CASE VI. *Given the three angles,* or A, B and C. (Fig. 9). We have three methods of finding the half sides:

1st. By the sines, (36).

2d. By the cosines, (38).

3d. By the tangents, (39).

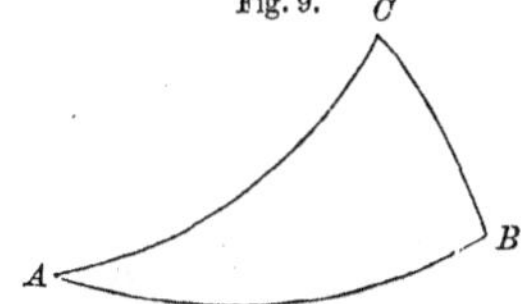

Fig. 9.

The computations are conducted precisely in the same form as those of the preceding case.

EXAMPLE.

Given $A = 120°$, $B = 130°$, $C = 80°$; find c.

Ans. $c = 41° 44' 14''.6$

99. The formulæ (39) may be arranged for convenient use in the same manner as the corresponding formulæ of the preceding case, Art. 96.

100. CASE VI. Given A, B and C. *Second Solution.* We have, by (5),

$$\cos a = \frac{\cos A + \cos B \cos C}{\sin B \sin C}$$

which may be adapted for logarithms by an auxiliary, thus:

$$\left.\begin{aligned} \cos \phi &= \cos B \cos C \\ \cos a &= \frac{2 \cos \frac{1}{2}(A + \phi) \cos \frac{1}{2}(A - \phi)}{\sin B \sin C} \end{aligned}\right\} \quad (170)$$

or,

$$\left.\begin{aligned} \tan \phi &= \frac{\cos B \cos C}{\sin A} \\ \cos a &= \frac{\cos (A - \phi)}{\sin B \sin C \cos \phi} \end{aligned}\right\} \quad (171)$$

---

* See NOTE at the end of this chapter, p. 211, for the method of computing many of the general formulæ of spherical trigonometry directly, without the aid of auxiliary angles.

Solution of Oblique Spherical Triangles by Means of a Perpendicular.

101. All the cases of oblique spherical triangles may be solved by dividing the triangle into two right triangles by a perpendicular from one of the vertices to the opposite side, and solving these partial triangles by the methods of the preceding chapter. Bowditch has given two rules, based upon Napier's Rules, (Art. 46), by which the application of this method is facilitated.

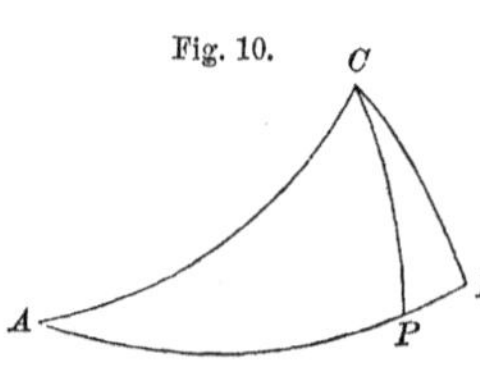

Fig. 10.

102. *Bowditch's Rules for Oblique Triangles.* "If in a spherical triangle, (Fig. 10), two right triangles are formed by a perpendicular let fall from one of its vertices upon the opposite side; and if, in the two right triangles, the middle parts are so taken that the perpendicular is an adjacent part in both of them; then

The sines of the middle parts in the two triangles are proportional to the tangents of the adjacent parts.

But if the perpendicular is an opposite part in both the triangles, then

The sines of the middle parts are proportional to the cosines of the opposite parts.

To prove which rules, let M denote the middle part in one of the right triangles, A an adjacent part, and O an opposite part. Also, let m denote the middle part in the other triangle, a an adjacent part, and o an opposite part; and let p denote the perpendicular.

First. If the perpendicular is an adjacent part in both triangles, we have, by Napier's Rules, (Art. 46,)

$$\sin M = \tan A \tan p$$
$$\sin m = \tan a \tan p$$

whence

$$\frac{\sin M}{\sin m} = \frac{\tan A \tan p}{\tan a \tan p} = \frac{\tan A}{\tan a}$$

or

$$\sin M : \sin m = \tan A : \tan a$$

Secondly. If the perpendicular is an opposite part in both triangles, we have, by Napier's Rules

$$\sin M = \cos O \cos p$$
$$\sin m = \cos o \cos p$$

whence

$$\frac{\sin M}{\sin m} = \frac{\cos O \cos p}{\cos o \cos p} = \frac{\cos O}{\cos o}$$

or

$$\sin M : \sin m = \cos O : \cos o\text{''}\ *$$

We proceed to solve the six cases of spherical triangles with the aid of a perpendicular. It will be seen, however, that Bowditch's Rules are applicable but in the first four cases.

---

* Peirce's Spherical Trigonometry, Art. 44.

103. Case I. Given b, c and A. Let the perpendicular CP, Fig. 10, be drawn from C, (that is, in such a manner as to put two given parts in one of the right triangles). Then the right triangle ACP gives, by Napier's Rules, if we put $AP = \phi$,

$$\tan \phi = \tan b \cos A \qquad (172)$$

then taking co. b and co. a as middle parts in the two triangles, $AP = \phi$ and $BP = c - \phi$* are the opposite parts, whence, by Bowditch's Rules,

$$\cos \phi : \cos (c - \phi) = \cos b : \cos a$$

whence

$$\cos a = \frac{\cos (c - \phi) \cos b}{\cos \phi} \qquad (173)$$

Again, taking AP and PB as middle parts, co. A and co. B are adjacent parts, whence, by Bowditch's Rules,

$$\sin \phi : \sin (c - \phi) = \cot A : \cot B$$

whence

$$\cot B = \frac{\sin (c - \phi) \cot A}{\sin \phi} \qquad (174)$$

and the formulæ (172), (173), (174), agree entirely with (122) and (123).

The triangle BCP gives as a check

$$\tan a \cos B = \tan (c - \phi) \qquad (175)$$

which agrees with (124).

By drawing the perpendicular from B, we may in the same manner obtain the formulæ (125).

The angle C may be found by the proportion

$$\sin a : \sin c = \sin A : \sin C$$

or if C has been found by means of a perpendicular from B, B may be found by a similar proportion, as in Art. 72; and the quadrant in which the angle is to be taken must be determined by the principles of Art. 69.

104. Case II. Given A, C and b. Let the perpendicular be drawn as before, Fig. 10, and let

$$ACP = \vartheta, \qquad BCP = C - \vartheta \dagger$$

then, by Napier's Rules,

$$\cot \vartheta = \tan A \cos b \qquad (176)$$

and by Bowditch's Rules, taking co. A and co. B as middle parts, and therefore co. ACP and co. BCP as opposite parts,

$$\sin \vartheta : \sin (C - \vartheta) = \cos A : \cos B$$

whence

$$\cos B = \frac{\sin (C - \vartheta) \cos A}{\sin \vartheta} \qquad (177)$$

---

* If AP should exceed AB, (that is, if the perpendicular should fall without the triangle), BP would be equal to $AP - AB = \phi - c$, and the solution could be modified accordingly. But the true results will always be obtained by regarding BP as negative; that is, by still taking $BP = c - \phi$ and attending to the signs of all the terms as already exemplified, p. 182.

† If $ACP > ACB$, $BCP = C - ACP$ will become negative, but the true results are still found by attending to the signs, as already shown, p. 187.

Again, taking co. $A\,C\,P$ and co. $B\,C\,P$ as middle parts, and therefore co. b and co. a as adjacent parts, Bowditch's Rules give

$$\cos \vartheta : \cos (C - \vartheta) = \cot b : \cot a$$

whence

$$\cot a = \frac{\cos (C - \vartheta) \cot b}{\cos \vartheta} \tag{178}$$

and (176), (177), (178), agree entirely with (134) and (135)

The triangle $B\,C\,P$ gives

$$\tan B \cos a = \cot (C - \vartheta) \tag{179}$$

which agrees with (136).

By drawing the perpendicular from A, we may in the same manner obtain the formulæ (137).

The side c may be found from the proportion

$$\sin A : \sin C = \sin a : \sin c$$

and Art. 69; or c being found by means of a perpendicular from A, we may find a by a similar proportion.

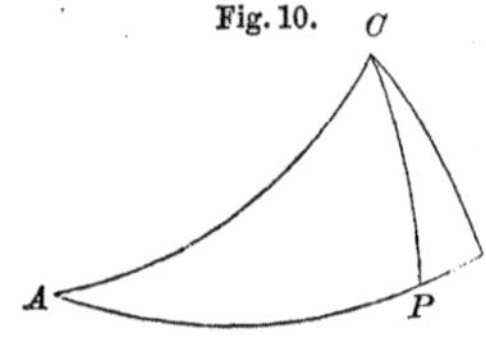

Fig. 10.

105. Case III. Given a, b and A. Let the perpendicular be drawn from C, Fig 10, as in the preceding cases, and let $A\,P = \phi$, $B\,P = \phi'$; then, by Napier's Rules,

$$\tan \phi = \tan b \cos A \tag{180}$$

and, by Bowditch's Rules,

$$\cos b : \cos a = \cos \phi : \cos \phi'$$

whence

$$\cos \phi' = \frac{\cos \phi \cos a}{\cos b} \tag{181}$$

and then

$$c = \phi + \phi'$$

In Art. 84, we have found, from analytical considerations, that this case admits of two solutions, and that the general expression for c is

$$c = \phi \pm \phi' \tag{182}$$

In fact, let us attempt to construct the triangle with the data a, b and A. Having constructed A equal to the given angle, and b equal to the adjacent side, Fig. 11, let

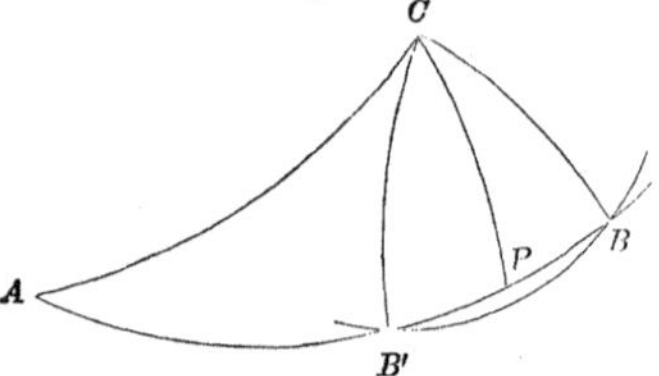

Fig. 11.

a small circle be described about C as a pole, with a (circular) radius $= a$; this circle intersects the great circle AB in two points, B and B', and both triangles, ACB and ACB' contain the same data a, b and A. If the perpendicular CP is drawn, we have $BP = B'P$, so that in one of the triangles, the side $c = AB = AP + PB = \phi + \phi'$, and in the other, $c = AB' = AP - B'P = \phi - \phi'$. If both points of intersection, B and B', fall on the same side of A, and within 180° of A, both solutions will be admissible.

To find C, let $ACP = \vartheta$, and $BCP = \vartheta'$, then by Napier's Rules,

$$\cot \vartheta = \tan A \cos b \tag{183}$$

and by Bowditch's Rules,

$$\cot b : \cot a = \cos \vartheta : \cos \vartheta'$$

whence

$$\cos \vartheta' = \cos \vartheta \tan b \cot a \tag{184}$$

and since in Fig. 11,

$$C = ACB = ACP + BCP = \vartheta + \vartheta', \text{ or } C = ACB' = ACP - B'CP = \vartheta - \vartheta',$$

we have

$$C = \vartheta \pm \vartheta' \tag{185}$$

and the formulæ (180), (181), (182), (183), (184), (185), agree entirely with (146) and (147).

After c was found, we might have found C from the proportion

$$\sin a : \sin c = \sin A : \sin C \tag{186}$$

and B is found from the proportion

$$\sin a : \sin b = \sin A : \sin B \tag{187}$$

The two values of B determined by (187), are both admissible when c has two values as above. It is also evident, from Fig. 11, that the two values of B are supplemental. To determine the corresponding values of c and B, we observe that, by Art. 49, the perpendicular CP is in the same quadrant with A and with CBP and $CB'P$, and therefore $CB'A$ is in a different quadrant from A. Hence, *that value of B which is in the same quadrant as A corresponds to the value of $c = \phi + \phi'$, and that value of B which is in a different quadrant from A corresponds to the value of $c = \phi - \phi'$*; which agrees with what is shown in Art. 84.

In computing (186), the two values of c must be employed successively, and the formula computed twice. At each computation we shall have two values of C found from the sine, one of which must be selected by Art. 69. But as the application of the principles of Art. 69 is tedious and embarrassing, it is better to find C by (184) and (185).

The formulæ (149), (150), (151), (152), for finding B, may easily be deduced by Napier's and Bowditch's Rules.

106. CASE IV. Given A, B and b. Let the perpendicular be drawn as before, Fig. 10, and let $AP = \phi$, $BP = \phi'$, then as before,

$$\tan \phi = \tan b \cos A \tag{188}$$

and by Bowditch's Rules,

$$\cot A : \cot B = \sin \phi : \sin \phi'$$

whence

$$\left.\begin{aligned} \sin \phi' &= \sin \phi \tan A \cot B \\ c &= \phi + \phi' \end{aligned}\right\} \tag{189}$$

which agree with (156). But ϕ' having two supplemental values determined by the sine, c has two values, as already explained in Art. 89.

Fig. 12.

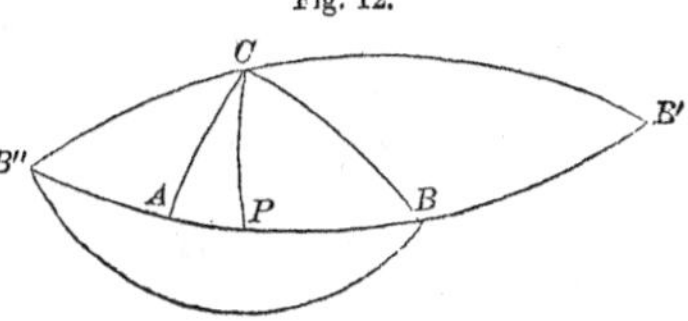

To show the same geometrically, let BP, Fig. 12, be the acute value of ϕ', and about C as a pole, let a small circle be described passing through B, and intersecting the great circle AB again in B''. Let $B''C$ be drawn, and produced to meet AB again in B', forming the lune $B''B'$. Then we have

$$B' = B'' = CBA$$

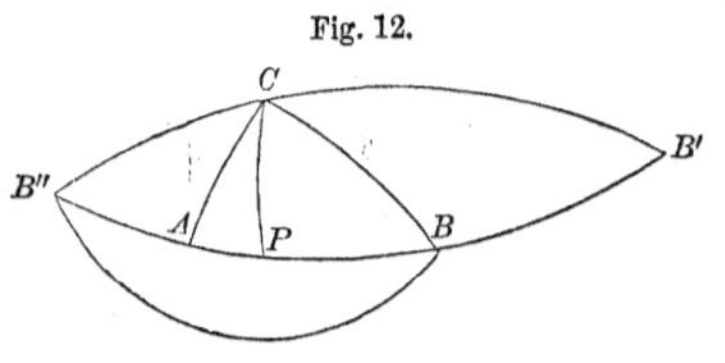

Fig. 12.

so that in both triangles, ACB and ACB', the value of the angle opposite the side b is the same, that is, both triangles contain the same data, A, B and b. Now

$$180° = B''B' = B''P + B'P = BP + B'P,$$

so that BP and $B'P$ are supplements of each other.

In the triangle ACB we have

$$c = \phi + \phi' = AP + BP$$

and in the triangle ACB' we have

$$c = \phi + \phi' = AP + B'P$$

and hence the two values of c are found by giving ϕ' its acute and obtuse values successively, as already shown analytically.

By Art. 49, CP must be in the same quadrant with A; hence, if B is in the same quadrant with A, P falls between A and B, as in the figure, and for the same reason, between A and B'. But if A and B were in different quadrants, both points, B and B', might fall between A and P. The two values of c would then be found by the formula

$$c = \phi - \phi'$$

ϕ' taking, successively, its acute and obtuse values. In that case, tan A and cot B would have opposite signs in (189), sin ϕ' would be negative, which would make ϕ' negative, so that the true results will be obtained, without reference to a diagram, by attending to the signs of the several terms, as already fully exemplified, p. 201.

To find C, let $ACP = \vartheta$, $BCP = \vartheta'$, then we have, as before,

$$\cot \vartheta = \tan A \cos b \qquad (190)$$

and by Bowditch's Rules

$$\cos A : \cos B = \sin \vartheta : \sin \vartheta'$$

whence

$$\left.\begin{aligned} \sin \vartheta' &= \frac{\sin \vartheta \cos B}{\cos A} \\ C &= \vartheta + \vartheta' \end{aligned}\right\} \qquad (191)$$

which agree with (157). It is evident from Fig. 12, that BCP and $B'CP$, are supplemental, and that the remarks above made with reference to ϕ' apply also to ϑ'.

After c was found, we might have found C by the proportion

$$\sin b : \sin c = \sin B : \sin C \qquad (192)$$

and a is found by the proportion

$$\sin B : \sin A = \sin b : \sin a \qquad (193)$$

The two values of a found by (193) are both admissible when c has two values. From Art. 50, it follows that when BP is acute, a must be in the same quadrant with CP, that is, (Art. 49), in the same quadrant with A; and when BP is obtuse, a must be in a different quadrant from A. That is, *that value of* a *which is in the same quadrant with* A, *belongs to the triangle in which* $\phi' < 90°$, *and that value of* a *which is in a different quadrant from* A, *belongs to the triangle in which* $\phi' > 90°$; which agrees with Art. 89.

The formulæ (159), (160), (161), and (162), for finding a may easily be deduced by Napier's and Bowditch's Rules.

107. CASE V. Given a, b and c. The perpendicular cannot be drawn, in this case, so that two of the given parts shall be in one triangle; nevertheless the case can be solved by means of a perpendicular. Let the perp. be drawn from any angle, as C, Fig. 13, and as before, put $AP = \phi$, $BP = \phi'$; then by Bowditch's Rules,

Fig. 13.

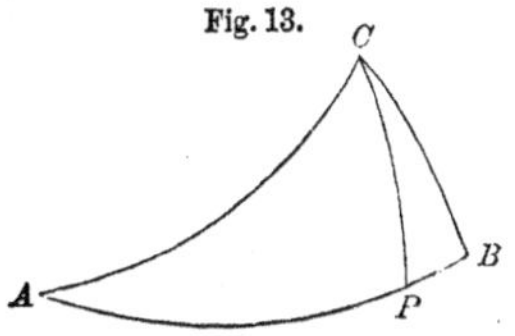

$$\cos\phi : \cos\phi' = \cos b : \cos a$$

whence

$$\frac{\cos\phi' - \cos\phi}{\cos\phi' + \cos\phi} = \frac{\cos a - \cos b}{\cos a + \cos b}$$

or, by Pl. Trig. (110),

$$\tan\tfrac{1}{2}(\phi + \phi')\tan\tfrac{1}{2}(\phi - \phi') = \tan\tfrac{1}{2}(b + a)\tan\tfrac{1}{2}(b - a)$$

whence, since $\phi + \phi' = c$,

$$\left.\begin{aligned}\tan\tfrac{1}{2}(\phi - \phi') &= \tan\tfrac{1}{2}(b + a)\tan\tfrac{1}{2}(b - a)\cot\tfrac{1}{2}c\\ \tfrac{1}{2}(\phi + \phi') &= \tfrac{1}{2}c\end{aligned}\right\} \quad (194)$$

which determine $\frac{1}{2}(\phi - \phi')$ and $\frac{1}{2}(\phi + \phi')$ whence ϕ and ϕ'. The angles A and B are then determined by Napier's Rules.

108. CASE VI. Given A, B and C. In Fig. 13, let $ACP = \vartheta$, $BCP = \vartheta'$; then, by Bowditch's Rules,

$$\sin\vartheta : \sin\vartheta' = \cos A : \cos B$$

whence

$$\frac{\sin\vartheta - \sin\vartheta'}{\sin\vartheta + \sin\vartheta'} = \frac{\cos A - \cos B}{\cos A + \cos B}$$

or, by Pl. Trig. (109) and (110),

$$\frac{\tan\tfrac{1}{2}(\vartheta - \vartheta')}{\tan\tfrac{1}{2}(\vartheta + \vartheta')} = \tan\tfrac{1}{2}(B + A)\tan\tfrac{1}{2}(B - A)$$

whence, since $\vartheta + \vartheta' = C$,

$$\left.\begin{aligned}\tan\tfrac{1}{2}(\vartheta - \vartheta') &= \tan\tfrac{1}{2}(B + A)\tan\tfrac{1}{2}(B - A)\tan\tfrac{1}{2}C\\ \tfrac{1}{2}(\vartheta + \vartheta') &= \tfrac{1}{2}C\end{aligned}\right\} \quad (195)$$

which determine $\frac{1}{2}(\vartheta - \vartheta')$ and $\frac{1}{2}(\vartheta + \vartheta')$ and therefore ϑ and ϑ'. The sides a and b are then found by Napier's Rules.

NOTE REFERRED TO ON PAGE 205.

Computation of Spherical Formulæ by the Gaussian Table.

The *Gaussian Table* is a table, first suggested by Gauss, for readily computing the logarithm of the sum or difference of two quantities, when the logarithms of these quantities are given.

If p and q are the two numbers whose logarithms are given, p being the greater number, (or $\log p$ the greater logarithm), we have, in the first place

$$p + q = q\left(1 + \frac{p}{q}\right) = p\left(1 + \frac{q}{p}\right)$$

If, then, we put $x = \dfrac{p}{q}$, we have

$$\log x = \log p - \log q$$

$$\log(p + q) = \log q + \log(1 + x)$$

or

$$\log(p + q) = \log p + \log\left(1 + \frac{1}{x}\right)$$

Downes's Table XXVIII., with the argument log x, the difference of the given logarithms, gives log $(1+x)$, which being added to log q, the *less* logarithm, gives the required log. sum, or log $(p+q)$. Table XXIX., with the argument log x, gives $\log\left(1+\frac{1}{x}\right)$ which, being added to log p, the *greater* logarithm, gives the required log. sum. Either table may, in general, be employed, but one or the other may be found more convenient in a particular application, and therefore both are given.

Again, we have

$$p-q=p\left(1-\frac{q}{p}\right)$$

so that, putting, as before, $x=\frac{p}{q}$, we have

$$\log x=\log p-\log q$$

$$\log(p-q)=\log p+\log\left(1-\frac{1}{x}\right)$$

Downes's Table XXX., with the argument, log x, gives $\log\left(1-\frac{1}{x}\right)$ which, being added to the *greater* logarithm, gives the required log. difference, or log $(p-q)$.

With these tables, then, we may readily compute any of the preceding formulæ which contain two terms in the second member, without the aid of auxiliary angles.

Examples.

1. Given $b=120°\ 30'\ 30''$, $c=70°\ 20'\ 20''$, $A=50°\ 10'\ 10''$; find a. (Same as Ex. 1. p. 182).

The formula is

$$\cos a=\cos b\cos c+\sin b\sin c\cos A$$

which will be thus computed:

$$\begin{array}{rl} \log\cos b & -\ 9{\cdot}70557 \\ \log\cos c & +\ 9{\cdot}52693 \\ \hline \log q & -\ 9{\cdot}23250 \\ \hline \log\sin b & +\ 9{\cdot}93529 \\ \log\sin c & +\ 9{\cdot}97391 \\ \log\cos A & +\ 9{\cdot}80654 \\ \hline \log p & +\ 9{\cdot}71574 \\ \hline \log p-\log q=\log x & -\ 0{\cdot}48324 \end{array}$$

The terms p and q have opposite signs, and although, by the formula, they are to be added (algebraically), an arithmetical difference is required. By marking the signs of all the quantities, as above, we shall always know whether a sum or difference is required by the sign before log x. In this case this sign being negative, we are to find a difference, and therefore, by Table XXX., we take

$$\begin{array}{rl} \log\left(1-\frac{1}{x}\right) & \quad 9{\cdot}82694 \\ \log p & +\ 9{\cdot}71574 \\ \hline \log\cos a & +\ 9{\cdot}54268 \\ a & =69°\ 34'\ 52'' \end{array}$$

2. With the same data, find B. The formula is

$$\cot B = \frac{\sin c \cot b - \cos c \cos A}{\sin A}$$

which must here be put under the form

$$\cot B = \frac{\sin c \cot b}{\sin A} - \cos c \cot A$$

and is thus computed:

log sin c	+ 9·97391
log cot b	— 9·77029
ar co log sin A	+ 0·11467
log p	— 9·85887
log (— cos c)	— 9·52693
log cot A	+ 9·92121
log q	— 9·44814
log p — log q = log x	+ 0·41073

where the sign before log x being positive, the tables for log. sum must be used. By Table XXVIII., we have

	log $(1 + x)$	0·55325
(which is to be added to the less log.)	log q	— 9·44814
	log cot B	— 0·00139

or, by Table XXIX., we have

	$\log\left(1 + \frac{1}{x}\right)$	0·14252
(which is to be added to the greater log.)	log p	— 9·85887
	log cot B	— 0·00139
	B =	135° 5′ 31″

In these isolated examples, the labor of computation is very little less than with the use of an auxiliary angle, as on p. 182; but the Gaussian Table has greatly the advantage when the same formula is to be repeatedly computed with successive values of one of the data while the others remain constant. Thus, in the first of the preceding examples, if successive values of a are to be found corresponding to successive values of A, while b and c are constant, log q will be constant, and log x will take successive values, corresponding to those of log cos A, so that after the first value of a is found the succeeding ones are rapidly obtained. On the other hand, as the auxiliary ϕ in the formulæ (122), depends upon A, the whole process would have to be repeated in finding each value of a.

For other forms of the Gaussian Table, see the original table, (to five places of decimals), by Gauss, published in Zach's *Monatliche Correspondenz*, Nov. 1812; Mattthiessen's, (to seven places), Altona, 1817; in Vega's *Sammlung mathematischer Tafeln*, (five places), Leipzig, 1840; Zech, (seven places), Leipzig, 1849; Shortrede's Collection of Tables, (seven places), Edinburgh, 1849; Gray's *Tables for the Computation of Life Contingencies*, (six places), London, 1849; Schumacher's *Hülfstafeln*, new ed. (four places).

Downes's Tables, the printing of which had commenced when the first edition of this work was published, have not yet appeared; but the above explanation will enable the reader to understand the principle of any Gaussian table.

CHAPTER IV.

SOLUTION OF THE GENERAL SPHERICAL TRIANGLE.

109. We have thus far, following the usual course, considered those spherical triangles only whose sides and angles are less than 180°. In the applications of this subject in astronomy, however, it is often necessary to consider triangles whose sides or angles exceed 180°. (For example, the right ascension of a heavenly body, admitting of all values from 0° to 360°, may be one part of such a triangle). We may, it is true, in such cases, always substitute another triangle whose parts are the supplements to 180° or 360° of those of the proposed triangle; but this mode, although very generally regarded as the simplest, is not really so in the cases alluded to. The construction of figures for discovering the supplemental triangles is often embarrassing and liable to mistake, while the solutions, when obtained, are mostly deficient in generality, and can only be regarded as solutions of the particular cases of a general problem. But if we proceed by a method that is as applicable when the parts of the triangle exceed as when they are less than 180°, we may investigate a problem under the simplest supposition of the values of these parts, and rely upon the generality of the method to cover all the particular cases.

110. We shall first endeavor, in an elementary manner, to give the student a conception of the nature of the general spherical triangle.

Fig. 14.

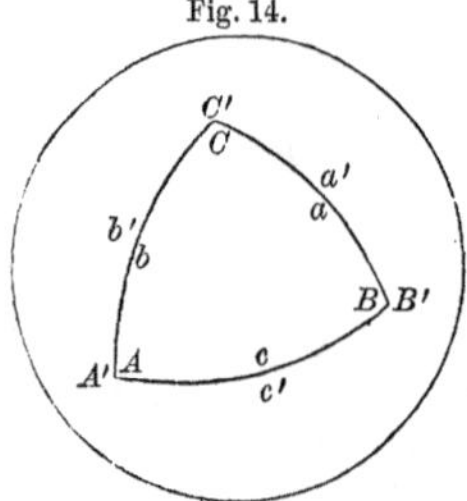

Let ABC, Fig. 14, be any spherical triangle whose parts are all less than 180°; then the remainder or complement of the sphere is also a spherical triangle whose sides are a, b and c, and whose angles are $360° - A$, $360° - B$, and $360° - C$. We shall distinguish these triangles from each other by means of accents, writing the letters *within* the triangle to which they respectively belong, as in Fig. 14. The sides are common, but when referred to as sides of $A'B'C'$, they will be denoted by a', b' and c'.

Again, one of the sides may exceed 180°, as the side a of the triangle ABC, Fig. 15. In this triangle, it is evident that we must have $A > 180°$, so long as B, C, b and c are each $< 180°$. In the triangle $A'B'C'$ we have $A' < 180°$, while $B' > 180°$, $C' > 180°$.

Fig. 15.

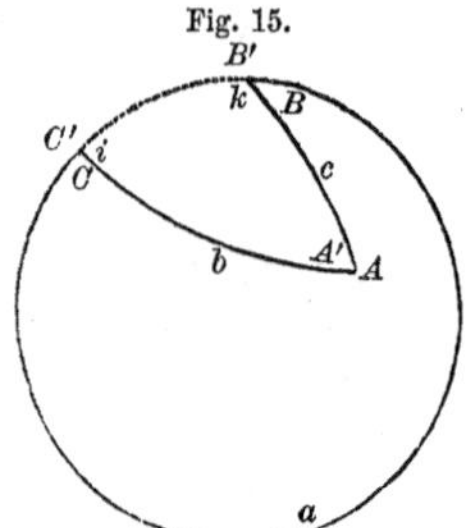

Fig. 16

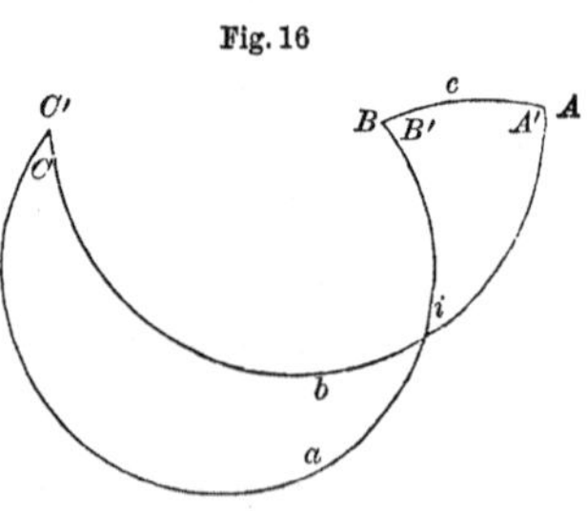

If we next suppose two of the sides to exceed 180°, as a and b, Fig 16, these sides intersecting in two points whose distance is 180°, the figure ceases to present the triangle as an enclosed surface, but it will presently appear that such triangles are solved by the same general methods that apply in other cases. To form a just conception of the triangle in this case, we may conceive Fig. 16 to be obtained from Fig. 15 by carrying the point A along the arc CA produced until it crosses the side a; the points A and B may then be joined either by an arc less than 180°, as in Fig. 16, or by its supplement to 360°, as in Fig. 17, in which last case every side exceeds 180°. In these figures, to avoid confusion, the point A is not placed in its true position according to perspective.

Fig. 17.

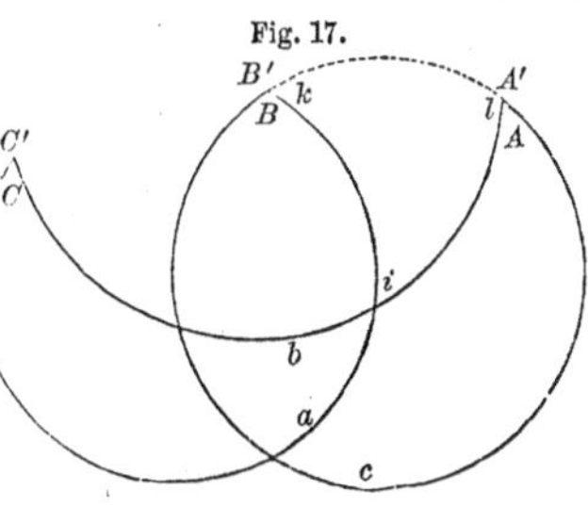

In each figure we have two triangles, whose sides are common, and whose angles are supplements to 360°. It will be easy to trace the two triangles signified by ABC and $A'B'C'$, by remarking that the letters in each case are all on the *same side* of the perimeter of the triangle.

We may go farther, and suppose the arc joining A and B to be a circumference + the arc AB, or any number of circumferences + AB; and similarly the angles may be supposed to be altogether unlimited; but since the relative positions of any three points of the sphere must be fully determined by arcs and angles less than 360°, nothing is gained by passing beyond this limit.

111. *All the formulæ of Chapter* I. *are applicable to the general spherical triangle.*

This proposition might be considered as established by the principle of Pl. Trig. Art. 49, but it is also very easily established by a continuation of the process of Spher. Trig. Art. 6, where the fundamental equation was shown to apply to all triangles whose parts are less than 180°.

It was proved in Art. 29, that all the equations of Chap. I. may be deduced from the fundamental one,

$$\cos a = \cos b \cos c + \sin b \sin c \cos A \qquad (\text{M})$$

We have then only to prove the generality of this single equation.

1st. Let all the sides be <180, but $A' > 180°$, Fig. 14. The formula being true for the triangle ABC, we have

$$\cos a = \cos b \cos c + \sin b \sin c \cos (360° - A')$$

or in the triangle $A'B'C'$, by Pl. Trig. (76),

$$\cos a' = \cos b' \cos c' + \sin b' \sin c' \cos A'$$

2d. Let $a > 180°$, Fig. 15, and produce a to complete the great circle. The triangles ABC and $A'B'C'$ are respectively the difference and sum of a hemisphere and the triangle $A'ik$, all of whose parts are $< 180°$. In the triangle $A'ik$ we have, in terms of the parts of ABC,

$$\cos (360° - a) = \cos b \cos c + \sin b \sin c \cos (360° - A)$$

and in terms of the parts of $A'B'C'$,

$$\cos (360° - a') = \cos b' \cos c' + \sin b' \sin c' \cos A'$$

both of which reduce to the form (M). But it is here necessary to show that the formula may also be applied to each of the other angles: thus the triangle $A'ik$ gives

$$\cos b = \cos (360° - a) \cos c + \sin (360° - a) \sin c \cos (180° - B)$$
$$\cos b' = \cos (360° - a') \cos c' + \sin (360° - a') \sin c' \cos (B' - 180°)$$

both of which reduce to the form (M).

3d. Let $a > 180°$, $b > 180°$, Fig. 18; these arcs intersect at i, and the triangle $A' B' i$ gives

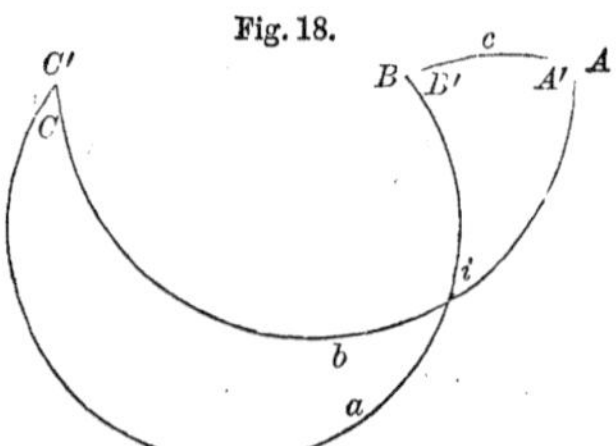

Fig. 18.

$$\cos (a - 180°) = \cos (b - 180°) \cos c + \sin (b - 180°) \sin c \cos (360° - A)$$

$$\cos (a' - 180°) = \cos (b' - 180°) \cos c' + \sin (b' - 180°) \sin c' \cos A'$$

which reduce to the form (M); and in the same way the formula applies to the angle B. We have also

$$\cos c = \cos (a - 180°) \cos (b - 180°) + \sin (a - 180°) \sin (b - 180°) \cos i$$

and since $\cos i = \cos C = \cos (360° - C') = \cos C'$, this also reduces to the form (M) for both $A B C$ and $A' B' C'$.

4th. Let $a > 180°$, $b > 180°$, $c > 180°$, Fig. 19; the side c being produced to complete the circle, the triangle $i k l$ gives

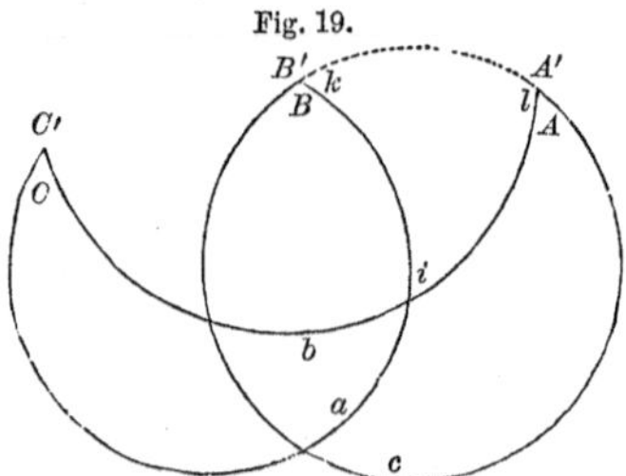

Fig. 19.

$$\cos (a - 180°) = \cos (b - 180°) \cos (360° - c) + \sin (b - 180°) \sin (360° - c) \cos l$$

and since $\cos l = \cos (180° - A) = -\cos A = \cos (A' - 180°) = -\cos A'$, this reduces to the form (M) for both $A B C$ and $A' B' C'$; and in the same way the formula applies to the angle B. We have also

$$\cos (360° - c) = \cos (a - 180°) \cos (b - 180°) + \sin (a - 180°) \sin (b - 180°) \cos i,$$

and since $\cos i = \cos C = \cos C'$, this reduces to the form (M) for both $A B C$ and $A' B' C'$.

The cases in which the angles or sides exceed 360° are included in the preceding, in consequence of Pl. Trig. Art. 45.

112. The preceding demonstration, though tedious, has the advantage of giving a definite conception of the figures which our formulæ represent. But perhaps the most satisfactory (as it is the most elegant) method, is to rest the demonstration of our fundamental equations themselves upon the principles of analytical geometry, and, for the sake of those who are acquainted with that subject, we add the following investigation:

Any point of the sphere may be referred by rectangular co-ordinates to three planes passing through the centre of the sphere at right angles to each other. Let O be the centre of the sphere, Fig. 20, and $A B C$ a spherical triangle upon its surface. Let one of the co-ordinate planes, as $X Y$, coincide with the great circle $A B$, and let the axis of X pass through B. If $C P$ be drawn perpendicular to the plane XY, and CP' and PP' to the axis OX, the co-ordinates of the point C are

$$x = OP', \qquad y = PP', \qquad z = CP$$

Fig. 20.

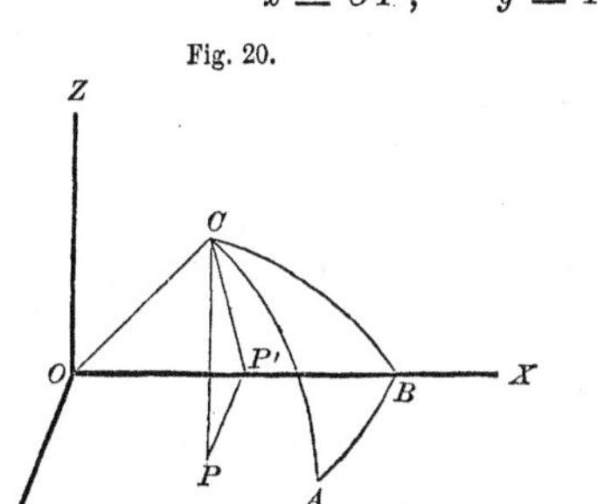

Fig. 21.

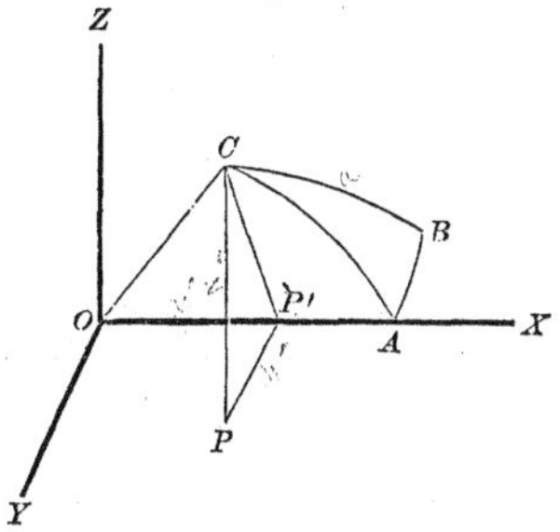

the values of which (OC being taken $= 1$) are

$$\begin{aligned} x &= \cos a \\ y &= \sin a \cos B \\ z &= \sin a \sin B \end{aligned}$$

If now the axis of X be made to pass through A, Fig. 21, without changing the position of the plane XY we shall have for x', y', z', the co-ordinates of C referred to the new axes,

$$\begin{aligned} x' &= \cos b \\ y' &= -\sin b \cos A \\ z' &= \sin b \sin A \end{aligned}$$

The axis of z being unchanged, the relations between x', y', and x, y, are expressed simply by the formulæ for the transformation of co-ordinates in a plane; the inclination of the new axes to the first is here expressed by c, and the formulæ of transformation are therefore

$$\begin{aligned} x &= x' \cos c - y' \sin c \\ y &= x' \sin c + y' \cos c \\ z &= z' \end{aligned}$$

substituting the values of the co-ordinates, we have at once the three following fundamental equations:

$$\left.\begin{aligned} \cos a &= \cos c \cos b + \sin c \sin b \cos A \\ \sin a \cos B &= \sin c \cos b - \cos c \sin b \cos A \\ \sin a \sin B &= \sin b \sin A \end{aligned}\right\} \quad (\text{N})$$

which are identical with (4), (6), and (3).

113. Having established the complete generality of our fundamental equations, we may now employ for the solution of the general triangle any of those deduced from them in Chap. I.

As a single trigonometric function is not sufficient to determine an unlimited angle or arc, (Pl. Trig. Art. 53), it becomes necessary in most cases to deduce expressions for both the sine and cosine of the required part.

It will be found that *all the six cases of the general triangle admit of two solutions, but that they all become determinate, when, in addition to the other data, the sign of the sine or cosine of one of the required parts is given.* In the practical applications in astronomy, it mostly happens that the conditions of the problem supply this sign.

114. CASE I. Given b, c and A. *First Solution;* when one of the remaining angles, as B, and the third side a are required. The relations between the given and required parts are

$$\left.\begin{aligned}\cos a &= \cos c \cos b + \sin c \sin b \cos A \\ \sin a \cos B &= \sin c \cos b - \cos c \sin b \cos A \\ \sin a \sin B &= \sin b \sin A\end{aligned}\right\} \quad (196)$$

The signs of the second members will be known from their computed numerical values; the sign of cos a is therefore known. If the sign of sin a is also given, the quadrant in which a must be taken will be known; the second and third equations will determine the sign of the sine and cosine of B, and therefore the quadrant in which B is to be taken.

In like manner, if the sign of either cos B or sin B is given, that of sin a becomes known, and the problem is determinate. If no conditions are atttached to the required parts, there must be two solutions.

The numerical solution will be conducted as follows: The values of the second members (or simply their logarithms) are to be separately computed, and their signs carefully noted; then the quotient of the 3d by the 2d (or the difference of their logs.) will give tan B, and hence B, which will be taken in the quadrant indicated by the signs of the sine and cosine. Then the 3d divided by sin B, or the 2d by cos B, will give sin a, which, agreeing with the value from the 1st equation, will serve to verify the correctness of the whole process.

This solution may be adapted for logarithms by the methods employed in the preceding chapter.

1st. Let k and ϕ be determined by the equations

$$\begin{aligned}k \sin \phi &= \sin b \cos A \\ k \cos \phi &= \cos b\end{aligned}$$

k being a positive number (Pl. Trig. Art. 174); then

$$\begin{aligned}\cos a &= k \cos (c - \phi) \\ \sin a \cos B &= k \sin (c - \phi) \\ \sin a \sin B &= \sin b \sin A\end{aligned} \quad (197)$$

2d. Eliminating k, and taking $\phi < 180°$, (Pl. Trig. Art. 174),

$$\left.\begin{aligned}\tan \phi &= \tan b \cos A \qquad (\phi < 180°) \\ \cos a &= \frac{\cos b}{\cos \phi} \cos (c - \phi) \\ \sin a \cos B &= \frac{\cos b}{\cos \phi} \sin (c - \phi) \\ \sin a \sin B &= \sin b \sin A\end{aligned}\right\} \quad (198)$$

3d. *If the quadrant in which* a *is to be taken is given,* we may give the preceding equations the following form:

$$\left.\begin{aligned}\tan \phi &= \tan b \cos A \qquad (\phi < 180°) \\ \tan a \cos B &= \tan (c - \phi) \\ \tan a \sin B &= \frac{\sin \phi \tan A}{\cos (c - \phi)}\end{aligned}\right\} \quad (199)$$

4th. If both a and b are less than 180°, as not unfrequently happens in the applications of this problem, let

$$m = \frac{k}{\sin b} \qquad\qquad n = \frac{\sin a}{k}$$

then m and n are both positive numbers (k being positive) and (197) gives

$$\left.\begin{aligned} m \sin \phi &= \cos A \\ m \cos \phi &= \cot b \\ n \sin B &= \sin \phi \tan A \\ n \cos B &= \sin (c - \phi) \\ \cot a &= \cot (c - \phi) \cos B \end{aligned}\right\} \qquad (200)$$

Check. We find

$$\left.\begin{aligned} \frac{\sin (c - \phi)}{\sin \phi} &= \frac{\sin a \cos B}{\sin b \cos A} = \frac{\tan A}{\tan B} \\ \frac{\cos (c - \phi)}{\cos \phi} &= \frac{\cos a}{\cos b} \end{aligned}\right\} \qquad (201)$$

besides which we may employ, in connection with (200), the equation $\sin a \sin B = \sin b \sin A$; or in connection with (197) or (198) the equation $\tan a \cos B = \tan (c - \phi)$. Or when (197) and (198) are employed, we may find a both by its sine and its cosine.

115. The angle C may be found in the same manner as B, interchanging B and C, b and c, in the preceding formulæ. But when B and C are both required, the Second Solution to be given presently is preferable.

Example.

Given $A = 261° 16'$ $b = 45° 54'$, $c = 138° 32'$, and $a < 180°$; to find a and B.

We shall first employ (197). The first column of the following computation, containing the symbols expressing the operations to be performed, should be prepared before opening the tables:

A	261° 16′
b	45° 54′
c	138° 32′
$\log \sin A$	— 9·9949352
$\log \cos A$	— 9·1813744
$\log \sin b$	+ 9·8562008
$\log \cos b = \log k \cos \phi$	+ 9·8425548
$\log \sin b \cos A = \log k \sin \phi$	— 9·0375752
$\log \tan \phi$	— 9·1950204
$\log \cos \phi$	+ 9·9947336
$\log k$	+ 9·8478212
ϕ	351° 5′ 42″·6
* $c - \phi$	147° 26′ 17″·4

* As $\phi > c$, we take $c = 138° 32' + 360°$, so that $c - \phi$ may be a positive angle; but it would be equally convenient to take $c - \phi = - 212° 33' 42''·6$.

$\log \sin (c - \phi)$	$+\ 9{\cdot}7309514$
$\log \cos (c - \phi)$	$-\ 9{\cdot}9257303$
$\log k \cos (c - \phi) = \log \cos a$	$-\ 9{\cdot}7735515$
a	$126^\circ\ 25'\ 6''{\cdot}6$
(1) $\log \sin b \sin A = \log \sin a \sin B$	$-\ 9{\cdot}8511360$
$\log k \sin (c - \phi) = \log \sin a \cos B$	$+\ 9{\cdot}5787726$
$\log \tan B$	$-\ 0{\cdot}2723634$
B	$298^\circ\ 6'\ 26''{\cdot}8$
$\log \sin a$	$+\ 9{\cdot}9056351$
$\log \sin B$	$-\ 9{\cdot}9455009$
(1) *Check.* $\log \sin a \sin B$	$-\ 9{\cdot}8511360$

If a were not limited, we should have two solutions, the second being $a = 233^\circ\ 34'\ 53''{\cdot}4$, $B = 118^\circ\ 6'\ 26''{\cdot}8$.

We shall next give the computation by (200), which is applicable to this example, since both a and b are less than 180°.

A	$261^\circ\ 16'$
b	$45^\circ\ 54'$
c	$138^\circ\ 32'$
$\log \cos A = \log m \sin \phi$	$-\ 9{\cdot}1813744$
$\log \cot b = \log m \cos \phi$	$+\ 9{\cdot}9863540$
$\log \tan \phi$	$-\ 9{\cdot}1950204$
ϕ	$351^\circ\ 5'\ 42''{\cdot}6$
$c - \phi$	$147^\circ\ 26'\ 17''{\cdot}4$
$\log \tan A$	$+\ 0{\cdot}8135608$
$\log \sin \phi$	$-\ 9{\cdot}1897534$
$\log \tan A \sin \phi = \log n \sin B$	$-\ 0{\cdot}0033142$
$\log \sin (c - \phi) = \log n \cos B$	$+\ 9{\cdot}7309514$
$\log \tan B$	$-\ 0{\cdot}2723628$
B	$298^\circ\ 6'\ 26''{\cdot}9$
$\log \cos B$	$+\ 9{\cdot}6731379$
$\log \cot (c - \phi)$	$-\ 0{\cdot}1947789$
$\log \cos B \cot (c - \phi) = \log \cot a$	$-\ 9{\cdot}8679168$
a	$126^\circ\ 25'\ 6''{\cdot}7$

116. Case I. Given b, c and A. *Second Solution;* when the two angles B and C, or when all the remaining parts are required. We have, by Gauss's Equations (44),

$$\left.\begin{aligned}
\cos \tfrac{1}{2} a \sin \tfrac{1}{2} (B + C) &= \cos \tfrac{1}{2} (b - c) \cos \tfrac{1}{2} A \\
\cos \tfrac{1}{2} a \cos \tfrac{1}{2} (B + C) &= \cos \tfrac{1}{2} (b + c) \sin \tfrac{1}{2} A \\
\sin \tfrac{1}{2} a \sin \tfrac{1}{2} (B - C) &= \sin \tfrac{1}{2} (b - c) \cos \tfrac{1}{2} A \\
\sin \tfrac{1}{2} a \cos \tfrac{1}{2} (B - C) &= \sin \tfrac{1}{2} (b + c) \sin \tfrac{1}{2} A
\end{aligned}\right\} \quad (202)$$

From the first two we deduce $\frac{1}{2} (B + C)$ and $\cos \frac{1}{2} a$, and from the second two $\frac{1}{2} (B - C)$ and $\sin \frac{1}{2} a$, whence B, C and a. The problem becomes determinate, as before; that is, when a is limited by one of the conditions

$$a < 180^\circ, \quad \text{or } a > 180^\circ$$

for then the signs of both $\cos \frac{1}{2} a$ and $\sin \frac{1}{2} a$ will be known.*

---

* By Art. 27, Gauss's Equations may also be taken with the negative sign when the triangle is unlimited, as in the group (45), but the same final results are obtained from either (44) or (45). See note at the end of this chapter, p. 227.

EXAMPLE.

Same as in Art. 115. $a < 180°$.

A	261° 16′
b	45° 54′
c	138° 32′
$\frac{1}{2}(b-c)$	46° 19′
$\frac{1}{2}(b+c)$	92° 13′
$\frac{1}{2}A$	130° 38′

$\log d = \log\cos\frac{1}{2}(b-c)$	$+$ 9·8392719
$\log e = \log\sin\frac{1}{2}(b-c)$	$-$ 9·8592393
$\log f = \log\cos\frac{1}{2}(b+c)$	$-$ 8·5874694
$\log g = \log\sin\frac{1}{2}(b+c)$	$+$ 9·9996749
$\log\cos\frac{1}{2}A$	$-$ 9·8137250
$\log\sin\frac{1}{2}A$	$+$ 9·8801803

$\log d\cos\frac{1}{2}A = \log\cos\frac{1}{2}a\sin\frac{1}{2}(B+C)$	$-$ 9·6529969
$\log f\sin\frac{1}{2}A = \log\cos\frac{1}{2}a\cos\frac{1}{2}(B+C)$	$-$ 8·4676497
$\log\tan\frac{1}{2}(B+C)$	$+$ 1·1853472
$\frac{1}{2}(B+C)$	266° 15′ 58″·0
$\log\sin\frac{1}{2}(B+C)$	$-$ 9·9990771
$\log\cos\frac{1}{2}a$	$+$ 9·6539198
$\frac{1}{2}a$	63° 12′ 33″·3
$\log e\cos\frac{1}{2}A = \log\sin\frac{1}{2}a\sin\frac{1}{2}(B-C)$	$+$ 9·6729643
$\log g\sin\frac{1}{2}A = \log\sin\frac{1}{2}a\cos\frac{1}{2}(B-C)$	$+$ 9·8798552
$\log\tan\frac{1}{2}(B-C)$	$+$ 9·7931091
$\frac{1}{2}(B-C)$	31° 50′ 28″·7

Verification.

$\log\sin\frac{1}{2}(B-C)$	$+$ 9·7222788
$\log\sin\frac{1}{2}a$	$+$ 9·9506855
$\frac{1}{2}a$	63° 12′ 33″·3

Ans. $\begin{cases} B = 298°\ 6'\ 26''\cdot7 \\ C = 234°\ 25'\ 29''\cdot3 \\ a = 126°\ 25'\ 6''\cdot6 \end{cases}$

117. If a only is required, we may find it by one of the methods of Arts. 75 and 76; and if the sign of $\sin a$ is given, the solution is determinate. If the sign of $\sin B$ or of $\sin C$ is given, we find that of $\sin a$ by inspecting the equation

$$\sin a = \frac{\sin A \sin b}{\sin B} = \frac{\sin A \sin c}{\sin C}$$

118. CASE II. Given A, C and b. *First Solution;* when the third angle B, and one of the remaining sides (as a) are required.

The general relations between the given and required parts are

$$\left.\begin{aligned} \cos B &= -\cos C\cos A + \sin C\sin A\cos b \\ \sin B\cos a &= \sin C\cos A + \cos C\sin A\cos b \\ \sin B\sin a &= \sin A\sin b \end{aligned}\right\} \quad (203)$$

which are solved in the same manner as (196). The problem is determinate when the sign of either $\sin B$, $\cos a$, or $\sin a$ is given.

Adapting (203) for logarithms, we find

1st.

$$\left.\begin{aligned} {}^{*}k \sin \vartheta &= \cos A \qquad (k \text{ positive}) \\ k \cos \vartheta &= \sin A \cos b \\ \cos B &= k \sin (C - \vartheta) \\ \sin B \cos a &= k \cos (C - \vartheta) \\ \sin B \sin a &= \sin A \sin b \end{aligned}\right\} \quad (204)$$

2d.

$$\left.\begin{aligned} \cot \vartheta &= \tan A \cos b \qquad (\vartheta < 180^\circ) \\ \cos B &= \frac{\cos A}{\sin \vartheta} \sin (C - \vartheta) \\ \sin B \cos a &= \frac{\cos A}{\sin \vartheta} \cos (C - \vartheta) \\ \sin B \sin a &= \sin A \sin b \end{aligned}\right\} \quad (205)$$

3d. *When the quadrant in which B is to be taken is given:*

$$\left.\begin{aligned} \cot \vartheta &= \tan A \cos b \qquad (\vartheta < 180^\circ) \\ \tan B \cos a &= \cot (C - \vartheta) \\ \tan B \sin a &= \frac{\tan b \cos \vartheta}{\sin (C - \vartheta)} \end{aligned}\right\} \quad (206)$$

4th. *When A and B are both less than* 180°, let

$$p = \frac{k}{\sin A} \qquad\qquad q = \frac{\sin B}{k}$$

then p and q are positive numbers, and we have from (204),

$$\left.\begin{aligned} p \sin \vartheta &= \cot A \\ p \cos \vartheta &= \cos b \\ q \sin a &= \tan b \cos \vartheta \\ q \cos a &= \cos (C - \vartheta) \\ \cot B &= \tan (C - \vartheta) \cos a \end{aligned}\right\} \quad (207)$$

Checks. We have

$$\left.\begin{aligned} \frac{\cos (C - \vartheta)}{\cos \vartheta} &= \frac{\sin B \cos a}{\sin A \cos b} = \frac{\tan b}{\tan a} \\ \frac{\sin (C - \vartheta)}{\sin \vartheta} &= \frac{\cos B}{\cos A} \end{aligned}\right\} \quad (208)$$

---

* The same factor k is used here and in (197), although the auxiliaries ϕ and ϑ are different. To show that k has the same value in (197) and (204), let the squares of the equations

$$k \sin \phi = \sin b \cos A \qquad\qquad k \cos \phi = \cos b$$

be added; we find

$$k^2 (\sin^2 \phi + \cos^2 \phi) = k^2 = \cos^2 b + \sin^2 b \cos^2 A = 1 - \sin^2 b \sin^2 A$$

and in the same manner, from the equations

$$k \sin \vartheta = \cos A \qquad\qquad k \cos \vartheta = \sin A \cos b$$

we find

$$k^2 = 1 - \sin^2 b \sin^2 A$$

and therefore, in both cases, $k = \sqrt{(1 - \sin^2 b \sin^2 A)}$

besides which we may employ, with (207), the equation $\sin B \sin a = \sin A \sin b$; or with (204) and (205), the equation $\tan B \cos a = \cot (C - S)$. Also, when (204) or (205) is employed, we may find B both by its sine and its cosine.

These formulæ are computed in the same manner as those of preceding case.

119. CASE II. Given A, C and b. *Second Solution;* when the two sides, a and c, or all the remaining parts, are required. We employ Gauss's Equations in the following form:

$$\left.\begin{aligned} \sin \tfrac{1}{2} B \sin \tfrac{1}{2} (a + c) &= \cos \tfrac{1}{2} (A - C) \sin \tfrac{1}{2} b \\ \sin \tfrac{1}{2} B \cos \tfrac{1}{2} (a + c) &= \cos \tfrac{1}{2} (A + C) \cos \tfrac{1}{2} b \\ \cos \tfrac{1}{2} B \sin \tfrac{1}{2} (a - c) &= \sin \tfrac{1}{2} (A - C) \sin \tfrac{1}{2} b \\ \cos \tfrac{1}{2} B \cos \tfrac{1}{2} (a - c) &= \sin \tfrac{1}{2} (A + C) \cos \tfrac{1}{2} b \end{aligned}\right\} \quad (209)$$

which are solved in the same manner as (202).

EXAMPLE.

Given $A = 121° 36' 19''\cdot 8$, $C = 42° 15' 13''\cdot 7$, $b = 40° 0' 10''$, and $B > 180°$.

By (209).

b	$40° \ 0' 10''\cdot 0$
A	$121° 36' 19''\cdot 8$
C	$42° 15' 13''\cdot 7$
$\frac{1}{2}(A - C)$	$39° 40' 33''\cdot 0$
$\frac{1}{2}(A + C)$	$81° 55' 46''\cdot 7$
$\frac{1}{2} b$	$20° \ 0' \ 5''\cdot 0$
$\log d = \log \cos \frac{1}{2}(A - C)$	$+ 9\cdot 8863038$
$\log e = \log \sin \frac{1}{2}(A - C)$	$+ 9\cdot 8051224$
$\log f = \log \cos \frac{1}{2}(A + C)$	$+ 9\cdot 1473326$
$\log g = \log \sin \frac{1}{2}(A + C)$	$+ 9\cdot 9956775$
$\log \sin \frac{1}{2} b$	$+ 9\cdot 5340806$
$\log \cos \frac{1}{2} b$	$+ 9\cdot 9729820$
$\log d \sin \frac{1}{2} b = \log \sin \frac{1}{2} B \sin \frac{1}{2}(a + c)$	$+ 9\cdot 4203844$
$\log f \cos \frac{1}{2} b = \log \sin \frac{1}{2} B \cos \frac{1}{2}(a + c)$	$+ 9\cdot 1203146$
$\log \tan \frac{1}{2}(a + c)$	$+ 0\cdot 3000698$
$\frac{1}{2}(a + c)$	$63° 23' 3''\cdot 3$
$\log \sin \frac{1}{2}(a + c)$	$+ 9\cdot 9513526$
$\log \sin \frac{1}{2} B$	$+ 9\cdot 4690318$
$\frac{1}{2} B$	$162° 52' 28''\cdot 6$
*$\log (- e \sin \frac{1}{2} b) = \log (- \cos \frac{1}{2} B) \sin \frac{1}{2}(a - c)$	$- 9\cdot 3392030$
*$\log (- g \cos \frac{1}{2} b) = \log (- \cos \frac{1}{2} B) \cos \frac{1}{2}(a - c)$	$- 9\cdot 9686595$
$\log \tan \frac{1}{2}(a - c)$	$+ 9\cdot 3705435$
$\frac{1}{2}(a - c)$	$193° 12' 32''\cdot 9$

$$\dagger Ans. \begin{cases} a = 256° 35' 36''\cdot 2 \\ c = 230° 10' 30''\cdot 4 \\ B = 325° 44' 57''\cdot 2 \end{cases}$$

* The sign of each of these factors is changed because $B > 180°$, and $\cos \frac{1}{2} B$ is negative.

† It was necessary to increase $\frac{1}{2}(a + c)$ by 360°, to obtain c. The corresponding value of b would be 616° 35′ 36″·2. See note at the end of this chapter, p. 227.

120. When B only is required, we may employ the methods of Arts. 81 and 82, which are determinate when the sign of $\sin B$ is given; or when that of either $\sin a$ or $\sin c$ is given, since we may then find that of $\sin B$ by inspecting the equations

$$\sin B = \frac{\sin A \sin b}{\sin a} = \frac{\sin C \sin b}{\sin c}$$

121. CASE III. Given a, b and A. *First Solution;* when the three remaining parts B, C, and c are all required.

We find B by the equation

$$\sin B = \frac{\sin A \sin b}{\sin a} \tag{210}$$

which is determinate when the sign of $\cos B$ is given.

Then, to find C, we have

$$-\cos C \cos A + \sin C \sin A \cos b = \cos B$$
$$\sin C \cos A + \cos C \sin A \cos b = \sin B \cos a$$

which have already been employed and adapted for logarithms in Art. 118. If we denote the auxiliary by ϑ, and put $C - \vartheta = \vartheta'$, we find, from (204),

$$\left.\begin{aligned} k \sin \vartheta &= \cos A \qquad (k \text{ positive}) \\ k \cos \vartheta &= \sin A \cos b \\ k \sin \vartheta' &= \cos B \\ k \cos \vartheta' &= \sin B \cos a \\ C &= \vartheta + \vartheta' \end{aligned}\right\} \tag{211}$$

To find c, we have

$$\cos c \cos b + \sin c \sin b \cos A = \cos a$$
$$\sin c \cos b - \cos c \sin b \cos A = \sin a \cos B$$

which have already been employed and adapted for logarithms in Art. 113. If we denote the auxiliary by ϕ, and put $c - \phi = \phi'$, we find, from (197),

$$\left.\begin{aligned} k \sin \phi &= \sin b \cos A \qquad (k \text{ positive}) \\ k \cos \phi &= \cos b \\ k \sin \phi' &= \sin a \cos B \\ k \cos \phi' &= \cos a \\ c &= \phi + \phi' \end{aligned}\right\} \tag{212}$$

Checks. We have

$$\left.\begin{aligned} \frac{\sin \vartheta}{\sin \vartheta'} &= \frac{\cos A}{\cos B} \qquad & \frac{\cos \vartheta}{\cos \vartheta'} &= \frac{\tan a}{\tan b} \\ \frac{\sin \phi}{\sin \phi'} &= \frac{\tan B}{\tan A} \qquad & \frac{\cos \phi}{\cos \phi'} &= \frac{\cos b}{\cos a} \end{aligned}\right\} \tag{213}$$

One of which may be used as a check when either C or c has been alone computed.*

When both C and c have been found, the obvious check is

$$\frac{\sin C}{\sin c} = \frac{\sin A}{\sin a} \tag{214}$$

---

* The following relations deserve a passing notice:

$$\frac{\sin \phi \cos \vartheta'}{\cos \phi' \sin \vartheta} = \sin b \sin B \qquad \frac{\sin \phi' \cos \vartheta}{\cos \phi \sin \vartheta'} = \sin a \sin A$$

$$\frac{\tan \phi \tan \phi'}{\tan \vartheta \tan \vartheta'} = \sin^2 a \sin^2 B = \sin^2 b \sin^2 A$$

$$\frac{\sin 2\phi \sin 2\vartheta'}{\sin 2\vartheta \sin 2\phi'} = \frac{\sin^2 b}{\sin^2 a}$$

EXAMPLE.

Given $a = 126° 25' 6''.6$, $b = 138° 32' 0''$, $A = 261° 16' 0''$, and $\cos B$ negative.

	a	$126° 25' 6''.6$
	b	$138° 32' 0''.0$
	A	$261° 16' 0''.0$
By (210),	$\log \sin a$	$+ 9.9056351$
	$\log \sin b$	$+ 9.8209788$
	$\log \sin A$	$- 9.9949352$
	$\log \sin B$	$- 9.9102789$
	B	$234° 25' 29''.3$
By (211),	$\log \cos b$	$- 9.8746795$
	$\log \sin A \cos b = \log k \cos \vartheta$	$+ 9.8696147$
	$\log \cos A = \log k \sin \vartheta$	$- 9.1813744$
	$\log \tan \vartheta$	$- 9.3117597$
	ϑ	$348° 24' 53''.0$
	$\log \cos a$	$- 9.7735515$
	$\log \sin B \cos a = \log k \cos \vartheta'$	$+ 9.6838304$
	$\log \cos B = \log k \sin \vartheta'$	$- 9.7647520$
	$\log \tan \vartheta'$	$- 0.0809216$
	ϑ'	$309° 41' 33''.7$
	$\vartheta + \vartheta' = C$	$298° 6' 26''.7$
By (212),	$\log \sin b \cos A = \log k \sin \phi$	$- 9.0023532$
	$\log \cos b = \log k \cos \phi$	$- 9.8746795$
	$\log \tan \phi$	$+ 9.1276737$
	ϕ	$187° 38' 31''.3$
	$\log \sin a \cos B = \log k \sin \phi'$	$- 9.6703871$
	$\log \cos a = \log k \cos \phi'$	$- 9.7735515$
	$\log \tan \phi'$	$+ 9.8968356$
	ϕ'	$218° 15' 28''.6$
	$\phi + \phi' = c$	$45° 53' 59''.9$
	$\log \sin C$	$- 9.9455010$
	$\log \sin c$	$+ 9.8562006$
	$\log \left(\frac{\sin C}{\sin c}\right)$	$- 0.0893004$
	Check. $\log \left(\frac{\sin A}{\sin a}\right)$	$- 0.0893001$

In this example, both $\vartheta + \vartheta'$ and $\phi + \phi'$ exceed 360°, and consequently we have to deduct 360° from each of them. We might have avoided this, however, by taking $\vartheta' = - 50° 18' 26''.3$, $\phi' = - 141° 44' 31''.4$.

122. CASE III. Given a, b and A. *Second Solution;* when C and c are required without finding B.

We have only to eliminate B from the fourth equation of (211) by means of (210), and then (omitting the third equation) determine ϑ' by its cosine, observing, however, to take it so that $\sin \vartheta'$ shall have the sign of $\cos B$, which sign is supposed to be given. The formulæ for finding C thus become

$$\left.\begin{aligned} k \sin \vartheta &= \cos A \qquad (k \text{ positive}) \\ k \cos \vartheta &= \sin A \cos b \\ \cos \vartheta' &= \cos \vartheta \cot a \tan b \\ (\vartheta' < 180^\circ &\text{ with the sign of } \cos B) \\ C &= \vartheta + \vartheta' \end{aligned}\right\} \quad (215)$$

To find c, we observe that $\sin \phi'$ has the sign of $\sin a \cos B$, so that we have the following formulæ:

$$\left.\begin{aligned} k \sin \phi &= \sin b \cos A \qquad (k \text{ positive}) \\ k \cos \phi &= \cos b \\ \cos \phi' &= \frac{\cos \phi \cos a}{\cos b} \\ (\phi' < 180^\circ &\text{ with the sign of } \sin a \cos B) \\ c &= \phi + \phi' \end{aligned}\right\} \quad (216)$$

Check. The equation (214).

123. Case IV. Given A, B and b. *First Solution;* when the three remaining parts a, c and C are all required.

We find a by the equation

$$\sin a = \frac{\sin A \sin b}{\sin B} \quad (217)$$

which is determinate when the sign of $\cos a$ is given. The remainder of the solution is by (211) and (212).

124. Case IV. Given A, B and b. *Second Solution;* when c and C are required, without finding a.

We easily find, from (211),

$$\left.\begin{aligned} k \sin \vartheta &= \cos A \qquad (k \text{ positive}) \\ k \cos \vartheta &= \sin A \cos b \\ \sin \vartheta' &= \frac{\sin \vartheta \cos B}{\cos A} \\ (\cos \vartheta' &\text{ and } \sin B \cos a \text{ to have the same sign}) \\ C &= \vartheta + \vartheta' \end{aligned}\right\} \quad (218)$$

And from (212),

$$\left.\begin{aligned} k \sin \phi &= \sin b \cos A \qquad (k \text{ positive}) \\ k \cos \phi &= \cos b \\ \sin \phi' &= \sin \phi \tan A \cot B \\ (\cos \phi' &\text{ and } \cos a \text{ to have the same sign}) \\ c &= \phi + \phi' \end{aligned}\right\} \quad (219)$$

Check. The equation (214)

125. Case V. Given a, b and c. The formula

$$\cos A = \frac{\cos a - \cos b \cos c}{\sin b \sin c} \quad (220)$$

determines A when the sign of $\sin A$ is known. If the sign of $\sin B$ or of $\sin C$ is given, that of $\sin A$ becomes known by the equation

$$\frac{\sin A}{\sin a} = \frac{\sin B}{\sin b} = \frac{\sin C}{\sin c}$$

The formulæ (31), (33), (34), may be used, each of which will become determinate when the sign of either sin A, sin B, or sin C is known.

126. CASE VI. Given A, B and C. The formula

$$\cos a = \frac{\cos A + \cos B \cos C}{\sin B \sin C} \tag{221}$$

determines a when the sign of sin a is given. If the sign of sin b or of sin c is given, that of sin a becomes known by the equation

$$\frac{\sin a}{\sin A} = \frac{\sin b}{\sin B} = \frac{\sin c}{\sin C}$$

The formulæ (36), (38), (39), may be used, each of which will be determinate when the sign of either sin a, sin b, or sin c is known.

NOTE UPON GAUSS'S EQUATIONS.

In the unlimited spherical triangle, we may consider any part, as a, to have an infinite number of values, viz. a, $a + 360°$, $a + 720°$, &c., expressed generally by the formula $a + 2 n \pi$, n being any whole number or zero; and since

$$\sin a = \sin (a + 2 n \pi) \qquad \cos a = \cos (a + 2 n \pi)$$

all those equations of Chap. I. that involve only sin a and cos a will not be changed by the substitution of $a + 2 n \pi$ for a. A similar substitution may be made for each of the parts, or for all of them, at the same time, so that there is an infinite series of triangles to which these equations are applicable.

But the substitution of $a + 360°$ for a, in Gauss's Equations, (202), will change the sign of all of them, since

$$\sin \tfrac{1}{2} (a + 360°) = - \sin \tfrac{1}{2} a \qquad \cos \tfrac{1}{2} (a + 360°) = - \cos \tfrac{1}{2} a$$

while the substitution of $a + 720°$ for a will not change their sign, since

$$\sin \tfrac{1}{2} (a + 720°) = \sin \tfrac{1}{2} a \qquad \cos \tfrac{1}{2} (a + 720°) = \cos \tfrac{1}{2} a$$

In general, their sign is changed by the substitution of $a + (4 n + 2) \pi$ for a, and it is not changed by the substitution of $a + 4 n \pi$. The same results follow like substitutions for each of the parts. It follows that these equations taken only with the positive sign, do not include all the triangles of the infinite series above spoken of, and that they are complete only when taken with the double sign, and expressed in two distinct groups, as (44) and (45) of Art. 27.

In practice, however, we may take them with the positive sign only; for they will then give at least *one* of the triangles of the series, from which all the others, (and particularly that whose parts are less than 360°), may be directly deduced by the application of 360°.*

This will be illustrated by the example of Art. 119, p. 223; we there find

$$\tfrac{1}{2} (a + c) = 63° \, 23' \, 3''\cdot 3$$
$$\tfrac{1}{2} (a - c) = 193° \, 12' \, 32''\cdot 9$$

or rather, since $\tfrac{1}{2} (a + c)$ should be greater than $\tfrac{1}{2} (a - c)$,

$$\tfrac{1}{2} (a + c) = 423° \, 23' \, 3''\cdot 3$$
$$\tfrac{1}{2} (a - c) = 193° \, 12' \, 32''\cdot 9$$

---

* Gauss (*Theoria Motus Corp. Cœl.* Art. 54) recommends the use of the positive sign only, observing that any side or angle may be diminished or increased by 360°, as the case may require, but confines himself to the statement of this practical precept, without explaining the grounds upon which it rests.

whence

$$a = 616^\circ\ 35'\ 36''\cdot 2$$
$$c = 230^\circ\ 10'\ 30''\cdot 4$$

which is the proper solution of the equations taken with the positive sign. If now we deduct 360° from a, and take, as on p. 223,

$$a = 256^\circ\ 35'\ 36''\cdot 6$$
$$c = 230^\circ\ 10'\ 30''\cdot 4$$

we have the solution that would have been obtained by taking the negative sign in all the equations; for we now have

$$\tfrac{1}{2}(a + c) = 243^\circ\ 23'\ \ 3''\cdot 3$$
$$\tfrac{1}{2}(a - c) = 13^\circ\ 12'\ 32''\cdot 9$$

which, differing from the former values by 180°, must change the sign of all the equations.

I have given some further particulars respecting unlimited spherical triangles, and a fuller discussion of Gauss's Equations, in an essay which the reader will find in the Astronomical Journal, Vol. I., published at Cambridge, Mass.

CHAPTER V.

AREA OF A SPHERICAL TRIANGLE.

127. *Given the three angles of a spherical triangle, to compute the area.*

This problem is solved in geometry, where it is proved that the surface of a spherical triangle is measured by the excess of the sum of its three angles over two right angles, by which is meant, that *the area is as many times the area of the tri-rectangular triangle as there are right angles in the excess of the sum of the angles over two right angles.*

To express this analytically, let

r = radius of the sphere
T = surface of the tri-rectangular triangle
$= \frac{1}{8}$ surface of a sphere $= \frac{1}{2}\pi r^2$
$2S = A + B + C$
K = area of the triangle ABC.

Also, let the angles A, B and C be expressed in the unit of Art. 11, that is, let A, B, C denote the arcs which measure the angles in a circle whose radius is unity. The right angle expressed in the same unit is $\frac{\pi}{2}$, therefore the number of right angles in $2S$ is

$$2S \div \frac{\pi}{2} = \frac{4S}{\pi}$$

and we have, according to the above theorem of geometry,

$$K = T \times \left(\frac{4S}{\pi} - 2\right) = \frac{2T}{\pi}(2S - \pi)$$

or
$$K = r^2(2S - \pi) \tag{222}$$

and if the radius of the sphere is taken = 1

$$K = 2S - \pi \tag{223}$$

128. In a plane triangle the sum of the angles is equal to π, and in a spherical triangle the sum *exceeds* π by K; hence this quantity, K, is commonly called the *spherical excess.*

129. *Given the three sides, to find the area.*

By (223), we have

$$\left.\begin{aligned} \sin\tfrac{1}{2}K &= \sin\left(S-\frac{\pi}{2}\right) = -\cos S \\ \cos\tfrac{1}{2}K &= \cos\left(S-\frac{\pi}{2}\right) = \sin S \\ \tan\tfrac{1}{2}K &= -\cot S \end{aligned}\right\} \quad (224)$$

in which we have only to substitute the values of $\cos S$, $\sin S$, and $\cot S$, given in Art. 35, to obtain the required solution. We find, $[s=\frac{1}{2}(a+b+c)]$,

$$\sin\tfrac{1}{2}K = \frac{\sqrt{[\sin s \sin(s-a)\sin(s-b)\sin(s-c)]}}{2\cos\frac{1}{2}a\cos\frac{1}{2}b\cos\frac{1}{2}c} \quad (225)$$

$$\cos\tfrac{1}{2}K = \frac{\cos a+\cos b+\cos c+1}{4\cos\frac{1}{2}a\cos\frac{1}{2}b\cos\frac{1}{2}c}$$

$$= \frac{\cos^2\frac{1}{2}a+\cos^2\frac{1}{2}b+\cos^2\frac{1}{2}c-1}{2\cos\frac{1}{2}a\cos\frac{1}{2}b\cos\frac{1}{2}c} \quad (226)$$

The numerator of (225) being denoted by n, we find,

$$\cot\tfrac{1}{2}K = \frac{1+\cos a+\cos b+\cos c}{2n} \quad (227)$$

which is known as *De Gua's* formula.

Again, from the formulæ of Art. 36, since $1-\sin S = 2\sin^2\frac{1}{4}K$, $1+\sin S = 2\cos^2\frac{1}{4}K$, we find

$$\left.\begin{aligned} \sin\tfrac{1}{4}K &= \sqrt{\left[\frac{\sin\frac{1}{2}s\sin\frac{1}{2}(s-a)\sin\frac{1}{2}(s-b)\sin\frac{1}{2}(s-c)}{\cos\frac{1}{2}a\cos\frac{1}{2}b\cos\frac{1}{2}c}\right]} \\ \cos\tfrac{1}{4}K &= \sqrt{\left[\frac{\cos\frac{1}{2}s\cos\frac{1}{2}(s-a)\cos\frac{1}{2}(s-b)\cos\frac{1}{2}(s-c)}{\cos\frac{1}{2}a\cos\frac{1}{2}b\cos\frac{1}{2}c}\right]} \\ \tan\tfrac{1}{4}K &= \sqrt{[\tan\tfrac{1}{2}s\tan\tfrac{1}{2}(s-a)\tan\tfrac{1}{2}(s-b)\tan\tfrac{1}{2}(s-c)]} \end{aligned}\right\} \quad (228)$$

the last of which is known as *Lhuillier's* formula.

130. *Given two sides and the included angle,* (or a, b and C) *to find the area.*

We have, from (224), by (77),

$$\cot\tfrac{1}{2}K = \frac{\cot\frac{1}{2}a\cot\frac{1}{2}b+\cos C}{\sin C}$$

or

$$\tan\tfrac{1}{2}K = \frac{\tan\frac{1}{2}a\tan\frac{1}{2}b\sin C}{1+\tan\frac{1}{2}a\tan\frac{1}{2}b\cos C} \quad (229)$$

131. If we admit more than three parts of the triangle into the expression of K, we have, by (56) and (71),

$$\left.\begin{aligned} \sin \tfrac{1}{2} K &= \frac{\sin \frac{1}{2} a \sin \frac{1}{2} b}{\cos \frac{1}{2} c} \sin C \\ \cos \tfrac{1}{2} K &= \frac{\cos \frac{1}{2} a \cos \frac{1}{2} b + \sin \frac{1}{2} a \sin \frac{1}{2} b \cos C}{\cos \frac{1}{2} c} \end{aligned}\right\} \quad (230)$$

the quotient of which gives (229).

132. Since there are always two triangles upon the surface of the sphere which have the same three sides, (Art. 110), the angles not being limited to values less than 180°, the formulæ (225), (226), (227) should give the areas of both of them, and their sum should be equal to the surface of the sphere $= 4\pi$. In fact, by (225), $\sin \frac{1}{2} K$ may be either positive or negative, while by (226) the cosine is fully determined, so that these formulæ give two values of $\frac{1}{2} K$ whose sum is 2π, and therefore two values of K, whose sum is 4π.

It follows that (225) alone is not sufficiently determinate when the triangle is unlimited, since it gives *four* solutions. The most convenient formula is therefore (228), for we must always have $\frac{1}{4} K < \pi$, and the double sign of the radical gives the two values of $\frac{1}{4} K$, one less and the other greater than $\frac{\pi}{2}$.

CHAPTER VI.

DIFFERENCES AND DIFFERENTIALS OF SPHERICAL TRIANGLES.

133. Two parts of a spherical triangle being constant, and a third receiving an increment, it is required to deduce the corresponding increments of the remaining three parts. As in plane triangles, (Pl. Trig. Chap. XII.), this will be effected by a comparison of two triangles having two parts in common. The triangle formed from the given one by applying the increments to the variable parts will be distinguished as *the derived triangle.*

We shall first consider the increments as *finite differences*, and give them the positive sign, (Pl. Trig. Art. 187).

134. Case I. A and c constant. The parts of ABC, Fig. 22, being A, c, B, C, a, b, those of the derived triangle ABC' are A, c, $B + \Delta B$, $C + \Delta C$, $a + \Delta a$, $b + \Delta b$; and the parts of the differential triangle BCC' are a, $a + \Delta a$, Δb, $180° - C$, $C + \Delta C$ and ΔB. We have, then, in BCC', by (3),

Fig. 22.

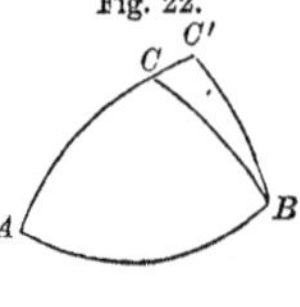

$$\frac{\sin \Delta b}{\sin \Delta B} = \frac{\sin (a + \Delta a)}{\sin C} = \frac{\sin a}{\sin (C + \Delta C)} \tag{231}$$

Also, in BCC', by (40), we have

$$\frac{\sin \frac{1}{2} (180° - C + C + \Delta C)}{\sin \frac{1}{2} (180° - C - C - \Delta C)} = \frac{\tan \frac{1}{2} \Delta b}{\tan \frac{1}{2} (a + \Delta a - a)}$$

whence

$$\frac{\tan \frac{1}{2} \Delta a}{\tan \frac{1}{2} \Delta b} = \frac{\cos (C + \frac{1}{2} \Delta C)}{\cos \frac{1}{2} \Delta C} \tag{232}$$

By (41) we find in a similar manner,

$$\frac{\tan \frac{1}{2} \Delta b}{\sin \frac{1}{2} \Delta C} = - \frac{\tan (a + \frac{1}{2} \Delta a)}{\sin (C + \frac{1}{2} \Delta C)} \tag{233}$$

By (42),

$$\frac{\sin \frac{1}{2} \Delta a}{\tan \frac{1}{2} \Delta B} = \frac{\sin (a + \frac{1}{2} \Delta a)}{\tan (C + \frac{1}{2} \Delta C)} \tag{234}$$

By (43),

$$\frac{\tan \frac{1}{2} \Delta C}{\tan \frac{1}{2} \Delta B} = -\frac{\cos (a + \frac{1}{2} \Delta a)}{\cos \frac{1}{2} \Delta a} \tag{235}$$

By combining (232) and (233),

$$\frac{\tan \frac{1}{2} \Delta a}{\tan \frac{1}{2} \Delta C} = -\frac{\tan (a + \frac{1}{2} \Delta a)}{\tan (C + \frac{1}{2} \Delta C)} \tag{236}$$

As these formulæ involve the increments in the second members, they are to be computed by successive approximations. (See Pl. Trig. Art. 201).

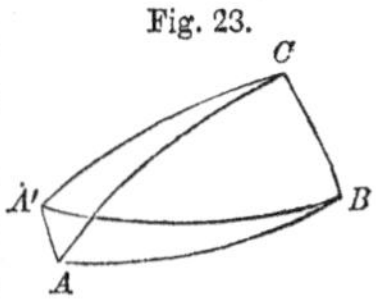

135. Case II. A and a constant. The given triangle being $A\,B\,C$, Fig. 23, the parts of the derived triangle $A'\,B\,C$ are A, a, $B + \Delta B$, $b + \Delta b$, $C + \Delta C$, $c + \Delta c$. Although the figure appears to show that the angle B is diminished, it is still proper to represent the angle $A'\,B\,C$ by $B + \Delta B$, to preserve uniformity in the algebraic signs of the increments; the essential signs being given by the equations of differences themselves. Hence we put the angle $A\,B\,A' = A\,B\,C - A'\,B\,C = B - (B + \Delta B) = -\Delta B$. Joining $A\,A'$ we have in $B\,A\,A'$ and $C\,A\,A'$, by (43),

Fig. 23.

$$\cos (c + \tfrac{1}{2} \Delta c) : \cos \tfrac{1}{2} \Delta c = -\cot \tfrac{1}{2} \Delta B : \tan \tfrac{1}{2} (A'A\,B + A\,A'B)$$

$$\cos (b + \tfrac{1}{2} \Delta b) : \cos \tfrac{1}{2} \Delta b = \cot \tfrac{1}{2} \Delta C : \tan \tfrac{1}{2} (A'A\,C + A\,A'C)$$

but since A is constant, or $B\,A\,C = B\,A'\,C$, we find that the fourth terms of these proportions are equal; whence

$$\frac{\tan \frac{1}{2} \Delta B}{\tan \frac{1}{2} \Delta C} = -\frac{\cos (b + \frac{1}{2} \Delta b) \cos \frac{1}{2} \Delta c}{\cos (c + \frac{1}{2} \Delta c) \cos \frac{1}{2} \Delta b} \tag{237}$$

In the polar triangle of $A\,B\,C$, the constants are still an angle and its opposite side, and the preceding equation applied to this polar triangle (by Art. 8) gives

$$\frac{\tan \frac{1}{2} \Delta b}{\tan \frac{1}{2} \Delta c} = -\frac{\cos (B + \frac{1}{2} \Delta B) \cos \frac{1}{2} \Delta C}{\cos (C + \frac{1}{2} \Delta C) \cos \frac{1}{2} \Delta B} \tag{238}$$

In $A\,B\,C$ and $A'B\,C$ we have

$$\sin a \sin B = \sin A \sin b$$

$$\sin a \sin (B + \Delta B) = \sin A \sin (b + \Delta b)$$

the difference and sum of which give

$$\sin a \cos (B + \tfrac{1}{2}\Delta B) \sin \tfrac{1}{2}\Delta B = \sin A \cos (b + \tfrac{1}{2}\Delta b) \sin \tfrac{1}{2}\Delta b$$
$$\sin a \sin (B + \tfrac{1}{2}\Delta B) \cos \tfrac{1}{2}\Delta B = \sin A \sin (b + \tfrac{1}{2}\Delta b) \cos \tfrac{1}{2}\Delta b$$

from which, by division, we find

$$\frac{\tan \frac{1}{2}\Delta b}{\tan \frac{1}{2}\Delta B} = \frac{\tan (b + \frac{1}{2}\Delta b)}{\tan (B + \frac{1}{2}\Delta B)} \qquad (239)$$

and in the same manner

$$\frac{\tan \frac{1}{2}\Delta c}{\tan \frac{1}{2}\Delta C} = \frac{\tan (c + \frac{1}{2}\Delta c)}{\tan (C + \frac{1}{2}\Delta C)} \qquad (240)$$

The product of (237) and (239) gives

$$\frac{\sin \frac{1}{2}\Delta b}{\tan \frac{1}{2}\Delta C} = -\frac{\sin (b + \frac{1}{2}\Delta b) \cos \frac{1}{2}\Delta c}{\cos (c + \frac{1}{2}\Delta c) \tan (B + \frac{1}{2}\Delta B)} \qquad (241)$$

whence also*

$$\frac{\sin \frac{1}{2}\Delta c}{\tan \frac{1}{2}\Delta B} = -\frac{\sin (c + \frac{1}{2}\Delta c) \cos \frac{1}{2}\Delta b}{\cos (b + \frac{1}{2}\Delta b) \tan (C + \frac{1}{2}\Delta C)} \qquad (242)$$

Fig. 24.

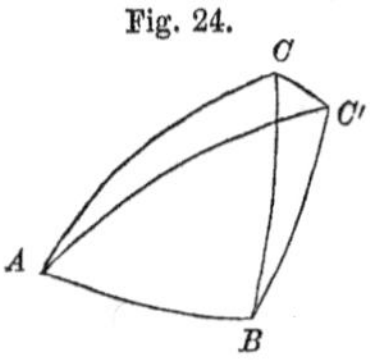

136. Case III. b and c constant. The given triangle being $A\,B\,C$, Fig. 24, the parts of the derived triangle $A\,B\,C'$ are b, c, $a + \Delta a$, $B + \Delta B$, $C + \Delta C$, $A + \Delta A$. Joining $C\,C'$ we have in $B\,C\,C'$, by (42),

$$\sin (a + \tfrac{1}{2}\Delta a) : \sin \tfrac{1}{2}\Delta a = \cot \tfrac{1}{2}\Delta B : \tan \tfrac{1}{2}(B\,C\,C' - B\,C'C)$$

But observing that $A\,C' = A\,C$, $A\,C\,C' = A\,C'C$, we have

$$B\,C\,C' = A\,C\,C' - C$$
$$B\,C'C = A\,C'C + C + \Delta C$$
$$\tfrac{1}{2}(B\,C\,C' - B\,C'C) = -(C + \tfrac{1}{2}\Delta C)$$

and the above proportion gives, therefore,

$$\frac{\sin \frac{1}{2}\Delta a}{\tan \frac{1}{2}\Delta B} = -\frac{\sin (a + \frac{1}{2}\Delta a)}{\cot (C + \frac{1}{2}\Delta C)} \qquad (243)$$

---

* The equations (239), (240), (241), and (242), contain each two factors less than the corresponding equations given by *Cagnoli.*

In the same manner we should find

$$\frac{\sin \frac{1}{2} \Delta a}{\tan \frac{1}{2} \Delta C} = -\frac{\sin (a + \frac{1}{2} \Delta a)}{\cot (B + \frac{1}{2} \Delta B)} \tag{244}$$

The quotient of (243) and (244) gives

$$\frac{\tan \frac{1}{2} \Delta B}{\tan \frac{1}{2} \Delta C} = \frac{\tan (B + \frac{1}{2} \Delta B)}{\tan (C + \frac{1}{2} \Delta C)} \tag{245}$$

In $A\,B\,C$ and $A\,B\,C'$, by (4), we have

$$\cos a = \cos b \cos c + \sin b \sin c \cos A$$

$$\cos (a + \Delta a) = \cos b \cos c + \sin b \sin c \cos (A + \Delta A)$$

the difference of which gives

$$\frac{\sin \frac{1}{2} \Delta a}{\sin \frac{1}{2} \Delta A} = \frac{\sin b \sin c \sin (A + \frac{1}{2} \Delta A)}{\sin (a + \frac{1}{2} \Delta a)} \tag{246}$$

The quotient of (243) divided by (246) gives

$$\frac{\sin \frac{1}{2} \Delta A}{\tan \frac{1}{2} \Delta B} = -\frac{\sin^2 (a + \frac{1}{2} \Delta a) \tan (C + \frac{1}{2} \Delta C)}{\sin b \sin c \sin (A + \frac{1}{2} \Delta A)} \tag{247}$$

and from (244) and (246), in the same manner,

$$\frac{\sin \frac{1}{2} \Delta A}{\tan \frac{1}{2} \Delta C} = -\frac{\sin^2 (a + \frac{1}{2} \Delta a) \tan (B + \frac{1}{2} \Delta B)}{\sin b \sin c \sin (A + \frac{1}{2} \Delta A)} \tag{248}$$

137. CASE IV. B and C constant. The equations of the preceding case (243 to 248), applied to the polar triangle, give

$$\frac{\sin \frac{1}{2} \Delta A}{\tan \frac{1}{2} \Delta b} = \frac{\sin (A + \frac{1}{2} \Delta A)}{\cot (c + \frac{1}{2} \Delta c)} \tag{249}$$

$$\frac{\sin \frac{1}{2} \Delta A}{\tan \frac{1}{2} \Delta c} = \frac{\sin (A + \frac{1}{2} \Delta A)}{\cot (b + \frac{1}{2} \Delta b)} \tag{250}$$

$$\frac{\tan \frac{1}{2} \Delta b}{\tan \frac{1}{2} \Delta c} = \frac{\tan (b + \frac{1}{2} \Delta b)}{\tan (c + \frac{1}{2} \Delta c)} \tag{251}$$

$$\frac{\sin \frac{1}{2} \Delta A}{\sin \frac{1}{2} \Delta a} = \frac{\sin B \sin C \sin (a + \frac{1}{2} \Delta a)}{\sin (A + \frac{1}{2} \Delta A)} \tag{252}$$

$$\frac{\sin \frac{1}{2} \Delta a}{\tan \frac{1}{2} \Delta b} = \frac{\sin^2 (A + \frac{1}{2} \Delta A) \tan (c + \frac{1}{2} \Delta c)}{\sin B \sin C \sin (a + \frac{1}{2} \Delta a)} \tag{253}$$

$$\frac{\sin \frac{1}{2} \Delta A}{\tan \frac{1}{2} \Delta c} = \frac{\sin^2 (A + \frac{1}{2} \Delta A) \tan (b + \frac{1}{2} \Delta b)}{\sin B \sin C \sin (a + \frac{1}{2} \Delta a)} \tag{254}$$

Finite Differences of Spherical Right Triangles.

138. All the preceding equations are, of course, applicable to right triangles, or to quadrantal triangles, and in some cases they assume simpler forms. Thus in Case I., if the variable $C = 90°$, (231) and (232) become

$$\sin \Delta b = \sin (a + \Delta a) \sin \Delta B$$

$$\tan \tfrac{1}{2} \Delta a = - \tan \tfrac{1}{2} \Delta b \tan \tfrac{1}{2} \Delta C$$

and similar modifications take place in other cases.

139. When one of the constants is 90°, the preceding equations do not generally assume any simpler forms, but they may be transformed so as to involve the *same variables in both members*, which is generally desirable in their practical applications.*

The method that we shall follow is so simple that it will be unnecessary to repeat it in every case. A single example will suffice to explain it.

Let C (= 90°) and b be the constants; to find the relation of Δc and ΔB, we have between the two variables and the constant b, the equations

$$\sin B = \sin b \operatorname{cosec} c$$

$$\sin (B + \Delta B) = \sin b \operatorname{cosec} (c + \Delta c)$$

the difference and sum of which, by Pl. Trig. (105), (106), (131), and (132), are

$$2 \cos (B + \tfrac{1}{2} \Delta B) \sin \tfrac{1}{2} \Delta B = - \frac{2 \sin b \cos (c + \frac{1}{2} \Delta c) \sin \frac{1}{2} \Delta c}{\sin c \sin (c + \Delta c)}$$

$$2 \sin (B + \tfrac{1}{2} \Delta B) \cos \tfrac{1}{2} \Delta B = \frac{2 \sin b \sin (c + \frac{1}{2} \Delta c) \cos \frac{1}{2} \Delta c}{\sin c \sin (c + \Delta c)}$$

---

* Cagnoli gives these equations reduced so as to involve the same variables in both members; but in almost every instance his formulæ involve two factors more than are necessary, and are far less simple and convenient than those here given.

and the quotient of these is

$$\frac{\tan \frac{1}{2} \Delta B}{\tan (B + \frac{1}{2} \Delta B)} = - \frac{\tan \frac{1}{2} \Delta c}{\tan (c + \frac{1}{2} \Delta c)}$$

which gives the first equation of the following article. This process always eliminates the constant, and is applicable in every case.

When the equation to be differenced involves cosines, we employ Pl. Trig. (107) and (108); if tangents, (115) and (116); if cotangents, (122); if secants, (129) and (130). The results are as follows:

140. Case I. $C = 90°$ and b constant.

$$\frac{\tan\frac{1}{2}\Delta c}{\tan\frac{1}{2}\Delta B} = -\frac{\tan(c + \frac{1}{2}\Delta c)}{\tan(B + \frac{1}{2}\Delta B)} \qquad \frac{\tan\frac{1}{2}\Delta c}{\tan\frac{1}{2}\Delta a} = -\frac{\cot(c + \frac{1}{2}\Delta c)}{\cot(a + \frac{1}{2}\Delta a)} \tag{255}$$

$$\frac{\sin \Delta a}{\sin \Delta A} = \frac{\sin(2a + \Delta a)}{\sin(2A + \Delta A)} \qquad \frac{\tan\frac{1}{2}\Delta a}{\sin \Delta B} = -\frac{\tan(a + \frac{1}{2}\Delta a)}{\sin(2B + \Delta B)} \tag{256}$$

$$\frac{\sin \Delta c}{\tan\frac{1}{2}\Delta A} = \frac{\sin(2c + \Delta c)}{\cot(A + \frac{1}{2}\Delta A)} \qquad \frac{\tan\frac{1}{2}\Delta A}{\tan\frac{1}{2}\Delta B} = -\frac{\tan(A + \frac{1}{2}\Delta A)}{\cot(B + \frac{1}{2}\Delta B)} \tag{257}$$

141. Case II. $C = 90°$ and c constant.

$$\frac{\sin \Delta A}{\sin \Delta B} = -\frac{\sin(2A + \Delta A)}{\sin(2B + \Delta B)} \qquad \frac{\tan\frac{1}{2}\Delta a}{\tan\frac{1}{2}\Delta b} = -\frac{\cot(a + \frac{1}{2}\Delta a)}{\cot(b + \frac{1}{2}\Delta b)} \tag{258}$$

$$\frac{\tan\frac{1}{2}\Delta a}{\tan\frac{1}{2}\Delta A} = \frac{\tan(a + \frac{1}{2}\Delta a)}{\tan(A + \frac{1}{2}\Delta A)} \qquad \frac{\tan\frac{1}{2}\Delta b}{\tan\frac{1}{2}\Delta B} = \frac{\tan(b + \frac{1}{2}\Delta b)}{\tan(B + \frac{1}{2}\Delta B)} \tag{259}$$

$$\frac{\sin \Delta a}{\tan\frac{1}{2}\Delta B} = -\frac{\sin(2a + \Delta a)}{\cot(B + \frac{1}{2}\Delta B)} \qquad \frac{\sin \Delta b}{\tan\frac{1}{2}\Delta A} = -\frac{\sin(2b + \Delta b)}{\cot(A + \frac{1}{2}\Delta A)} \tag{260}$$

142. Case III. $C = 90°$ and A constant.

$$\frac{\tan\frac{1}{2}\Delta c}{\tan\frac{1}{2}\Delta a} = \frac{\tan(c + \frac{1}{2}\Delta c)}{\tan(a + \frac{1}{2}\Delta a)} \qquad \frac{\sin \Delta a}{\tan\frac{1}{2}\Delta b} = \frac{\sin(2a + \Delta a)}{\tan(b + \frac{1}{2}\Delta b)} \tag{261}$$

$$\frac{\tan\frac{1}{2}\Delta c}{\sin \Delta B} = \frac{\cot(c + \frac{1}{2}\Delta c)}{\sin(2B + \Delta B)} \qquad \frac{\tan\frac{1}{2}\Delta b}{\tan\frac{1}{2}\Delta B} = \frac{\cot(b + \frac{1}{2}\Delta b)}{\cot(B + \frac{1}{2}\Delta B)} \tag{262}$$

$$\frac{\sin \Delta c}{\sin \Delta b} = \frac{\sin(2c + \Delta c)}{\sin(2b + \Delta b)} \qquad \frac{\tan\frac{1}{2}\Delta a}{\tan\frac{1}{2}\Delta B} = \frac{\cot(a + \frac{1}{2}\Delta a)}{\tan(B + \frac{1}{2}\Delta B)} \tag{263}$$

143. If a constant side is 90°, the equations of finite differences for the triangle may be obtained by applying the preceding equations to the polar triangle.

Differential Variations of Spherical Oblique Triangles.

144. To obtain the differential variations, we have only to make the increments infinitely small in the equations of finite differences, observing the principles of Pl. Trig. Art. 192. Or we may differentiate the equations of spherical triangles directly, employing the differentials of the trigonometric functions given in Pl. Trig. Art. 192. For example, A and c being constant, to find the relation of $d\,a$ and $d\,B$, we have

$$\sin A \sin c = \sin a \sin C$$

the differential of which is

$$0 = \sin a\, d \sin C + \sin C\, d \sin a$$

$$= \sin a \cos C\, d\, C + \cos a \sin C\, d\, a$$

$$\frac{d\,a}{d C} = -\frac{\tan a}{\tan C}$$

and to find the relation of $d\,a$ and $d\,b$, we have

$$\cos a = \cos b \cos c + \sin b \sin c \cos A$$

$$-\sin a\, d\, a = -\sin b \cos c\, d\, b + \cos b \sin c \cos A\, d\, b$$

$$\frac{d\,a}{d\,b} = \frac{\sin b \cos c - \cos b \sin c \cos A}{\sin a}$$

or by (7),

$$\frac{d\,a}{d\,b} = \cos C$$

results which agree with those found from (236) and (232), by making Δa, Δb and ΔC infinitely small. By either method, then, the following equations may be readily verified.

145. Case I. A and c constant.

$$\frac{d\,a}{d\,C} = -\frac{\tan a}{\tan C} \qquad \frac{d\,b}{d\,B} = \frac{\sin a}{\sin C} \tag{264}$$

$$\frac{d\,a}{d\,b} = \cos C \qquad \frac{d\,b}{d\,C} = -\frac{\tan a}{\sin C} \tag{265}$$

$$\frac{d\,a}{d\,B} = \frac{\sin a}{\tan C} \qquad \frac{d\,C}{d\,B} = -\cos a \tag{266}$$

146. CASE II. A and a constant.

$$\frac{dB}{dC} = -\frac{\cos b}{\cos c} \qquad \frac{db}{dc} = -\frac{\cos B}{\cos C} \tag{267}$$

$$\frac{db}{dB} = \frac{\tan b}{\tan B} \qquad \frac{dc}{dC} = \frac{\tan c}{\tan C} \tag{268}$$

$$\frac{db}{dC} = -\frac{\sin b}{\cos c \tan B} \qquad \frac{dc}{dB} = -\frac{\sin c}{\cos b \tan C} \tag{269}$$

147. CASE III. b and c constant.

$$\frac{da}{dB} = -\sin a \tan C \qquad \frac{da}{dC} = -\sin a \tan B \tag{270}$$

$$\frac{dB}{dC} = \frac{\tan B}{\tan C} \qquad \frac{da}{dA} = \sin b \sin C \tag{271}$$

$$\frac{dA}{dB} = -\frac{\sin A}{\sin B \cos C} \qquad \frac{dA}{dC} = -\frac{\sin A}{\sin C \cos B} \tag{272}$$

148. CASE IV. B and C constant.

$$\frac{dA}{db} = \sin A \tan c \qquad \frac{dA}{dc} = \sin A \tan b \tag{273}$$

$$\frac{db}{dc} = \frac{\tan b}{\tan c} \qquad \frac{dA}{da} = \sin B \sin c \tag{274}$$

$$\frac{da}{db} = \frac{\sin a}{\sin b \cos c} \qquad \frac{da}{dc} = \frac{\sin a}{\sin c \cos b} \tag{275}$$

DIFFERENTIAL VARIATIONS OF SPHERICAL RIGHT TRIANGLES.

The preceding may also be used for right triangles; but it may be desirable to have the same variables in both members, as in the following formulæ derived from those of Arts. 140, 141, and 142:

149. CASE I. $C = 90°$ and b constant.

$$\frac{dc}{dB} = -\frac{\tan c}{\tan B} \qquad \frac{dc}{da} = \frac{\cot c}{\cot a} \tag{276}$$

$$\frac{da}{dA} = \frac{\sin 2a}{\sin 2A} \qquad \frac{da}{dB} = -\frac{2 \tan a}{\sin 2B} \tag{277}$$

$$\frac{dc}{dA} = \frac{\sin 2c}{2 \cot A} \qquad \frac{dA}{dB} = -\frac{\tan A}{\cot B} \tag{278}$$

150. CASE II. $C = 90°$ and c constant.

$$\frac{dA}{dB} = -\frac{\sin 2A}{\sin 2B} \qquad \frac{da}{db} = -\frac{\cot a}{\cot b} \tag{279}$$

$$\frac{da}{dA} = \frac{\tan a}{\tan A} \qquad \frac{db}{dB} = \frac{\tan b}{\tan B} \tag{280}$$

$$\frac{da}{dB} = -\frac{\sin 2a}{2\cot B} \qquad \frac{db}{dA} = -\frac{\sin 2b}{2\cot A} \tag{281}$$

151. CASE III. $C = 90°$ and A constant.

$$\frac{dc}{da} = \frac{\tan c}{\tan a} \qquad \frac{da}{db} = \frac{\sin 2a}{2\tan b} \tag{282}$$

$$\frac{dc}{dB} = \frac{2\cot c}{\sin 2B} \qquad \frac{db}{dB} = \frac{\cot b}{\cot B} \tag{283}$$

$$\frac{dc}{db} = \frac{\sin 2c}{\sin 2b} \qquad \frac{da}{dB} = \frac{\cot a}{\tan B} \tag{284}$$

152. The differential variations are often employed for approximate results, instead of the equations of finite differences, when the increments are very small. The remarks of Pl. Trig. Art. 203, apply here also, but it is not necessary to introduce the radius in seconds, since all the parts of a spherical triangle are expressed in the same unit.

DIFFERENTIAL VARIATIONS OF SPHERICAL TRIANGLES WHEN ALL THE PARTS ARE VARIABLE.

153. Let the equation

$$\cos a = \cos b \cos c + \sin b \sin c \cos A$$

be differentiated, all the parts being variable; we find

$$\begin{aligned}\sin a\, da = &(\sin b \cos c - \cos b \sin c \cos A)\, db \\ &+ (\sin c \cos b - \cos c \sin b \cos A)\, dc \\ &+ \sin b \sin c \sin A\, dA\end{aligned}$$

Dividing by $\sin a$, this becomes, by (7) and (3),

$$da = \cos C\, db + \cos B\, dc + \sin b \sin C\, dA \tag{285}$$

and in the same manner from the 2d and 3d equations of (4) we find

$$db = \cos A\, dc + \cos C\, da + \sin c \sin A\, dB \tag{286}$$

$$dc = \cos B\, da + \cos A\, db + \sin a \sin B\, dC \tag{287}$$

From these three equations, any three of the six differentials da, db, dc, dA, dB, dC, being given, the other three may be determined by the usual processes of elimination.

If any one of the parts be supposed constant, its differential will become zero, and these equations will assume simpler forms. If two of the parts be supposed constant, we can easily deduce all the equations of Arts. 145, 146, 147 and 148.

CHAPTER VII.

APPROXIMATE SOLUTION OF SPHERICAL TRIANGLES IN CERTAIN CASES.

154. WHEN some of the parts of the triangle are small, or nearly 90°, or nearly 180°, approximate solutions may be employed with advantage. These are generally found by means of series.

155. *In a spherical right triangle* (the right angle being C), *given A and c, to find b.* We have

$$\tan b = \cos A \tan c \tag{288}$$

which is of the form in Pl. Trig. (493), and may therefore be developed by (495) and (496) by putting $x = b$, $y = c$, $p = \cos A$, whence

$$q = \frac{p-1}{p+1} = -\frac{1-\cos A}{1+\cos A} = -\tan^2 \tfrac{1}{2} A$$

and (495) and (496) become, [taking $n = 0$ in (495), and $n = 1$ in (496)],

$$b = \quad c - \tan^2 \tfrac{1}{2} A \sin 2c + \tfrac{1}{2} \tan^4 \tfrac{1}{2} A \sin 4c - \&c. \tag{289}$$

$$b = \pi - c + \cot^2 \tfrac{1}{2} A \sin 2c - \tfrac{1}{2} \cot^4 \tfrac{1}{2} A \sin 4c + \&c. \tag{290}$$

If A is small, $\cos A$ is nearly equal to unity, and b exceeds c by a small quantity which is approximately found by one or more terms of the series (289).

If A is nearly 180°, or $\cos A$ nearly $= -1$, b exceeds $\pi - c$ by a small quantity, which is found by (290).

For examples of the mode of computation, see Pl. Trig. Art. 255.

156. Although these solutions are termed *approximate*, it must not be inferred that they are *less accurate* in practice than the direct solution of (288) by the tables; for the logarithmic tables are themselves only approximate, and the neglect of the higher powers in such series as (289) and (290) may involve a less theoretical error than the similar neglect of the higher powers in the series by which the tables are computed. In the examples of Pl. Trig. Art. 255, the thousandths of a second were found with accuracy, which could not have been effected by a direct solution with less than eight decimal places in the logarithms.

These considerations lead to the frequent employment of approximate solutions in astronomy.

157. If A and b are given, to find c, we have

$$\tan c = \sec A \tan b$$

which is reduced to Pl. Trig. (493), by putting $x = c$, $y = b$, $p = \sec A$,

$$q = \frac{\sec A - 1}{\sec A + 1} = \frac{1-\cos A}{1+\cos A} = \tan^2 \tfrac{1}{2} A$$

and the series will be

$$c = \quad b + \tan^2 \tfrac{1}{2} A \sin 2b + \tfrac{1}{2} \tan^4 \tfrac{1}{2} A \sin 4b + \&c. \tag{291}$$

$$c = \pi - b - \cot^2 \tfrac{1}{2} A \sin 2b - \tfrac{1}{2} \cot^4 \tfrac{1}{2} A \sin 4b - \&c. \tag{292}$$

158. Similar solutions apply to the equations of right triangles,

$$\tan a = \sin b \tan A$$

$$\cot B = \cos c \tan A$$

the last being solved under the form

$$\tan (90^\circ - B) = \cos c \tan A$$

We may also compute, in the same manner, the auxiliaries ϕ and ϑ in (122) and (134), so frequently employed in the solutions of oblique triangles.

159. *In a right spherical triangle, given c and A, to find a, when A is nearly* 90°.

We have

$$\sin a = \sin A \sin c \tag{293}$$

from which we deduce

$$\tan \tfrac{1}{2} (c - a) = \tan^2 (45^\circ - \tfrac{1}{2} A) \tan \tfrac{1}{2} (c + a) \tag{294}$$

From this we may find $c - a$, which is supposed very small, by successive approximations. For a first approximation, let $a = c$ in the second member, and find thence the value of $c - a$ and of a; for a second approximation substitute in the second member the value of a just found; and so on until two successive values agree as nearly as may be desired.

EXAMPLE.

Given $A = 89^\circ$, $c = 87^\circ$; find a.

Here $45^\circ - \frac{1}{2} A = 0^\circ\ 30'$, and for the first approximation $\frac{1}{2} (c + a) = 87^\circ$.

$\log \tan \frac{1}{2} (c + a)$	1·28060
$\log \tan^2 (45^\circ - \frac{1}{2} A)$	5·88172
ar co log sin 1″	5·31443
$\log \frac{1}{2} (c - a)$	2·47675

$\frac{1}{2} (c - a) = 299''\!\cdot\!74$

$a = 87^\circ - 9'\ 59''\!\cdot\!48 = 86^\circ\ 50'\ 0''\!\cdot\!52$

	2D APPROX.	3D APPROX.	4TH APPROX.
$\frac{1}{2} (c + a)$	86° 55′ 0″	86° 55′ 8″	86° 55′ 8″·17
$\log \tan \frac{1}{2} (c + a)$	1·26868	1·26899	1·26900
$\log \dfrac{\tan^2 (45^\circ - \frac{1}{2} A)}{\sin 1''}$	1·19615	1·19615	1·19615
$\log \frac{1}{2} (c - a)$	2·46483	2·46514	2·46515
$\frac{1}{2} (c - a)$	291″·63	291″·83	291″·84
$c - a$	9′ 43″·26	9′ 43″·66	9′ 43″·68
a	86° 50′ 16″·74	86° 50′ 16″·34	86° 50′ 16″·32

The direct solution of (293) gives $a = 86^\circ\ 50'\ 16''$, but cannot give the fractions of a second without tables of more than seven figure logs. We have given this problem, however, not so much on account of its particular utility, as for the purpose of introducing the method of approximation to which it leads, and which is often employed.

The process here explained may obviously be applied to any equation of the form

$$\sin x = m \sin y$$

when m is nearly equal to unity.

160. *In a spherical oblique triangle, given two sides and the included angle, to find the other angles and side by series.*

If a, b and C are the data, to find c, we have

$$\cos c = \cos a \cos b + \sin a \sin b \cos C$$

Substituting half arcs,

$$\sin^2 \tfrac{1}{2} c = \sin^2 \tfrac{1}{2} a \cos^2 \tfrac{1}{2} b + \cos^2 \tfrac{1}{2} a \sin^2 \tfrac{1}{2} b$$
$$- 2 \sin \tfrac{1}{2} a \cos \tfrac{1}{2} b \cos \tfrac{1}{2} a \sin \tfrac{1}{2} b \cos C$$

which is of the form Pl. Trig. (507), and may be developed by (508) by substituting $\sin \frac{1}{2} c$ for c, $\sin \frac{1}{2} a \cos \frac{1}{2} b$ for a, and $\cos \frac{1}{2} a \sin \frac{1}{2} b$ for b; so that (508) becomes

$$\log \sin \tfrac{1}{2} c = \log \cos \tfrac{1}{2} a \sin \tfrac{1}{2} b - M \left[\frac{\tan \frac{1}{2} a}{\tan \frac{1}{2} b} \cos C + \left(\frac{\tan \frac{1}{2} a}{\tan \frac{1}{2} b} \right)^2 \frac{\cos 2C}{2} + \&c. \right] \quad (295)$$

To find A and B, we have,

$$\tan \tfrac{1}{2} (A + B) = \frac{\cos \frac{1}{2} (a - b)}{\cos \frac{1}{2} (a + b)} \cot \tfrac{1}{2} C$$

$$\tan \tfrac{1}{2} (A - B) = \frac{\sin \frac{1}{2} (a - b)}{\sin \frac{1}{2} (a + b)} \cot \tfrac{1}{2} C$$

whence

$$\tan [\tfrac{1}{2} \pi - \tfrac{1}{2} (A + B)] = \left(\frac{\cot \frac{1}{2} a - \tan \frac{1}{2} b}{\cot \frac{1}{2} a + \tan \frac{1}{2} b} \right) \tan \tfrac{1}{2} C$$

$$\tan [\tfrac{1}{2} \pi - \tfrac{1}{2} (A - B)] = \left(\frac{\tan \frac{1}{2} a + \tan \frac{1}{2} b}{\tan \frac{1}{2} a - \tan \frac{1}{2} b} \right) \tan \tfrac{1}{2} C$$

Comparing these equations with Pl. Trig. (493), and developing by (495), $n = 0$,

$$\tfrac{1}{2} \pi - \tfrac{1}{2} (A + B) = \tfrac{1}{2} C - \frac{\tan \frac{1}{2} b}{\cot \frac{1}{2} a} \sin C + \left(\frac{\tan \frac{1}{2} b}{\cot \frac{1}{2} a} \right)^2 \frac{\sin 2C}{2} - \&c. \quad (296)$$

$$\tfrac{1}{2} \pi - \tfrac{1}{2} (A - B) = \tfrac{1}{2} C + \frac{\tan \frac{1}{2} b}{\tan \frac{1}{2} a} \sin C + \left(\frac{\tan \frac{1}{2} b}{\tan \frac{1}{2} a} \right)^2 \frac{\sin 2C}{2} + \&c. \quad (297)$$

If we develop by (496), we find

$$\tfrac{1}{2} \pi - \tfrac{1}{2} (A + B) = - \tfrac{1}{2} C + \left(\frac{\cot \frac{1}{2} a}{\tan \frac{1}{2} b} \right) \sin C - \left(\frac{\cot \frac{1}{2} a}{\tan \frac{1}{2} b} \right)^2 \frac{\sin 2C}{2} + \&c. \quad (298)$$

$$\tfrac{1}{2} \pi - \tfrac{1}{2} (A - B) = - \tfrac{1}{2} C - \left(\frac{\tan \frac{1}{2} a}{\tan \frac{1}{2} b} \right) \sin C - \left(\frac{\tan \frac{1}{2} a}{\tan \frac{1}{2} b} \right)^2 \frac{\sin 2C}{2} - \&c. \quad (299)$$

from which a selection will be made in any particular case, according to the convergency of the series. The terms of the series are in arc, and must be reduced to seconds, by dividing by $\sin 1''$.

This solution may be applied to the case where two angles and the included side are the data, by means of the polar triangle.

161. *To express the area of a spherical triangle in series.*

Comparing (229) with Pl. Trig. (500), and developing by (502), we find

$$\tfrac{1}{2} K = \tan \tfrac{1}{2} a \tan \tfrac{1}{2} b \sin C - \tfrac{1}{2} \tan^2 \tfrac{1}{2} a \tan^2 \tfrac{1}{2} b \sin 2C + \&c. \quad (300)$$

162. LEGENDRE'S THEOREM. *If the sides of a spherical triangle are very small compared with the radius of the sphere, and a plane triangle be formed whose sides are equal to those of the spherical triangle, then each angle of the plane triangle is equal to the corresponding angle of the spherical triangle* minus *one-third of the spherical excess.*

Let a, b and c be the sides of the spherical triangle expressed in arc, the radius of the sphere being unity; and let A', B' and C' be the angles of the plane triangle whose sides are a, b and c. Then we have, in the spherical triangle,

$$\cos A = \frac{\cos a - \cos b \cos c}{\sin b \sin c}$$

Substitute in the second member of this, the values of $\cos a$, &c., in series, by Pl. Trig. (405) and (406), neglecting only powers above the fourth, viz.

$$\cos a = 1 - \tfrac{1}{2} a^2 + \tfrac{1}{24} a^4$$

$$\cos b = 1 - \tfrac{1}{2} b^2 + \tfrac{1}{24} b^4 \qquad \sin b = b - \tfrac{1}{6} b^3$$

$$\cos c = 1 - \tfrac{1}{2} c^2 + \tfrac{1}{24} c^4 \qquad \sin c = c - \tfrac{1}{6} c^3$$

we find

$$\cos A = \frac{\tfrac{1}{2}(b^2 + c^2 - a^2) + \tfrac{1}{24}(a^4 - b^4 - c^4 - 6\, b^2 c^2)}{b\, c\, [1 - \tfrac{1}{6}(b^2 + c^2)]}$$

Multiplying the numerator and denominator by $1 + \tfrac{1}{6}(b^2 + c^2)$, and neglecting terms of a higher order than the fourth, as before, we have

$$\cos A = \frac{b^2 + c^2 - a^2}{2\, b\, c} + \frac{a^4 + b^4 + c^4 - 2\, a^2 b^2 - 2\, a^2 c^2 - 2\, b^2 c^2}{24\, b\, c}$$

which, by Pl. Trig. (225) and (239), becomes

$$\cos A = \cos A' - \tfrac{1}{6}\, b\, c \sin^2 A'$$

Let $A = A' + x$, then since x is small, we may put $\cos x = 1$, so that, by Pl. Trig. (38),

$$\cos A = \cos A' - x \sin A'$$

whence

$$x = \tfrac{1}{6}\, b\, c \sin A$$

But $\tfrac{1}{2}\, b\, c \sin A$ = area of the plane triangle = very nearly area of the spherical triangle = K, whence

$$x = \tfrac{1}{3} K \qquad A' = A - \tfrac{1}{3} K$$

The same reasoning applies to each of the other angles, so that

$$B' = B - \tfrac{1}{3} K \qquad C' = C - \tfrac{1}{3} K$$

which proves the theorem.

163. This theorem is applied in geodetical surveying, and is found to be sufficiently accurate for triangles whose sides are considerably greater than 1°. It is to be remembered that the sides are to be expressed in arc; and if they are given in feet (for example), they must be reduced to arc by dividing by the radius in feet, or, which is equivalent, the area must be divided by the square of this radius. If then r = radius of the earth in units of any kind, a, b and c the sides of the triangle in units of the same kind, and k the area of the plane triangle, we shall have K in seconds, by the equation

$$K = \frac{k}{r^2 \sin 1''}$$

EXAMPLE.

In a triangle upon the earth's surface, given $b = 183496{\cdot}2$ feet, $c = 156122{\cdot}1$ feet, and $A = 48^\circ\, 4'\, 32''{\cdot}35$; to find the remaining parts.

We have $k = \frac{1}{2} b c \sin A$, and the mean value of $r = 20888780$ feet. Hence

$\log b$	5·26363
$\log c$	5·19346
$\log \sin A$	9·87159
ar co $\log 2 r^2 \sin 1''$	0·37356
$K = 5''{\cdot}04$ $\quad$ $\log K$	0·70224

It is evident that great accuracy in the value of r and of the other data is not required in computing K. We now have $\frac{1}{3} K = 1''{\cdot}68$, $A' = 48° \; 4' \; 30''{\cdot}67$, and by solving the plane triangle with the data A', b and c, we find

$$a = 140580{\cdot}0 \text{ feet} \qquad B' = 76° \; 12' \; 22''{\cdot}19 \qquad C' = 55° \; 43' \; 7''{\cdot}13$$

Adding $\frac{1}{3} K$ to each of these angles, the angles of the spherical triangle are

$$B = 76° \; 12' \; 23''{\cdot}87 \qquad C = 55° \; 43' \; 8''{\cdot}81.$$

For further details respecting geodetical triangles, and for the methods of solving spheroidal triangles, special works upon geodesy must be consulted, such as Legendre's *Analyse des Triangles tracés sur la surface d'une sphéroide;* Puissant's *Traité de Géodésie;* Puissant's *Nouvel essai de trigonométrie sphéroidique;* Fischer's *Lehrbuch der höheren Geodäsie;* various papers by Gauss, Bessel, &c.

164. *To solve a spherical triangle when two of its sides are nearly* 90°.

If a and b are nearly 90°, c and C are nearly equal, and it will be expedient to compute the small quantity $C - c$ by an approximate method. We have, by (25),

$$\sin^2 \tfrac{1}{2} c = \sin^2 \tfrac{1}{2} (a + b) \sin^2 \tfrac{1}{2} C + \sin^2 \tfrac{1}{2} (a - b) \cos^2 \tfrac{1}{2} C$$

and by Pl. Trig.

$$\sin^2 \tfrac{1}{2} C = [\sin^2 \tfrac{1}{2} (a + b) + \cos^2 \tfrac{1}{2} (a + b)] \sin^2 \tfrac{1}{2} C$$

the difference of which equations is

$$\sin \tfrac{1}{2} (C + c) \sin \tfrac{1}{2} (C - c) = \cos^2 \tfrac{1}{2} (a + b) \sin^2 \tfrac{1}{2} C - \sin^2 \tfrac{1}{2} (a - b) \cos^2 \tfrac{1}{2} C$$

Let

$$a' = 90° - a \qquad b' = 90° - b$$

a' and b' being very small: also, since C and c are nearly equal, put

$$\tfrac{1}{2} (C + c) = C$$

then the above equation becomes

$$\sin C \sin \tfrac{1}{2} (C - c) = \sin^2 \tfrac{1}{2} (a' + b') \sin^2 \tfrac{1}{2} C - \sin^2 \tfrac{1}{2} (a' - b') \cos^2 \tfrac{1}{2} C$$

Dividing by $\sin C = 2 \sin \frac{1}{2} C \cos \frac{1}{2} C$, and substituting the arcs $\frac{1}{2} (C - c)$, $\frac{1}{2} (a' + b')$, $\frac{1}{2} (a' - b')$, for their sines, we find

$$C - c = \sin 1'' \left[\left(\frac{a' + b'}{2}\right)^2 \tan \tfrac{1}{2} C - \left(\frac{a' - b'}{2}\right)^2 \cot \tfrac{1}{2} C \right] \qquad (301)$$

which is the required approximate formula for the case when a', b' and C are given to find c.

If a', b' and c are given, to find C, we may exchange C for c in the second member, whence

$$C - c = \sin 1'' \left[\left(\frac{a' + b'}{2}\right)^2 \tan \tfrac{1}{2} c - \left(\frac{a' - b'}{2}\right)^2 \cot \tfrac{1}{2} c \right] \qquad (302)$$

CHAPTER VIII.

MISCELLANEOUS PROBLEMS OF SPHERICAL TRIGONOMETRY.

165. *In a given spherical triangle, to find the perpendicular from one of the angles upon the opposite side.*

Fig. 25.

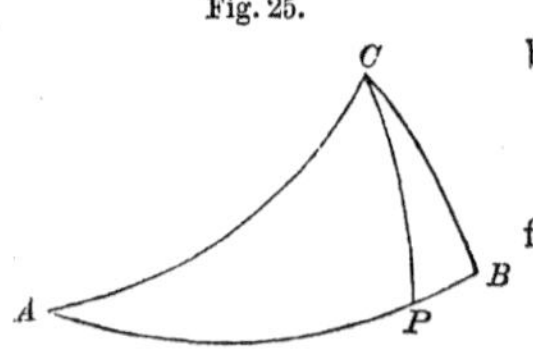

Denoting the perpendicular upon the side c (Fig. 25) by p, we have

$$\sin p = \sin b \sin A \tag{303}$$

If the three sides or the three angles are given, we find by (48), or (51), and (303),

$$\sin p = \frac{2\,n}{\sin c} = \frac{2\,N}{\sin C} \tag{304}$$

in which n and N are given by (47) and (50).

If we admit more than three parts of the triangle into the expression of p, we have, by (55), (56), and (303),

$$\sin p = \frac{2 \sin \frac{1}{2} A \sin \frac{1}{2} B}{\cos \frac{1}{2} C} \sin s = -\frac{2 \cos \frac{1}{2} a \cos \frac{1}{2} b}{\sin \frac{1}{2} c} \cos S \tag{305}$$

166. *To find the radius of the circle described about a given spherical triangle.*

Fig. 26.

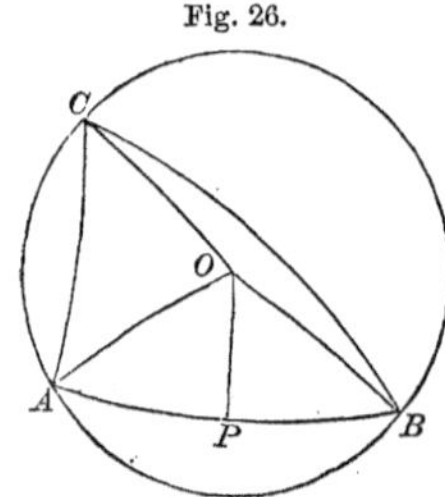

The radius here understood is the arc $O A = O B = O C$, Fig. 26, drawn from the pole of the small circle $A B C$ to either of the angles. Let

$$O A B = O B A = x$$

then

$$C = O C A + O C B = O A C + O B C$$
$$= A - x + B - x$$
$$x = \tfrac{1}{2} (A + B - C) = S - C$$

putting $S = \frac{1}{2} (A + B + C)$.

The triangle $A O B$ being isosceles, the perpendicular $O P$ bisects the side c, therefore if $O A = R$, we have

$$\tan R = \frac{\tan \frac{1}{2} c}{\cos x} = \frac{\tan \frac{1}{2} c}{\cos (S - C)} \tag{306}$$

or, by (70),

$$\tan R = \frac{2 \sin \frac{1}{2} a \sin \frac{1}{2} b \sin \frac{1}{2} c}{n} \tag{307}$$

By applying the principles of Art. 37, this will give the corresponding formulæ of Pl. Trig. (285).

167. From (69) and (70) we find

$$\cos (S - C) = -\cos S \cot \tfrac{1}{2} a \cot \tfrac{1}{2} b$$

by which (306) is reduced to

$$\tan R = \frac{\tan \frac{1}{2} a \tan \frac{1}{2} b \tan \frac{1}{2} c}{-\cos S} \tag{308}$$

Also, by the last equation of (56), (306) becomes

$$\tan R = \frac{\sin \frac{1}{2} c}{\cos \frac{1}{2} a \cos \frac{1}{2} b \sin C} \tag{309}$$

168. Substituting in (306) for $\tan \frac{1}{2} c$ by (39),

$$\tan R = \sqrt{\left(\frac{-\cos S}{\cos (S-A) \cos (S-B) \cos (S-C)}\right)}$$

or, by (50),

$$\tan R = \frac{-\cos S}{N} \tag{310}$$

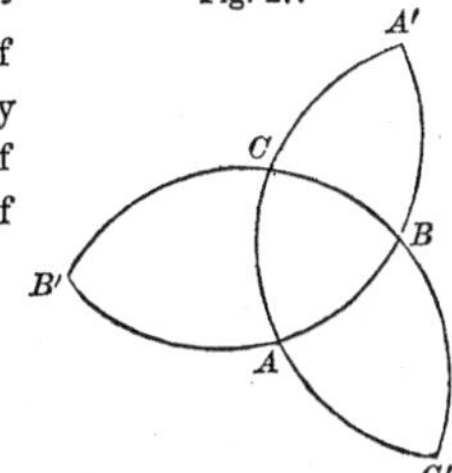

Fig. 27.

169. Let the sides of the triangle $A\,B\,C$, Fig. 27, be produced to meet in A', B', and C'; and denote the radii of the circles circumscribed about $A'\,B\,C$, $B'A\,C$, $C'A\,B$ by R', R'', R''' respectively. Then if $2\,S'$ denote the sum of the angles of $A'\,B\,C$, (A, B and C being the angles of $A\,B\,C$),

$$2\,S' = 2\,\pi - B - C + A$$

$$S' - A' = \pi - \tfrac{1}{2}(A + B + C) = \pi - S$$

so that (306) applied to $A'\,B\,C$ gives

$$\tan R' = \frac{\tan \frac{1}{2} a}{\cos (S' - A')} = \frac{\tan \frac{1}{2} a}{-\cos S}$$

and in like manner

$$\tan R'' = \frac{\tan \frac{1}{2} b}{-\cos S}$$

$$\tan R''' = \frac{\tan \frac{1}{2} c}{-\cos S} \tag{311}$$

Substituting for $\tan \frac{1}{2} a$, &c., by (39), or for $\cos S$ by (69),

$$\left.\begin{aligned} \tan R' &= \frac{\cos (S-A)}{N} = \frac{2 \sin \frac{1}{2} a \cos \frac{1}{2} b \cos \frac{1}{2} c}{n} \\ \tan R'' &= \frac{\cos (S-B)}{N} = \frac{2 \cos \frac{1}{2} a \sin \frac{1}{2} b \cos \frac{1}{2} c}{n} \\ \tan R''' &= \frac{\cos (S-C)}{N} = \frac{2 \cos \frac{1}{2} a \cos \frac{1}{2} b \sin \frac{1}{2} c}{n} \end{aligned}\right\} \tag{312}$$

170. Combining (310) with (312), we find the relation

$$\cot R \cot R' \cot R'' \cot R''' = N^2 \tag{313}$$

If this be multiplied successively by the squares of (310) and (312), we obtain

$$\left.\begin{aligned} \tan R \cot R' \cot R'' \cot R''' &= \cos^2 S \\ \cot R \tan R' \cot R'' \cot R''' &= \cos^2 (S-A) \\ \cot R \cot R' \tan R'' \cot R''' &= \cos^2 (S-B) \\ \cot R \cot R' \cot R'' \tan R''' &= \cos^2 (S-C) \end{aligned}\right\} \tag{314}$$

171. Again, from (310) and (312) we find

$$-\tan R + \tan R' + \tan R'' + \tan R'''$$
$$= \frac{\cos S + \cos (S - A) + \cos (S - B) + \cos (S - C)}{N}$$
$$= \frac{2 \cos \frac{1}{2} A \cos \frac{1}{2} (B + C) + 2 \cos \frac{1}{2} A \cos \frac{1}{2} (B - C)}{N}$$

whence

$$-\tan R + \tan R' + \tan R'' + \tan R''' = \frac{4 \cos \frac{1}{2} A \cos \frac{1}{2} B \cos \frac{1}{2} C}{N} \quad (315)$$

We shall find in a similar manner

$$\left.\begin{aligned} \tan R - \tan R' + \tan R'' + \tan R''' &= \frac{4 \cos \frac{1}{2} A \sin \frac{1}{2} B \sin \frac{1}{2} C}{N} \\ \tan R + \tan R' - \tan R'' + \tan R''' &= \frac{4 \sin \frac{1}{2} A \cos \frac{1}{2} B \sin \frac{1}{2} C}{N} \\ \tan R + \tan R' + \tan R'' - \tan R''' &= \frac{4 \sin \frac{1}{2} A \sin \frac{1}{2} B \cos \frac{1}{2} C}{N} \end{aligned}\right\} \quad (316)$$

It is also easily shown that

$$\tan^2 R + \tan^2 R' + \tan^2 R'' + \tan^2 R''' = \frac{2 + 2 \cos A \cos B \cos C}{N^2} \quad (317)$$

172. *To find the radius of the circle inscribed in a given spherical triangle.*

Fig. 28.

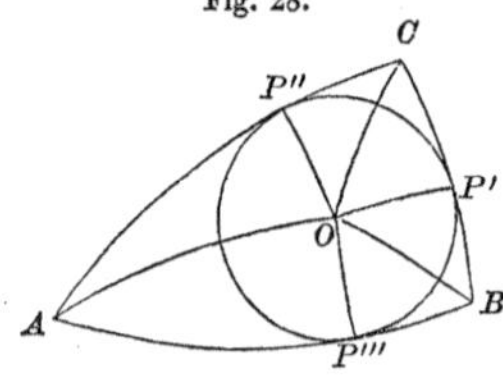

In Fig. 28, O being the pole of the required circle, draw OP', OP'' and OP''' to the points of contact, and join OA, OB. We have $OP'' = OP'''$ and the triangles AOP'' and AOP''' right-angled at P'' and P'''; hence

$$\sin OAP'' = \frac{\sin OP''}{\sin AO} = \frac{\sin OP'''}{\sin AO} = \sin OAP'''$$

therefore $OAP'' = OAP'''$, (for we cannot have $OAP'' = \pi - OAP'''$), and the pole of the inscribed circle is consequently found by the same construction as *in plano*, namely, by bisecting the angles of the triangle.

If then we put $s = \frac{1}{2}(a + b + c)$, and r = radius of the inscribed circle, we have

$$AP''' + BP' + CP' = AP''' + a = s, \qquad AP''' = s - a$$

and the right triangle AOP''' gives

$$\tan r = \sin (s - a) \tan \tfrac{1}{2} A \quad (318)$$

corresponding with the formula of Pl. Trig. (288).

Substituting, in (318), the value of $\tan \frac{1}{2} A$,

$$\tan r = \sqrt{\left(\frac{\sin (s - a) \sin (s - b) \sin (s - c)}{\sin s}\right)}$$

or

$$\tan r = \frac{n}{\sin s} \quad (319)$$

Substituting, in (318), the value of $\sin (s - a)$ given by (58),

$$\tan r = \frac{N}{2 \cos \frac{1}{2} A \cos \frac{1}{2} B \cos \frac{1}{2} C} \quad (320)$$

Also, by (51), we have $N = \frac{1}{2} \sin B \sin C \sin a$, which reduces (320) to

$$\tan r = \frac{\sin \frac{1}{2} B \sin \frac{1}{2} C}{\cos \frac{1}{2} A} \sin a \quad (321)$$

173. Let the radii of the circles inscribed in the three triangles $A'BC$, $B'AC$, $C'AB$ of Fig. 27, be r', r'' and r'''. Then if s' denote the half sum of the sides of $A'BC$, we have

$$2\,s' = 2\,\pi - b - c + a$$

$$s' - a = \pi - \tfrac{1}{2}\,(a + b + c) = \pi - s$$

so that (318) applied to the three triangles, gives

$$\left.\begin{aligned} \tan r' &= \sin s \tan \tfrac{1}{2} A \\ \tan r'' &= \sin s \tan \tfrac{1}{2} B \\ \tan r''' &= \sin s \tan \tfrac{1}{2} C \end{aligned}\right\} \quad (322)$$

Substituting in these the values of $\tan \frac{1}{2} A$, &c., or of $\sin s$,

$$\left.\begin{aligned} \tan r' &= \frac{n}{\sin (s - a)} = \frac{N}{2 \cos \frac{1}{2} A \sin \frac{1}{2} B \sin \frac{1}{2} C} \\ \tan r'' &= \frac{n}{\sin (s - b)} = \frac{N}{2 \sin \frac{1}{2} A \cos \frac{1}{2} B \sin \frac{1}{2} C} \\ \tan r''' &= \frac{n}{\sin (s - c)} = \frac{N}{2 \sin \frac{1}{2} A \sin \frac{1}{2} B \cos \frac{1}{2} C} \end{aligned}\right\} \quad (323)$$

Also, by (321),

$$\left.\begin{aligned} \tan r' &= \frac{\cos \frac{1}{2} B \cos \frac{1}{2} C}{\cos \frac{1}{2} A} \sin a \\ \tan r'' &= \frac{\cos \frac{1}{2} C \cos \frac{1}{2} A}{\cos \frac{1}{2} B} \sin b \\ \tan r''' &= \frac{\cos \frac{1}{2} A \cos \frac{1}{2} B}{\cos \frac{1}{2} C} \sin c \end{aligned}\right\} \quad (324)$$

174. The product of (319) and (323) gives

$$\tan r \tan r' \tan r'' \tan r''' = \frac{n^4}{n^2} = n^2 \quad (325)$$

whence, as in Art. 170,

$$\left.\begin{aligned} \cot r \tan r' \tan r'' \tan r''' &= \sin^2 s \\ \tan r \cot r' \tan r'' \tan r''' &= \sin^2 (s - a) \\ \tan r \tan r' \cot r'' \tan r''' &= \sin^2 (s - b) \\ \tan r \tan r' \tan r'' \cot r''' &= \sin^2 (s - c) \end{aligned}\right\} \quad (326)$$

175. We find from (319) and (323), as in Art. 171,

$$\left.\begin{aligned} -\cot r + \cot r' + \cot r'' + \cot r''' &= \frac{4 \sin \frac{1}{2} a \sin \frac{1}{2} b \sin \frac{1}{2} c}{n} \\ \cot r - \cot r' + \cot r'' + \cot r''' &= \frac{4 \sin \frac{1}{2} a \cos \frac{1}{2} b \cos \frac{1}{2} c}{n} \\ \cot r + \cot r' - \cot r'' + \cot r''' &= \frac{4 \cos \frac{1}{2} a \sin \frac{1}{2} b \cos \frac{1}{2} c}{n} \\ \cot r + \cot r' + \cot r'' - \cot r''' &= \frac{4 \cos \frac{1}{2} a \cos \frac{1}{2} b \sin \frac{1}{2} c}{n} \\ \cot^2 r + \cot^2 r' + \cot^2 r'' + \cot^2 r''' &= \frac{2 - 2 \cos a \cos b \cos c}{n^2} \end{aligned}\right\} \quad (327)$$

176. From (309) and (321), we find

$$\frac{\tan r}{\tan R} = 4 \sin \tfrac{1}{2} A \sin \tfrac{1}{2} B \sin \tfrac{1}{2} C \cos \tfrac{1}{2} a \cos \tfrac{1}{2} b \cos \tfrac{1}{2} c \tag{328}$$

From (307) and the first of (327),

$$-\cot r + \cot r' + \cot r'' + \cot r''' = 2 \tan R \tag{329}$$

From (315) and (320),

$$-\tan R + \tan R' + \tan R'' + \tan R''' = 2 \cot r \tag{330}$$

and other similar relations are found by comparing (312) with (327), and (316) with (323).

177. The following relations are also worth remarking.

If p is the perpendicular from C upon c,

$$\left.\begin{aligned} \tan R \sin p &= \frac{2 \sin \frac{1}{2} a \sin \frac{1}{2} b}{\cos \frac{1}{2} c} \\ \cot r \sin p &= \frac{2 \cos \frac{1}{2} A \cos \frac{1}{2} B}{\sin \frac{1}{2} C} \end{aligned}\right\} \tag{331}$$

178. *The pole of the circle inscribed in a spherical triangle is also the pole of the circle circumscribed about the polar triangle; and the radii of these circles are complements of each other.*

The arcs bisecting the angles of a given triangle will evidently bisect the sides of the polar triangle, and will be perpendicular to those sides respectively; the common intersection of these arcs is therefore at once the pole of the circle inscribed in the first and circumscribed about the second.

Again, if we join the angular points of the polar triangle with this common pole, the arcs thus drawn, being produced to meet the sides of the first triangle, are perpendicular to those sides, and therefore pass through the points of contact of the inscribed circle. Each of these arcs $= 90°$, and is at the same time the sum of the two radii of the circles in question.

This latter property is also obvious from the analytical expressions of the two radii. By means of it, we might have deduced all the formulæ for the inscribed from those for the circumscribed circle, or vice versa.

179. *To find the arc joining the poles of the circles inscribed in, and circumscribed about a given spherical triangle.**

Fig. 29.

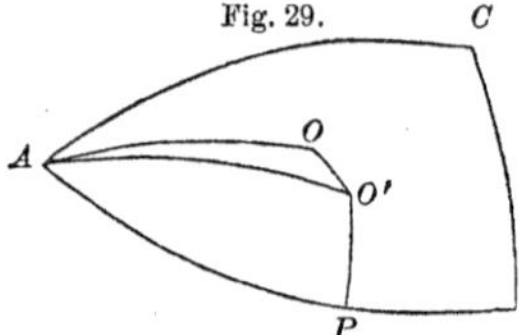

Let O be the pole of the circumscribed circle, Fig. 29, and O' that of the inscribed circle. Put $OO' = D$; then

$$\cos D = \cos AO \cos AO' + \sin AO \sin AO' \cos OAO'$$

By Art. 166, we have $OAB = S - C$, whence

$$OAO' = S - C - \tfrac{1}{2} A = \tfrac{1}{2} (B - C)$$

We have also

$$\cos AO' = \cos O'P \cos AP = \cos r \cos (s - a)$$

$$\sin AO' = \frac{\sin O'P}{\sin O'AP} = \frac{\sin r}{\sin \frac{1}{2} A}$$

* Hymer's Spherical Trigonometry.

Therefore,

$$\frac{\cos D}{\cos R \sin r} = \cot r \cos (s - a) + \tan R \frac{\cos \frac{1}{2}(B - C)}{\sin \frac{1}{2} A}$$

Substituting by (319), (307), and (44),

$$\frac{\cos D}{\cos R \sin r} = \frac{\sin s \cos (s - a) + 2 \sin \frac{1}{2} b \sin \frac{1}{2} c \sin \frac{1}{2}(b + c)}{n}$$

$$= \frac{\sin a + \sin b + \sin c}{2 n}$$

whence, by (53),

$$\left(\frac{\cos D}{\cos R \sin r}\right)^2 - 1 = \frac{1 + \sin a \sin b + \sin a \sin c + \sin b \sin c - \cos a \cos b \cos c}{2 n^2}$$

by Pl. Trig. (179), $$= \left(\frac{\sin s + 2 \sin \frac{1}{2} a \sin \frac{1}{2} b \sin \frac{1}{2} c}{n}\right)^2$$

$$= (\cot r + \tan R)^2$$

$$\cos^2 D = \cos^2 (R - r) + \cos^2 R \sin^2 r$$

$$\sin^2 D = \sin^2 (R - r) - \cos^2 R \sin^2 r \qquad (332)$$

If the inscribed circle is inscribed in $A'BC$, Fig. 27, and its radius $= r'$, we have, by a similar process,

$$\sin^2 D' = \sin^2 (R + r') - \cos^2 R \sin^2 r' \qquad (333)$$

180. *To find the equilateral spherical triangle inscribed in a given circle.*

If $R =$ radius of the given circle, and $A =$ one of the angles of the equilateral triangle, we have, by (310), and Pl. Trig. Art. 76,

$$\tan^2 R = \frac{-\cos \frac{3}{2} A}{\cos^3 \frac{1}{2} A} = \frac{3 \cos \frac{1}{2} A - 4 \cos^3 \frac{1}{2} A}{\cos^3 \frac{1}{2} A}$$

whence $$\cos \tfrac{1}{2} A = \sqrt{\left(\frac{3}{4 + \tan^2 R}\right)} \qquad (334)$$

181. *To find the equilateral spherical triangle circumscribed about a given circle.*

If $r =$ radius of the given circle, and $a =$ one of the sides of the triangle, we find

$$\sin \tfrac{1}{2} a = \sqrt{\left(\frac{3}{4 + \cot^2 r}\right)} \qquad (335)$$

182. *Given the base and area of a spherical triangle, to find the locus of the vertex.*

Fig. 30.

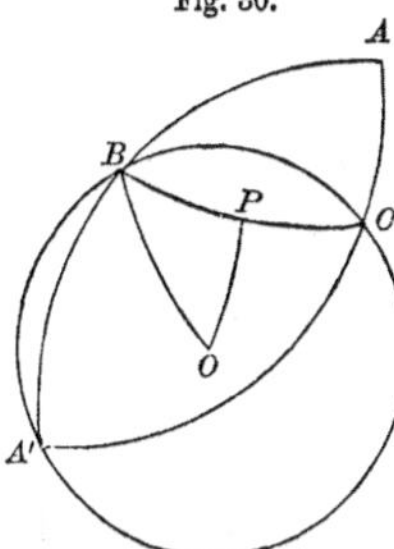

Let $a =$ the given base, and $K =$ area of ABC, Fig. 30. Produced AB and AC to meet in A'. Let O be the pole of the circle described about $A'BC$. The radius of this circle is given by the first equation of (311), which, by (224) becomes

$$\tan R' = \frac{\tan \frac{1}{2} a}{\sin \frac{1}{2} K} \qquad (336)$$

The second member of this equation, being constant for all the triangles of the same base a, and the same area K, shows that R' is also constant, and consequently, that the point A' is always found upon the circumference of the same small circle $A'BC$. But A and A' being the extremities of the same diameter of the sphere, A is also found upon a small circle, equal and parallel to the circle $A'BC$.

The perpendicular distance (p') of O from the base BC, is found by the equation

$$\cos p' = \frac{\cos R'}{\cos \frac{1}{2} a}$$

and the pole of the locus of A is in the same perpendicular, at a distance from $BC = \pi - p' = p$, whence

$$\cos p = -\frac{\cos R'}{\cos \frac{1}{2} a} \tag{337}$$

The equations (336) and (337) determine the radius and position of the pole of the required locus, which may therefore be constructed.

This elegant proposition is due to *Lexell.*

183. *To find the angle between the chords of two sides of a spherical triangle.*

Fig. 31.

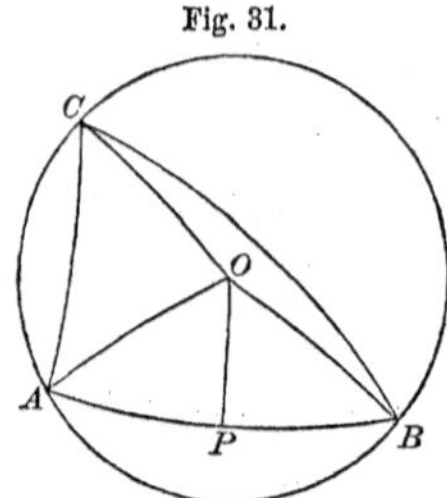

In Fig. 31, O being the center of the circumscribed circle, the angle between the chords of the sides AC and BC is half the spherical angle AOB. If, then,

$$C_1 = \text{angle between the chords of } a \text{ and } b$$

we have

$$\cos C_1 = \cos AOP = \sin OAP \cos AP$$

or, by Art. 166,

$$\cos C_1 = \sin (S - C) \cos \tfrac{1}{2} c \tag{338}$$

By (72) this becomes

$$\cos C_1 = \sin \tfrac{1}{2} a \sin \tfrac{1}{2} b + \cos a \cos \tfrac{1}{2} b \cos C \tag{339}$$

184. The preceding problem is employed for geodetical triangles, in which C_1 differs very little from C, in which case it is expedient to compute the small difference $C - C_1 = x$. We easily reduce (339) to the following:

$$\cos C_1 = \cos \tfrac{1}{2}(a-b) \cos^2 \tfrac{1}{2} C - \cos \tfrac{1}{2}(a+b) \sin^2 \tfrac{1}{2} C$$
$$= \cos^2 \tfrac{1}{2} C - 2 \sin^2 \tfrac{1}{4}(a-b) \cos^2 \tfrac{1}{2} C - \sin^2 \tfrac{1}{2} C + 2 \sin^2 \tfrac{1}{4}(a+b) \sin^2 \tfrac{1}{2} C$$

Subtracting $\cos C = \cos^2 \frac{1}{2} C - \sin^2 \frac{1}{2} C$, we have,

$$\sin \tfrac{1}{2}(C + C_1) \sin \tfrac{1}{2}(C - C_1) = \sin^2 \tfrac{1}{4}(a+b) \sin^2 \tfrac{1}{2} C - \sin^2 \tfrac{1}{4}(a-b) \cos^2 \tfrac{1}{2} C$$

or approximately, taking

$$\sin \tfrac{1}{2}(C + C_1) = \sin C = 2 \sin \tfrac{1}{2} C \cos \tfrac{1}{2} C$$

and

$$\sin \tfrac{1}{2}(C - C_1) = \tfrac{1}{2} x \sin 1''$$

x being expressed in seconds,

$$x = \frac{1}{\sin 1''} \sin^2 \tfrac{1}{4}(a + b) \tan \tfrac{1}{2} C - \frac{1}{\sin 1''} \sin^2 \tfrac{1}{4}(a - b) \cot \tfrac{1}{2} C \tag{340}$$

185. *If a great circle* (DE, Fig. 32) *bisect the base of a spherical triangle at right angles, any great circle* (FG), *perpendicular to it, divides the sides* (AC, BC) *into segments whose sines are proportional;* that is,

$$\sin FA : \sin FC = \sin GB : \sin GC \tag{341}$$

Fig. 32.

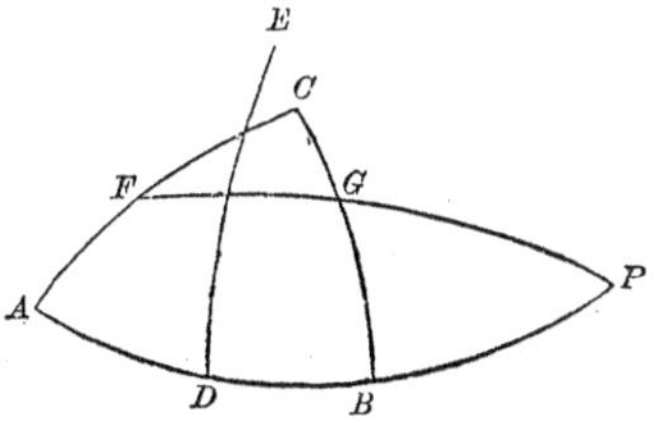

Let P be the pole of ED, ($DP = 90°$), and PGF any great circle drawn through P, and therefore perpendicular to DE. Then, since

$$PB + PA = 2\,PD = 180°$$

we have, by (3),

$$\sin F \sin FA = \sin P \sin PA = \sin P \sin PB$$
$$= \sin G \sin GB$$
$$\sin F \sin FC = \sin G \sin GC$$

whence, by division, the theorem (341). The arc FG is analogous to the parallel to the base in plane triangles.

186. *If two arcs of great circles,* (AB, CD, Fig. 33), *terminated by any circle, intersect, the products of the tangents of the semi-segments are equal to one another;* that is,

$$\tan \tfrac{1}{2} AE \tan \tfrac{1}{2} EB = \tan \tfrac{1}{2} CE \tan \tfrac{1}{2} ED \tag{342}$$

Let P be the pole of the circle $DACB$. Join PE and draw the perpendiculars PF, PG, bisecting the arcs AB and CD. Then we have

Fig. 33.

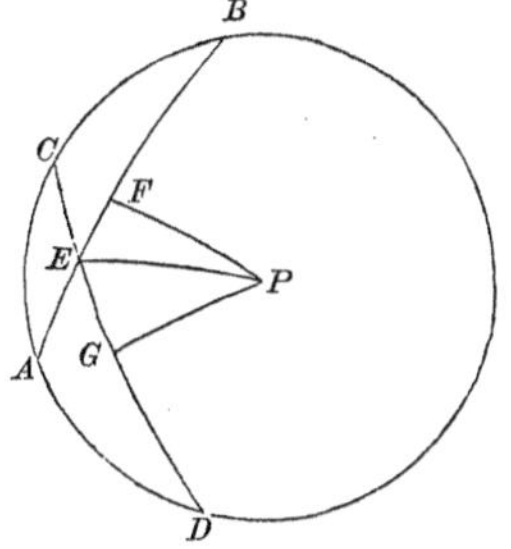

$$\frac{\cos FE}{\cos FB} = \frac{\cos PE}{\cos PB} = \frac{\cos PE}{\cos PD} = \frac{\cos GE}{\cos GD}$$

$$\frac{\cos FE - \cos FB}{\cos FE + \cos FB} = \frac{\cos GE - \cos GD}{\cos GE + \cos GD}$$

which, by Pl. Trig. (110), gives (342).

187. *If three arcs be drawn from the angles of a spherical triangle through the same point, to meet the opposite sides, the products of the sines of the alternate segments of the sides will be equal.*

Thus, in Fig. 34, we shall have

$$\sin AB' \sin CA' \sin BC' = \sin CB' \sin BA' \sin AC' \tag{343}$$

For we easily find

$$\frac{\sin AB'}{\sin CB'} = \frac{\sin AP}{\sin CP} \cdot \frac{\sin APB'}{\sin CPB'}$$

$$\frac{\sin CA'}{\sin BA'} = \frac{\sin CP}{\sin BP} \cdot \frac{\sin CPA'}{\sin BPA'}$$

$$\frac{\sin BC'}{\sin AC'} = \frac{\sin BP}{\sin AP} \cdot \frac{\sin BPC'}{\sin APC'}$$

Fig. 34.

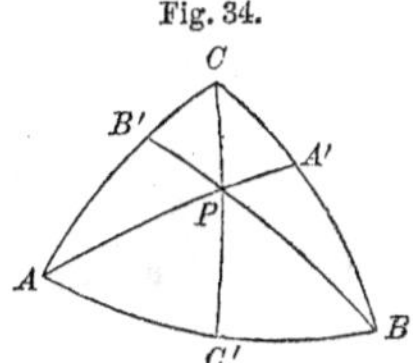

Multiplying these equations together, the product of the second members is unity, whence (343).

The same property is easily extended to the segments of the angles.

188. It follows, that when three arcs are drawn from the three angles, so as to satisfy the condition (343), they must intersect in the same point. This occurs in the same cases as in plane triangles, that is, when the angles are bisected; when the sides are bisected; when the three arcs are drawn from the angles to the points of

contact of the inscribed circle; and when the three arcs are the three perpendiculars upon the sides.

The first three of these cases are obvious. To prove the last, if A', B' and C', Fig. 34, are right angles, we have

$$\frac{\cos AB'}{\cos CB'}\cdot\frac{\cos CA'}{\cos BA'}\cdot\frac{\cos BC'}{\cos AC'}=\frac{\cos AB}{\cos CB}\cdot\frac{\cos CA}{\cos BA}\cdot\frac{\cos BC}{\cos AC}=1$$

whence $\cos AB' \cos CA' \cos BC' = \cos CB' \cos BA' \cos AC'$

and in the same manner we find

$$\tan AB' \tan CA' \tan BC' = \tan CB' \tan BA' \tan AC'$$

The product of these two equations gives the condition (343), and therefore the perpendiculars intersect in the same point.

189. *To find the arc drawn from any angle of a spherical triangle to a given point in the opposite side.*

In the triangle PAA'', Fig. 35, let PA' be drawn; we have

Fig. 35.

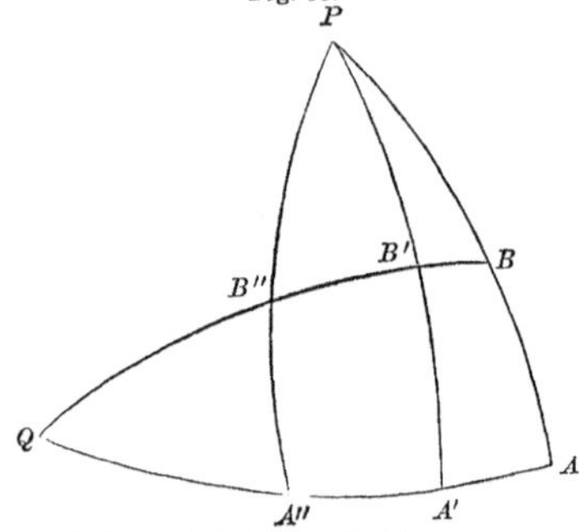

$$\begin{aligned}\cos PA' \sin AA'' &= \cos PA' \sin (AA' + A'A'')\\ &= \cos PA' \cos A'A'' \sin AA' + \cos PA' \cos AA' \sin A'A''\end{aligned}$$

But in the triangles PAA', $PA'A''$ we have, by (4),

$$\cos PA' \cos AA' = \cos PA - \sin PA' \sin AA' \cos PA'A$$
$$\cos PA' \cos A'A'' = \cos PA'' + \sin PA' \sin A'A'' \cos PA'A$$

which substituted above give

$$\cos PA' \sin AA'' = \cos PA \sin A'A'' + \cos PA'' \sin AA' \qquad (344)$$

which determines PA', the sides PA and PA'' and the segments of the side AA'' being given.

190. Let three arcs PA, PA', PA'', Fig. 35, passing through the same point P, be intersected by two others AA'' and BB'' whose intersection is Q; we have several symmetrical relations among the parts of the figure which find their application in astronomy.

Let the points A, A', A'' be given in position by their distances from Q, and put

$$\begin{array}{lll} AQ = \alpha & AB = \beta & PB = \gamma \\ A'Q = \alpha' & A'B' = \beta' & PB' = \gamma' \\ A''Q = \alpha'' & A''B'' = \beta'' & PB'' = \gamma'' \end{array}$$

By Pl. Trig. (171), we have

$$\sin \alpha \sin (\alpha' - \alpha'') + \sin \alpha' \sin (\alpha'' - \alpha) + \sin \alpha'' \sin (\alpha - \alpha') = 0$$

and in Fig. 35,

$$\sin \alpha = \frac{\sin \beta \sin B}{\sin Q} \qquad\qquad \sin B = \frac{\sin \gamma'' \sin B''}{\sin \gamma}$$

whence

$$\sin \alpha = \frac{\sin \beta}{\sin \gamma} \cdot \frac{\sin \gamma'' \sin B''}{\sin Q}$$

and similarly

$$\sin \alpha' = \frac{\sin \beta'}{\sin \gamma'} \cdot \frac{\sin \gamma'' \sin B''}{\sin Q}$$

$$\sin \alpha'' = \frac{\sin \beta''}{\sin \gamma''} \cdot \frac{\sin \gamma'' \sin B''}{\sin Q}$$

which, substituted above, give

$$\frac{\sin \beta}{\sin \gamma} \sin (\alpha' - \alpha'') + \frac{\sin \beta'}{\sin \gamma'} \sin (\alpha'' - \alpha) + \frac{\sin \beta''}{\sin \gamma''} \sin (\alpha - \alpha') = 0 \qquad (345$$

Again, if we express (344) in the notation of this article, it becomes

$$\cos (\beta + \gamma) \sin (\alpha' - \alpha'') + \cos (\beta' + \gamma') \sin (\alpha'' - \alpha) + \cos (\beta'' + \gamma'') \sin (\alpha - \alpha') = 0 \quad (346)$$

which, added to (345), gives

$$\frac{\sin(\beta+\gamma)}{\tan \gamma} \sin(\alpha' - \alpha'') + \frac{\sin(\beta'+\gamma')}{\tan \gamma'} \sin(\alpha'' - \alpha) + \frac{\sin(\beta''+\gamma'')}{\tan \gamma''} \sin(\alpha - \alpha') = 0 \quad (347)$$

191. If P is the pole of $A\,Q$, we have

$$\beta + \gamma = \beta' + \gamma' = \beta'' + \gamma'' = 90^\circ$$

and (345) and (347) both give

$$\tan \beta \sin (\alpha' - \alpha'') + \tan \beta' \sin (\alpha'' - \alpha) + \tan \beta'' \sin (\alpha - \alpha') = 0 \quad (348)$$

192. *To find the inclination of two adjacent faces of a regular polyhedron, and the radii of the inscribed and circumscribed spheres.*

Let C and E, Fig. 36, be the centres of two adjacent faces whose common edge is $A\,B$; O the centre of the inscribed and circumscribed spheres. Draw $O\,D$ bisecting $A\,B$ at right angles; draw $C\,D$, $E\,D$, which will also evidently be perpendicular to $A\,B$; and put

I = inclination of the faces = $C\,D\,E$
R = radius of the circumscribed sphere = $O\,A = O\,B$
r = radius of the inscribed sphere = $O\,C = O\,E$
a = one of the edges = $A\,B$
m = number of faces that form a solid angle
n = number of sides of a face

Fig. 36.

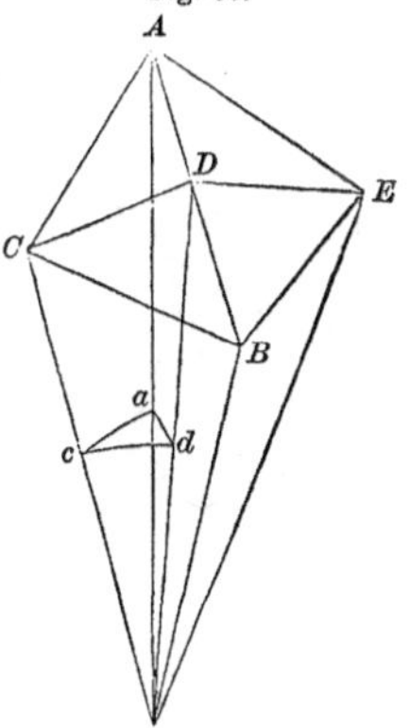

Suppose a sphere to be described about the centre O with any radius, and $c\,a\,d$ the triangle formed upon its surface by the planes $C\,O\,D$, $C\,O\,A$, $A\,O\,D$; this triangle is right-angled at d and gives

$$\cos c\,d = \frac{\cos c\,a\,d}{\sin a\,c\,d}$$

But $\cos c\,d = \cos C\,O\,D$, and

$$C\,O\,D = 90^\circ - C\,D\,O = \frac{\pi}{2} - \tfrac{1}{2} I$$

$$c\,a\,d = \tfrac{1}{2} \text{ angle of the planes } O\,A\,C \text{ and } O\,A\,E$$

$$= \frac{1}{2} \cdot \frac{2\pi}{m} = \frac{\pi}{m}$$

$$a\,c\,d = \tfrac{1}{2} A\,C\,B = \frac{1}{2} \cdot \frac{2\pi}{n} = \frac{\pi}{n}$$

therefore

$$\sin \tfrac{1}{2} I = \frac{\cos \frac{\pi}{m}}{\sin \frac{\pi}{n}} \tag{349}$$

Then from the triangles $O\,C\,D$, $A\,C\,D$, &c., we find

$$r = \frac{a}{2} \tan \tfrac{1}{2} I \cot \frac{\pi}{n} \tag{350}$$

$$r = R \cos a\,c = R \cot a\,c\,d \cot c\,a\,d = R \cot \frac{\pi}{n} \cot \frac{\pi}{m}$$

$$R = \frac{a}{2} \tan \tfrac{1}{2} I \tan \frac{\pi}{m} \tag{351}$$

193. *To find the surface and volume of a regular polyhedron.*

Let $f =$ number of faces of the polyhedron; $S =$ the surface, and $V =$ the volume; then the area of each face (the notation of the preceding article being continued) is equal to

$$\tfrac{1}{2} A\,B \times C\,D \times n = a^2 \cdot \frac{n}{4} \cot \frac{\pi}{n}$$

whence

$$S = a^2 \cdot \frac{n f}{4} \cot \frac{\pi}{n} \tag{352}$$

and since $V = S \times \frac{1}{3} r$,

$$V = a^3 \cdot \frac{n f}{24} \tan \tfrac{1}{2} I \cot^2 \frac{\pi}{n} \tag{353}$$

194. *To find the surface and volume of a parallelopiped, given the edges and their inclinations to each other.*

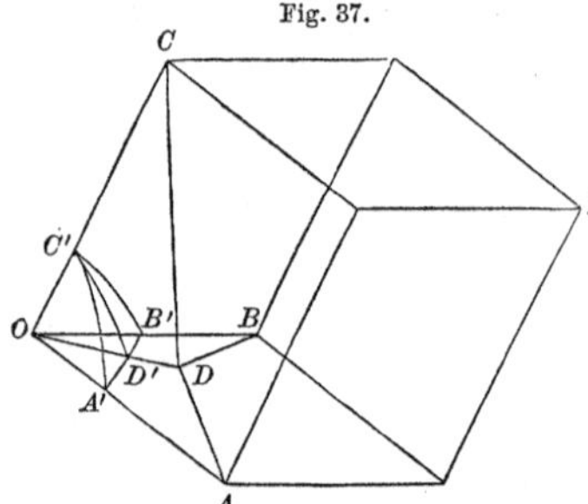

Fig. 37.

Let $O\,P$, Fig. 37, be a parallelopiped, whose edges $O\,A = a$, $O\,B = b$, $O\,C = c$, and their inclinations $B\,O\,C = \alpha$, $A\,O\,C = \beta$, $A\,O\,B = \gamma$, are given.

The area of any face, as $B\,C$, is found by the formula $b\,c \sin \alpha$, and therefore for the whole surface, we have

$$S = 2\,(b\,c \sin \alpha + a\,c \sin \beta + a\,b \sin \gamma) \tag{354}$$

To find the volume, let $C\,D$ be the altitude, then

$$V = \text{base } A\,B \times C\,D = a\,b \sin \gamma \times C\,D$$

Suppose a sphere to be described about O, whose intersections with the planes BOC, AOC, AOB and DOC are $B'C' = \alpha$, $A'C' = \beta$, $A'B' = \gamma$, and $C'D$. The triangle $A'C'D'$ is right-angled at D', whence

$$C\,D = c \sin C'D' = c \sin \beta \sin C'A'B'$$

or by (46), if $\sigma = \frac{1}{2}(\alpha + \beta + \gamma)$,

$$C\,D = \frac{2\,c}{\sin \gamma} \sqrt{[\sin \sigma \sin (\sigma - \alpha) \sin (\sigma - \beta) \sin (\sigma - \gamma)]}$$

whence

$$V = 2\,a\,b\,c \sqrt{[\sin \sigma \sin (\sigma - \alpha) \sin (\sigma - \beta) \sin (\sigma - \gamma)]} \tag{355}$$

THE END.

www.ingramcontent.com/pod-product-compliance
Lightning Source LLC
LaVergne TN
LVHW010249110826
845151LV00004B/1422

* 9 7 8 1 4 2 5 5 2 3 2 3 7 *